Springer Proceedings in Business and Economics

Springer Proceedings in Business and Economics brings the most current research presented at conferences and workshops to a global readership. The series features volumes (in electronic and print formats) of selected contributions from conferences in all areas of economics, business, management, and finance. In addition to an overall evaluation by the publisher of the topical interest, scientific quality, and timeliness of each volume, each contribution is refereed to standards comparable to those of leading journals, resulting in authoritative contributions to the respective fields. Springer's production and distribution infrastructure ensures rapid publication and wide circulation of the latest developments in the most compelling and promising areas of research today.

The editorial development of volumes may be managed using Springer Nature's innovative EquinOCS, a proven online conference proceedings submission, management and review system. This system is designed to ensure an efficient timeline for your publication, making Springer Proceedings in Business and Economics the premier series to publish your workshop or conference volume.

This book series is indexed in SCOPUS.

Göran Lindahl • Stefan Christoffer Gottlieb
Editors

SDGs in Construction Economics and Organization

The 11th Nordic Conference on Construction Economics and Organisation (CREON), May 18–20, 2022

Springer

CREON

Editors
Göran Lindahl
Architecture and Civil Engineering
Chalmers University of Technology
Göteborg, Sweden

Stefan Christoffer Gottlieb
Department of the Built Environment
Aalborg University
Copenhagen, Denmark

ISSN 2198-7246 ISSN 2198-7254 (electronic)
Springer Proceedings in Business and Economics
ISBN 978-3-031-25497-0 ISBN 978-3-031-25498-7 (eBook)
https://doi.org/10.1007/978-3-031-25498-7

This Springer imprint is published by the registered company Springer Nature Switzerland AG
The registered company address is: Gewerbestrasse 11, 6330 Cham, Switzerland

Preface

United Nations' Sustainable Development Goals (SDGs) have gained widespread attention across multiple societal sectors. In the context of construction and the built environment, governments are introducing new regulatory initiatives and client organisations incorporate the SDGs in their procurement and contracting policies. At the same time, market-based certification schemes are gaining momentum, and businesses are reformulating their business models and strategies incorporating SDGs and other approaches to sustainability. This development takes place against the backdrop of growing societal complexity and systemic interdependence, which entails that solutions increasingly rely on partnerships between disparate actors and actions across scales and levels to succeed.

At the 11th Nordic Conference on Construction Economics and Organisation that was held in Malmö and Copenhagen on 18–20 May 2022, scholars in the field of construction management were invited to contribute to sustainable action by addressing the overall question of how current developments challenge understandings in the academic field, and how our research may contribute to reaching the goals of society and businesses in the context of construction.

In response to our call, we received 37 contributions, which through a double-blind peer review process resulted in 25 high-quality papers being accepted for the presentation at the conference and published in these proceedings. The papers fall into six coherent themes that formed the basis for stimulating discussions at the conference about new perspectives and avenues for construction management informed research on sustainability in the construction industry. The themes covered include:

- Education and social sustainability
- Business processes and the circular economy
- Renovation and resource efficiency in the built environment
- Innovations and digitalisation for sustainable development
- Policy support for sustainability and the circular economy
- Production and procurement for sustainable development

Collectively, the papers explore and contribute to two main discussions within the field of construction management. First, how management theories can be applied to address societal grand challenges and contribute to application and development of the SDGs. This includes questions of how procurement practices can be developed to consider consequences of the circular economy and how new practices are developed to reduce the consumption of scarce resources. Second, papers also encompass reflexive studies of how global societal grand challenges, and the efforts to implement the SDGs, affect organisations and institutions related to the built environment. Examples of questions explored in this respect deal with how the new EU Taxonomy Regulation for sustainable financing impacts business processes and whether environmental regulation leads to greenwashing or deep sustainable change.

In responding to the call, the papers published in these proceedings explicitly address one or more of the UN SDGs, and it is our hope that the papers will be seen as a valuable starting-point and resource for further studies on the effect of, and approaches to implement, more sustainable business and policy practices in the built environment.

Gothenburg, Sweden Göran Lindahl
Copenhagen, Denmark Stefan Christoffer Gottlieb

Contents

Contributors

Mahgol Afshari Department of Civil and Environmental Engineering, Norwegian University of Science and Technology (NTNU), Trondheim, Norway

Rickard Andersson Department of Architecture and Civil Engineering, Chalmers University of Technology, Gothenburg, Sweden

Nikolaos Anevlavis Department of Management Engineering, Technical University of Denmark, Lyngby, Denmark

Radhlinah Aulin Division of Construction Management, Lund University, Lund, Sweden

Daniela Baer SINTEF, Trondheim, Norway

Alexander Olin Barfoed Department of Management Engineering, Technical University of Denmark, Lyngby, Denmark

Håvard Bergsdal SINTEF, Trondheim, Norway

Marco Alvise Bragadin Department of Architecture, University of Bologna, Bologna, Italy

Martine Buser Department of Architecture and Civil Engineering, Chalmers University of Technology, Gothenburg, Sweden

Tom B. J. Coenen Civil Engineering & Management, University of Twente, Enschede, The Netherlands

Anandasivakumar Ekambaram SINTEF, Trondheim, Norway

Ali El Chaabi Department of Management Engineering, Technical University of Denmark, Lyngby, Denmark

Atle Engebø Department of Civil and Environmental Engineering, Norwegian University of Science and Technology (NTNU), Trondheim, Norway

Marianne Forman Department of the Built Environment, Aalborg University, Copenhagen, Denmark

Nicolaj Frederiksen Department of the Built Environment, Aalborg University, Copenhagen, Denmark

Lasse M. Fredslund Department of the Built Environment, Aalborg University, Copenhagen, Denmark

Sophie Führer Group for Sustainability and Technology, ETH Zürich, Zürich, Switzerland

Stefan Christoffer Gottlieb Department of the Built Environment, Aalborg University, Copenhagen, Denmark

Neil Govender School of Construction Economics and Management, University of the Witwatersrand, Johannesburg, South Africa

Erik Gran SINTEF Digital, Oslo, Norway

Kim Haugbølle Department of the Built Environment, Aalborg University, Copenhagen, Denmark

Agnar Johansen Department of Civil and Environmental Engineering, Norwegian University of Science and Technology (NTNU), Trondheim, Norway

Keith Jones Anglia Ruskin University, Cambridge, UK

Kalle Kähkönen Faculty of Built Environment, Tampere University, Tampere, Finland

Dimosthenis Kifokeris Department of Architecture and Civil Engineering, Chalmers University of Technology, Gothenburg, Sweden

David King Anglia Ruskin University, Cambridge, UK

Lin Kjerulf Department of the Built Environment, Aalborg University, Copenhagen, Denmark

Ole Jonny Klakegg Department of Civil and Environmental Engineering, Norwegian University of Science and Technology (NTNU), Trondheim, Norway

Christian Koch Department of the Built Environment, Aalborg University, Copenhagen, Denmark

Riikka Kyrö Real Estate Science, Faculty of Engineering at Lund University (LTH), Lund, Sweden

Ola Lædre Department of Civil and Environmental Engineering, Norwegian University of Science and Technology (NTNU), Trondheim, Norway

Samuel Laryea School of Construction Economics and Management, University of the Witwatersrand, Johannesburg, South Africa

Irene Lill Department of Civil Engineering and Architecture, Tallinn University of Technology, Tallinn, Estonia

Claes Seehausen Lindtofte RGS Nordic, Copenhagen, Denmark

Jardar Lohne Department of Civil and Environmental Engineering, Norwegian University of Science and Technology (NTNU), Trondheim, Norway

Hanne Tønder Ludvigsen Department of Construction, University College Lillebaelt, Odense, Denmark

Rebecka Lundgren Real Estate Science, Faculty of Engineering at Lund University (LTH), Lund, Sweden

Javier Alonso Madrid ATANGA Arquitectura e Ingeniería, Madrid, Spain

Henrik Kryger Madsen Department of Construction, University College Lillebaelt, Odense, Denmark

Carlos Martinez Department of Building and Environmental Technology, Lund University, Lund, Sweden

Eero Nippala Tampere University of Applied Sciences, Built Environment and Bioeconomy, Tampere, Finland

Joseph Ofori-Kuragu Anglia Ruskin University, Cambridge, UK

Stefan Olander Construction Management, Faculty of Engineering at Lund University (LTH), Lund, Sweden

Theophilus Olowa Department of Civil Engineering and Architecture, Tallinn University of Technology, Tallinn, Estonia

Nils O. E. Olsson Department of Mechanical and Industrial Engineering, Norwegian University of Science and Technology (NTNU), Trondheim, Norway

Guillermo Sotorrío Ortega Eduardo Torroja Institute for Construction Science, Spanish National Research Council (CSIC), Madrid, Spain

Luca Pozzi Department of Architecture, University of Bologna, Bologna, Italy

Bintang Noor Prabowo Department of Civil and Environmental Engineering, Norwegian University of Science and Technology (NTNU), Trondheim, Norway

Mikkel Bruun Primdahl Department of the Built Environment, Aalborg University, Copenhagen, Denmark

Janak Raja Department of Management Engineering, Technical University of Denmark, Lyngby, Denmark

José Antonio Tenorio Rios Eduardo Torroja Institute for Construction Science, Spanish National Research Council (CSIC), Madrid, Spain

Thor Biering Rohdin Department of Management Engineering, Technical University of Denmark, Lyngby, Denmark

Lillian S. Rokseth SINTEF, Trondheim, Norway

Julian Andreas Moen Rygh Veidekke, Trondheim, Norway

Vera Rytter Division Civil SE South, NCC Sverige AB, Örebro, Sweden

Nina H. Sandberg SINTEF, Trondheim, Norway

Marit Støre-Valen Department of Civil and Environmental Engineering, Norwegian University of Science and Technology (NTNU), Trondheim, Norway

Rikard Sundling Department of Building and Environmental Technology, Lund University, Lund, Sweden

Henrik Szentes Construction Management, Lund University, Lund, Sweden

Alenka Temeljotov-Salaj Department of Civil and Environmental Engineering, Norwegian University of Science and Technology (NTNU), Trondheim, Norway

Christian Thuesen Department of Management Engineering, Technical University of Denmark, Lyngby, Denmark

Artur Tomczak Department of Civil and Environmental Engineering, Norwegian University of Science and Technology (NTNU), Trondheim, Norway

Kaleem Ullah Department of Civil Engineering and Architecture, Tallinn University of Technology, Tallinn, Estonia

Pål C. B. Vågbø SINTEF, Trondheim, Norway

Terttu Vainio VTT Technical Research Centre of Finland, Smart Energy and Built Environment, Espoo, Finland

Francesca Vergani Department of Building and Environmental Technology, Lund University, Lund, Sweden

Klaasjan Visscher Department of Technology, Policy & Society, University of Twente, Enschede, The Netherlands

Leentje Volker Civil Engineering & Management, University of Twente, Enschede, The Netherlands

Nadeeshani Wanigarathna Anglia Ruskin University, Cambridge, UK

Ron Watermeyer School of Construction Economics and Management, University of the Witwatersrand, Johannesburg, South Africa

Emlyn Witt Department of Civil Engineering and Architecture, Tallinn University of Technology, Tallinn, Estonia

Part I
Education and Social Sustainability

Chapter 1
Identifying Methods and Tools Toward More People-Friendly Environment: A Scoping Review

Mahgol Afshari, Alenka Temeljotov-Salaj, Agnar Johansen, and Jardar Lohne

Abstract Cities are contending with issues such as traffic congestion, air pollution, road accidents, and urban sprawl as the world's population grows at a rapid rate. Cycling and walking are nonmotorized modes that use no fossil fuel energy and require comparatively little infrastructure. They also have lower implementation and maintenance costs for users and governments than motorized forms of transport. Therefore, this study aims to identify methods and tools for more active mobility. The identification of approaches that can be used as incentives to increase walkability or bikeability in the Elgeseter district in the city of Trondheim has been done through a scoping literature review. The analysis is carried out according to the following research question: what can motivate citizens that commute to or travel inside the Elgeseter district to change their behavior toward more walking or biking? The findings are divided into four groups: active mobility advantages, bikeability motivators, walkability motivators, and active mobility barriers. Though almost all cities around the world are eager to address these issues, they will need integrated planning approaches that include everything from land use to city infrastructure design. Such approaches are necessary to encourage people to embrace green-sustainable modes of transportation as a lifestyle choice rather than a forced obligation. The study contributes to the knowledge about determinants that are important for encouraging commuters toward active mobility in the Elgeseter district.

Keywords Active mobility · People friendly environment · Walkability · Bikeability

M. Afshari (✉) · A. Temeljotov-Salaj · A. Johansen · J. Lohne
Department of Civil and Environmental Engineering, Norwegian University of Science and Technology (NTNU), Trondheim, Norway
e-mail: mahgol.afshari@ntnu.no; alenka.temeljotov-salaj@ntnu.no; a.johansen@ntnu.no; jardar.lohne@ntnu.no

G. Lindahl, S. C. Gottlieb (eds.), *SDGs in Construction Economics and Organization*, Springer Proceedings in Business and Economics,
https://doi.org/10.1007/978-3-031-25498-7_1

1.1 Introduction

It is an ongoing trend that people are moving into cities, and there is a focus on the development of smart cities in many countries in Europe (Collins et al., 2021). The location, design, and operation of a residential or commercial complex have an impact on how often people walk, bike, use public transportation, or drive, as well as whether their commuting experience is pleasant or unpleasant. Other factors can also influence people's travel behavior, such as geographical characteristics, cultural backgrounds, and awareness of traveling habits' effect on climate change. Active mobility including walking and cycling may result in cost savings, lower CO_2 emissions, less noise and air pollution, and less car congestion (Rabl & de Nazelle, 2012).

In accordance with the Trondheim city commitment to reduce greenhouse gas (GHG) emissions based on (Trondheim kommune., 2017), the city is attempting to overcome the effects of urbanization, city expansion, and highways as barriers to efficient collaboration in a part of the city called the Elgeseter district. In the area, the largest university and the largest hospital in Norway as well as many technologies and other companies are situated. There are different goals for the Elgeseter project, and the project is moving toward Sustainable Development Goals. Achieving zero emission, consolidation of sustainable lifestyles, supporting mental and societal health, moving toward innovation and development in an urban context, and achieving a systemic change toward a sustainable society are some of the targets of the project. Furthermore, improving active mobility in the Elgeseter district will achieve three Sustainable Development Goals at the same time which are good health and well-being, sustainable cities and communities, and climate action.

Elgeseter gate is an urban thoroughfare just south of Trondheim city center, between Professor Brochs gate in the south and Klostergata in the north. The road is a continuation of main road from the city center toward south. In this paper, the term Elgeseter district will be used to refer to the area surrounding Elgeseter gate. Figure 1.1 depicts the case study area, with the whole red circle representing Elgeseter gate and surrounding area; the red dotted circle includes the connecting routes to Elgeseter gate.

1.1.1 Scope of the Study

This paper reports on a scoping literature review focusing on identifying methods and tools that increase people-centric and active mobility that is relevant for the Elgeseter district. More specifically, the analysis ambitions to recognize incentives toward increasing walkability or bikeability in the Elgeseter district. Based on this, the research presented in this paper addresses the following main research question: *what can motivate citizens that commute to or travel inside the Elgeseter district to change their behavior toward more walking and biking?* In the upcoming chapter, a

Fig. 1.1 The focus area of the case study. (Map data: Google, 2022)

theoretical background about active mobility as part of developing a new town area is discussed according to previous studies.

1.2 Theoretical Background: Active Mobility as Part of New Town Development

In 122 nations around the world, more than 30% of adults were found to be physically inactive (Hallal et al., 2012). A considerable proportion of people in countries all over the world have adopted sedentary and physically inactive lifestyles (Van Dyck et al., 2013). This chapter provides a theoretical background on the relevance of physical activities in people's daily life, as well as how active mobility might meet this requirement. Furthermore, walking and cycling as two major types of active transportation are discussed, with bikeability being the more popular way of transportation among the general public. Finally, a summary of active mobility as part of the development of a new town area is presented.

Physical activity has been shown to increase emotion, sense of recognition, overall life quality, anxiety neurosis (Ohmatsu et al., 2014), and lower depression (Dunn et al., 2001). Regular physical activity can assist to reduce the risk of a variety of chronic diseases and their risk factors, thereby improving global public health. On the other hand, *active mobility* (i.e., walking/cycling to get from one place to another) can be done regularly, is cost-effective, and is an easily accessible form of physical activity. It is simple to incorporate into adults' daily lives; it may be an important contributor to meeting the daily physical activity guidelines for health (Mertens, 2016). Active mobility (also known as nonmotorized mobility) is critical to the development of efficient and equitable transportation networks as well as the transition to more sustainable communities (Victoria Transport Policy Institute, 2016).

Walking and cycling activities, among other sorts of physical activities, have recently gotten more attention from both civic and academic sectors to increase people's physical activity levels. Their popularity has been aided by a variety of factors. First, walking and cycling are suitable for people of all ages because they do not necessitate any special skills or equipment. Second, even though cycling is better for longer excursions, walking and cycling allow people to choose their preferred movement intensity. Finally, walking and cycling can assist people, particularly those from low-income groups, in breaking free from sedentary and inactive lifestyles (Brownson et al., 2000).

From the perspective of commuter cyclists, a city's cycling culture is the most important determinant in commuter riding levels (Sager, 2002). In various places throughout the world, policies and initiatives targeted at boosting the use, accessibility, and safety of cycling have increased during the previous decade (Pucher & Buehler, 2017). Furthermore, according to Waitt and Stanes (2022), barriers to "commuter cycling" as "stop-start" journeys filled with interruptions from traffic lights, crossing main roads, sharp corners, or pedestrians in the same lane as the bikers are important elements to be considered to increase bikeability in an area.

In sum, the natural and man-made environments, as well as individual and household characteristics, all have an impact on the decision to travel by bicycle (Heinen et al., 2009). Bicycling infrastructure can help reduce greenhouse gas emissions, while other measures that promote human-powered modes can help improve air and noise pollution. These advantages motivate towns to encourage greater riding, but doing so necessitates legislative changes that bring bicycling on par with other modes of transportation (Desjardins et al., 2021a, b).

During the research presented in the paper, walkability and bikeability motivators and barriers appear to be little analyzed in the literature. Not many papers are identified focusing solely on the methods for motivating people toward more active mobility. Therefore, the current paper tries to fill this knowledge gap by, firstly, identifying the methods and tools for increasing active mobility, and, secondly, by analyzing the factors which can affect citizens' commuting behavior. This latter concerns especially the factors that can motivate and encourage toward more walking and biking.

1.3 Methodology and Research Design

Scoping literature reviews are useful when the research intends to overview an existing body of literature within a specific field to find potential research gaps (Munn et al., 2018). The structure of the scoping review was inspired by the framework developed by Arksey and O'Malley (2005). Scoping reviews are also adequate methods when the research questions asked are broad and holistic without the intention of confirming or denying existing practices within the selected field (Arksey & O'Malley, 2005; Colquhoun et al., 2014; Munn et al., 2018).

1.3.1 Systematic Search

Four databases were chosen for the scoping literature review: Google Scholar, ScienceDirect, Web of Science, and Scopus. The study was limited to articles and books that have been published in the last 10 years. By defining "motivation AND commute AND walkability AND bikeability" as the main search string, 447 results appeared in GS, 694 results appeared in SD, and no results in WoS and Scopus. Based on the research question, the titles of the findings were read to select the most relevant literature for the topic. After transferring all 346 relevant articles based on their titles to Mendeley, five sets of duplicates have been found. So, for the next step, we started to read the abstracts of the 341 remaining documents and transfer the relevant ones to the comparison table. By reading abstracts, 53 final documents were chosen as the most relevant to the research topic. The framework for the scoping review performed in this paper is visualized in Fig. 1.2. In addition, two studies with pertinent data were added to the references after a particular search with the keyword "cycling commuter" in Google Scholar.

1.3.2 Search Procedure

Following the protocol of the scoping review, the steps are explained as follows (Fig. 1.3).

1. One research question is defined.
2. After several trials and errors, an initial search of relevant studies was conducted using available scientific databases with the following search string: "motivation AND commute AND walkability AND bikeability."

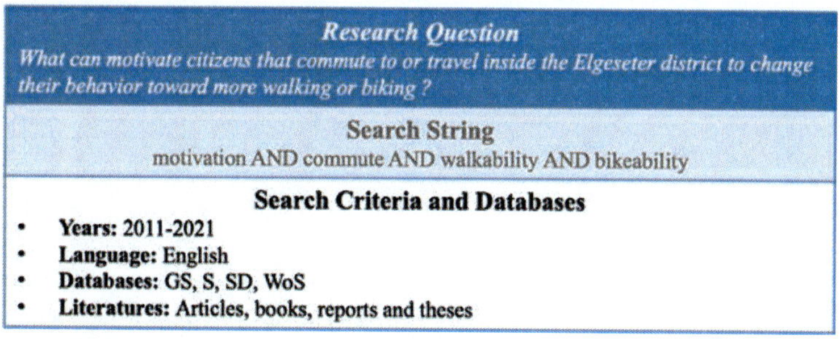

Fig. 1.2 Framework scoping study

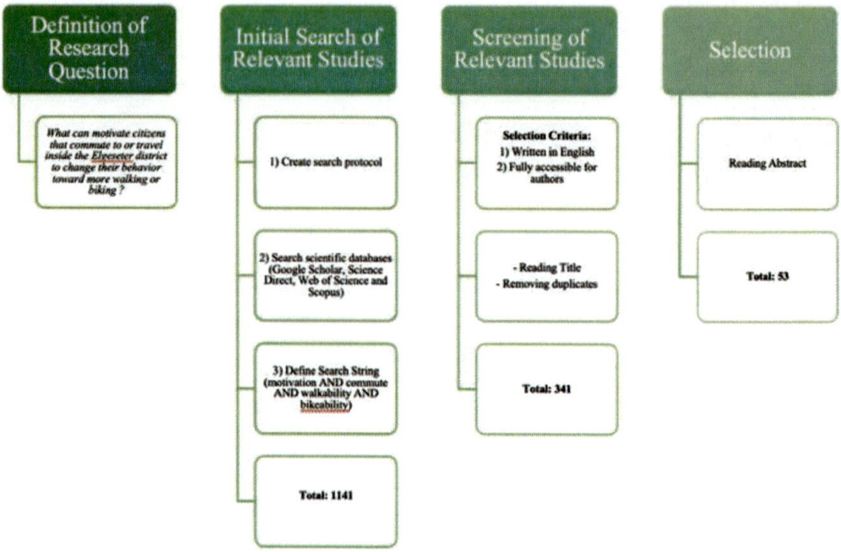

Fig. 1.3 Scoping review process as followed in the research presented in this paper

3. The selected databases were Google Scholar (GS), Scopus (S), ScienceDirect (SD), and Web of Science (WoS).
4. The language is limited to English, and the year of publication was set from 2011 to 2021.

In the upcoming chapter, the findings from the literature review are presented.

1.4 Results: Active Mobility, Bikeability, and Walkability Factors

In this chapter, the results are presented in two different categories. In the first section, the focus is on descriptive analysis, and in the second section, an overview of the overall findings from the literature is presented.

1.4.1 Descriptive Analysis

In this section, the descriptive findings from review are analyzed from two different aspects. In the first part, the number of publications of the examined papers during the last 10 years is provided, and in the second part, the top journals are identified.

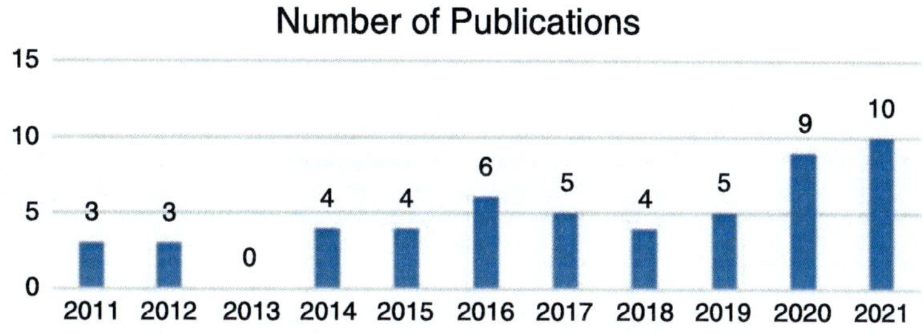

Fig. 1.4 The number of publication trends between 2011 and 2021

1.4.1.1 Number of Publications

In general, there was a notable increase in the number of publications connected to active mobility in 2020 and 2021, with nine and ten publications, respectively. Between 2011 and 2019, the number of papers published ranged from three to six, except for 2013, when no relevant papers were found. Figure 1.4 depicts the number of publication trends from 2011 to 2021.

1.4.1.2 Top Journals of the Examined Papers

The number of papers in the most prestigious journals for the examined 53 papers is depicted in Fig. 1.5.

"Transportation Research Part A: Policy and Practice" and "Environmental Research and Public Health," each with four papers, were the top journals in the field of study of this report, according to the analysis. In addition to them, three other journals, each with two papers, were active in this instance. The remaining papers originate from grey papers, other journals, conference proceedings, and publishers, with each having less than two papers.

By presenting the descriptive analysis in this section, the main focus of the next part will be on four different sets of findings from the literature.

1.4.2 Findings from the Literature

In order to identify approaches to increase active mobility, the main findings from the scoping literature review are categorized into four groups. First, the benefits of active mobility are presented. Then, bikeability and walkability motivators are discussed respectively, and finally, the barriers of active mobility are mentioned.

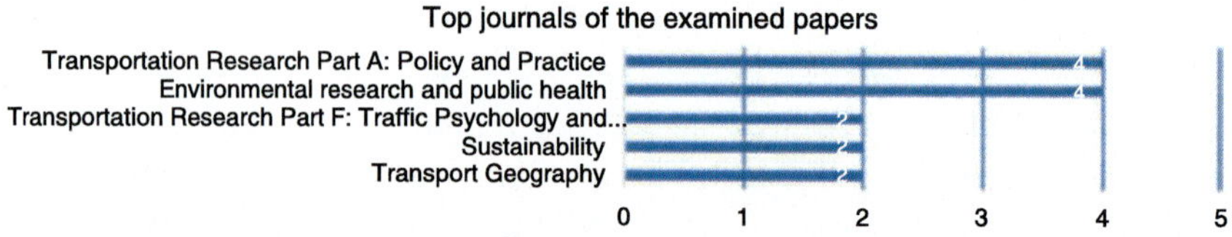

Fig. 1.5 Top journals of the examined paper

1.4.2.1 Active Mobility Advantages

Active mobility modalities are a low-cost means of commuting with a low environmental impact. Because of their low cost, flexibility, beneficial physical and psychological health impacts, and zero emissions, active modes (such as walking and bicycling) are deemed green, economic, equitable, and convenient (Gan et al., 2018). Walking or cycling as an alternative to motorized transportation for everyday journeys is example of active mobility modes. Based on previous studies, each of these alternatives is beneficial for the communities, and they have many advantages for the people, societies, and environment.

Physical activity can benefit people physiologically by having a favorable impact on their mental health, in addition to enhancing their physical health. Therefore, active mobility, which is linked to health, physical activity, and the prevention of chronic diseases, is increasingly being included in transportation and urban planning studies looking for alternatives to motorized transportation (Arbab et al., 2020a). Cycling as one of the active mobility modes has been shown to reduce the incidence of obesity, increase cardiovascular fitness, and reduce the risk of heart disease, diabetes, high blood pressure, and a variety of cancer-related side effects (Oja et al., 2011).

Kim and Dumitrescu (2016) believe that bicycling is critical for creating a city with sustainable development by lowering pollution from motorized vehicle emissions, improving inhabitants' health and physical fitness, and, most critically, minimizing road traffic accidents. As a result, promoting bikeability and walkability as a mode of transportation can help communities become more sustainable and livable.

1.4.2.2 Bikeability Motivators

Changes in travel behavior have been demonstrated to be one of the most effective ways to reduce greenhouse gas (GHG) emissions in transportation. Based on this fact, cycling, in particular, is becoming increasingly popular as a non-automobile means of transportation. Therefore, the main focus of this section is introducing some incentives which can lead to an increase in the rate of biking between people.

According to prior research, there are a variety of bikeability motivators that can encourage people to choose riding as a mode of transportation. Winters et al. (2010) present that in Vancouver, Canada, different sorts of bicyclists, both existing and potential, rated "routes with magnificent scenery" as a top motivator, slightly higher than routes with divided bicycle tracks or a flat slope. In another research, Heesch et al. (2012), compared biking incentives between men and women and mentioned that women were significantly more motivated by fun and enjoyment, getting fresh air, incorporating physical activity into a busy lifestyle, confidence in their cycling abilities, seeing other people cycle, encouragement from others, convenient or inexpensive mode of transportation, and environmental concerns than men.

According to Dill and McNeil (2013), protected bike lanes, known as "gold standard" bike lanes, are perceived to be safer than their non-protected counterparts because they use a barrier to separate cyclists from motorists. This sense of security, or comfort, could be critical in drawing more bicycles to the roads. In another study, Habib et al. (2014) indicated that people who have a greater perception of a city's bikeability and a low level of safety awareness are more likely to pedal for utilitarian reasons. It is also important to consider the quality of the urban environment.

While many research studies are discussing bikeability motivators without focusing on the specific areas, some other researchers present their findings based on different case studies in different geographical locations. For example, based on research in Brisbane, Australia, shorter distances to destinations, such as a commercial district with jobs and a river with bicycle routes, enhanced the likelihood of riding (Heesch et al., 2015). According to another study, bicyclists in Seattle, Washington, choose short, flat routes with well-connected amenities on highways with low traffic speeds. Their research discovered higher variation in preferences for views along routes with mixed land use, street trees, illumination, and city elements (Chen et al., 2018).

1.4.2.3 Walkability Motivators

A neighborhood's walkability is a measure of how walkable it is considered to be for people that walk in the district daily. The availability or absence of footpaths, sidewalks, or other pedestrian rights-of-way, traffic and road conditions, land-use patterns, building accessibility, and safety are all factors that can influence people's decision to walk as their primary means of transportation. According to Hess et al. (1999), in more walkable communities that have a higher density and a diversified land-use mix, there is a higher use of active modes and transit. Safe accessibility, such as strengthening personal security and improving transportation safety, and physical setting, such as boosting comfort level and providing supporting facilities, can be also some incentives toward active mobility (Arbab et al., 2020b).

Hillnhutter (2022) and Vukmirovic and Gavrilovic (2020) approached the stimulators in the urban environment, which influence the experience of walking (non-monotone environment, not boring streetscape, green features, artistic elements, gathering places, good visibility, safety). Alfonzo et al. (2008) believe that sidewalks' width and quality, benches, and crosswalks all had a beneficial impact on the number of pedestrians and/or the amount of time they spent walking. In other words, well-designed green street facilities contributed to more attractive walking environments (Adkins et al., 2012). Moreover, the likelihood of preferring to walk for both access and egress trips was positively and significantly linked with enough perceived walking amenities and comfortable walking space (Wu et al., 2018). Zhang and Mu (2020) also mentioned that if it's busy, dark, or hazardous, people will avoid walking. While strolling, pedestrians often consider additional facilities such as a water fountain, a restroom, and shade.

1.4.2.4 Active Mobility Barriers

Identifying the constraints that prevent individuals from walking or cycling to their destinations is the first step toward promoting active mobility. Greater distance, increased household income, and increased car ownership are consistently related to lower rates of active mobility among the factors that cannot be controlled for. According to Pucher and Buehler (2006), bicycle journeys are less common in low-density areas, as there are fewer places that can be visited in a short amount of time. Elgeseter district can be described as a low-density area. There is the potential for some restaurants, cafes, businesses, and perhaps a shopping mall to be built there, but currently, there are not enough places there to be visited or make the district attractive to walk or bike.

Ma et al. (2014) investigated active mobility barriers from an age standpoint. Ma et al. believe that younger individuals are more likely to bicycle. Older adults are less likely to ride a bike, which could be explained by the fact that as people get older, they become more concerned about safety and fear of being injured in an accident. However, Habib et al. (2014) explored cycling barriers from the perspective of gender. Based on his findings, women are more concerned about traffic and safety conditions, which is why they are less likely to cycle.

Based on Rojas López and Wong (2017), the most commonly reported walking difficulties in Singapore were distance limitations, sluggish transport speeds, and hot, wet weather. The need of carrying stuff (particularly for students) was also emphasized. Users, primarily younger users, stated that they must commute a significant distance to work or education. As a result, walking trips were frequently overlooked. Some people said they have to carry a lot of stuff to go to work (notebooks, lunch, paperwork, etc.), which makes walking more than a few blocks difficult.

1.5 Discussion

This study aimed to get a better knowledge of the advantages of active mobility for people and societies as well as the barriers that exist in the growth of walking and biking in the city as a genuine mode of transportation in such a constrained area. Moreover, this paper tries to identify some motivators toward more walkability and bikeability by using scoping literature review as the main research method to answer the research question: *what can motivate citizens that commute to or travel inside the Elgeseter district to change their behavior toward more walking and biking.* Here the results are discussed concerning the theoretical framework. In comparison to active mobility advantages and barriers and bikeability motivators, walkability motivators were discussed less in the literature.

Physical activities have been shown to help people's health by lowering the risk of becoming overweight or obese, as well as in the primary and secondary prevention of a variety of chronic illnesses (Warburton et al., 2006). Therefore, in recent years, numerous studies have attempted to discover various techniques for promoting physical activity in the general public, with a particular focus on active mobility as one of the most essential ways to improve an active lifestyle by utilizing walkability and bikeability as modes of commuting.

Trondheim municipality has aimed to introduce Trondheim as a model and a collaborative arena for green value creation and the development of a climate-friendly lifestyle. Furthermore, based on Trondheim kommune (2017), the municipality's goal is to reduce the greenhouse gas emissions by 80% before 2030, compared to the 1991 level. Therefore, improving active mobility in the Elgeseter area will simultaneously entail three Sustainable Development Goals: good health and well-being, sustainable cities and communities, and climate action.

According to Zhang (2016), although individuals care about the walking environment, the current metrics are insufficient in several ways. First, present methodologies do not take into account aspects of urban planning such as sidewalk quality, walking buffers, and other elements that impact people's walking behavior. Second, understanding the neighborhood's purpose and, more crucially, local people's preferences for the walking environment is vital for evaluating walkability. In a business center, a residential neighborhood, and a university campus, people have various walking requirements and expectations.

Inactivity and decreased physical activity/active transportation may be caused by poor sidewalk conditions, restricted access to recreational amenities such as parks, and a lack of local attractions (Arbab et al., 2020b). Moreover, areas with trees and green space are also associated with more bicycling. Currently, there is not enough green space in the Elgeseter district, and the area is mostly surrounded by old buildings which makes it a bit less attractive for the bikers. In other words, more walkable and bikeable communities may increase inhabitants' views toward active commute modes.

As unsafe paths discourage walking and biking, pedestrian safety is crucial to improving active transportation. One of the significant issues which make citizens less motivated to walk in the Elgeseter district is the lack of walking amenities in the area. In most places, the sidewalk is not divided from bicycling paths, and the sidewalks are either too wide or too narrow. As a walkable city is one with safe, accessible, and comfortable walkways, trails, and street crossings for people of all abilities, planners should emphasize constructing paths to connect residences with services and investing in more recreational facilities within walking distances in rural regions, where physical activity and active mobility alternatives are severely limited (Pavlick et al., 2020).

Finally, findings indicate that it's crucial to recognize that various users have distinct travel habits and requirements. As a result, measures to encourage walking and cycling should be tailored to the requirements of everybody, resulting in a greater number of prospective users.

1.6 Conclusions

Elgeseter district as one of the most important streets in Trondheim city is experiencing challenges such as increased traffic, toxic pollutants, and noise pollution. Increased active mobility in the area can help solve a lot of these problems. In accordance with Temeljotov-Salaj and Lindkvist (2021) to holistically approach the regeneration of urban spaces, the contribution to health and well-being is important, from both physical causes and symptoms of poor health, and the social, economic, and environmental components of individual community and overall well-being. In other words, based on the importance of considering walking and cycling as a way to improve the quality of life in cities, particularly Trondheim in Norway, with the added benefits of enhancing public and private health and lowering harmful emissions, traffic congestion, and noise associated with excessive automobility, the main focus of this paper is on looking for incentives that will motivate residents to change their behavior and choose walkability and bikeability as their preferred means of transportation.

According to the practical findings in this study walking and cycling for transportation ("active mobility") are usually thought to minimize CO_2 emissions by substituting for at least some motorized travel (de Nazelle et al., 2010). This is only one of the benefits of active mobility in Elgeseter gate. Active mobility may not only boost health as a source of physical activity, but it may also help achieve social and environmental goals such as promoting social cohesion and lowering CO_2 emissions by offsetting air pollution from motorized cars on such travels. So, to achieve these environmental goals and move toward greater sustainability, certain recommendations are made based on the research conducted by the authors of this paper, particularly for the Elgeseter district in Trondheim. Moreover, as one of the aims of UN SDG is to protect the planet, by improving active mobility, this goal will be more achievable.

It is important that urban area developments be rethought and reconfigured to improve traffic flow by including and supporting nonmotorized, less polluting modes of transportation such as cycling and walking. In other words, the main street just south of Trondheim city center, between Professor Brochs gate in the south and Klostergata in the north, must be adapted to include walkways, crossing junctions, and distinct cycling and pedestrian lanes with end-to-end connections. Secondly, to reduce dependency on unsustainable modes of transportation, Trondheim kommune has to construct a more inexpensive, accessible, and appealing transportation infrastructure that is available to the commuters of Elgeseter district at any time.

As a result of these insights, policymakers of Trondheim will be able to establish more effective policies for encouraging and developing active forms of transportation in the Elgeseter district. Therefore, it is critical for infrastructure and regulations to match existing and future users' expectations to provide an acceptable walking and bicycle transportation network service and entice people to utilize it. Furthermore, given the important public health, economic, and climatic implications of transportation behavior, for future work, researchers and funders should pay particular attention to finding motivators for active mobility more specifically.

References

Adkins, A., Dill, J., Luhr, G., & Neal, M. (2012). Unpacking walkability: Testing the influence of urban design features on perceptions of walking environment attractiveness. *Journal of Urban Design, 17*(4), 499–510.

Alfonzo, M. A., Boarnet, M. G., Day, K., Mcmillan, T., & Anderson, C. L. (2008). The relationship of neighbourhood built environment features and adult parents' walking. *Journal of Urban Design, 13*, 29.

Arbab, P., Martinez, J., Amer, S., & Pfeffer, K. (2020a). Toward active transport as a utilitarian and recreational form of sustainable urban mobility. *Adv. Intell. Syst. Comput., 1278*, 635–644.

Arbab, P., Schrenk, M., Popovich, V. V., Zeile, P., Elisei, P., Beyer, C., Ryser, J., Reicher, C., Çelik, C., Pfeffer, K., Martinez, J., & Amer, S. (2020b). *Active mobility as a response to physical inactivity in cities*. pp. 15–18.

Arksey, H., & O'Malley, L. (2005). Scoping studies: towards a methodological framework. *International Journal of Social Research Methodology, 8*(1), 19–32.

Brownson, R. C., Housemann, R. A., Brown, D. R., Jackson-Thompson, J., King, A. C., Malone, B. R., & Sallis, J. F. (2000). Promoting physical activity in rural communities: Walking trail access, use, and effects. *American Journal of Preventive Medicine, 18*(3), 235–241.

Chen, P., Shen, Q., & Childress, S. (2018). A GPS data-based analysis of built environment influences on bicyclist route preferences. *International Journal of Sustainable Transportation, 12*(3), 218–231.

Collins, D., Johansen, A., Kalsaas, B. T., Temeljotov Salaj, A., & Hamdy, M. (2021). Brought by degrees: A focus on the current indicators of lean 'smartness' in smart cities. *IGLC*, 167–176.

Colquhoun, H. L., et al. (2014, December). Scoping reviews: Time for clarity in definition, methods, and reporting. *Journal of Clinical Epidemiology, 67*(12), 1291–1294.

de Nazelle, A., Morton, B. J., Jerrett, M., & Crawford-Brown, D. (2010). Short trips: An opportunity for reducing mobile-source emissions? *Transportation Research Part D: Transport and Environment, 15*(8), 451–457.

Desjardins, E., Apatu, E., Razavi, S. D., Higgins, C. D., Scott, D. M., & Páez, A. (2021a). "Going through a little bit of growing pains": A qualitative study of the factors that influence the route choice of regular bicyclists in a developing cycling city. *Transportation Research Part F: Traffic Psychology and Behaviour, 81*, 431–444.

Desjardins, E., Higgins, C. D., Scott, D. M., Apatu, E., & Páez, A. (2021b). Correlates of bicycling trip flows in Hamilton, Ontario: Fastest, quietest, or balanced routes? *Transportation, 49*, 1–29.

Dill, J., & McNeil, N. (2013). Four types of cyclists?: Examination of typology for better understanding of bicycling behavior and potential. *Transportation Research Record, 2387*, 129.

Dunn, A. L., Trivedi, M. H., & O'Neal, H. A. (2001). Physical activity dose-response effects on outcomes of depression and anxiety. *Medicine and Science in Sports and Exercise, 33*(6 Suppl), S587.

Gan, Z., Feng, T., & Yang, M. (2018). Exploring the effects of car ownership and commuting on subjective well-being: A nationwide questionnaire study. *Sustainability, 11*(1), 84.
Habib, K. N., Mann, J., Mahmoud, M., & Weiss, A. (2014). Synopsis of bicycle demand in the City of Toronto: Investigating the effects of perception, consciousness and comfortability on the purpose of biking and bike ownership. *Transportation Research Part A: Policy and Practice, 70*, 67–80.
Hallal, P. C., Andersen, L. B., Bull, F. C., Guthold, R., Haskell, W., Ekelund, U., Alkandari, J. R., Bauman, A. E., Blair, S. N., Brownson, R. C., Craig, C. L., Goenka, S., Heath, G. W., Inoue, S., Kahlmeier, S., Katzmarzyk, P. T., Kohl, H. W., Lambert, E. V., Lee, I. M., et al. (2012). Global physical activity levels: Surveillance progress, pitfalls, and prospects. *The Lancet, 380*(9838), 247–257.
Heesch, K. C., Sahlqvist, S., & Garrard, J. (2012). Gender differences in recreational and transport cycling: A cross-sectional mixed-methods comparison of cycling patterns, motivators, and constraints. *International Journal of Behavioral Nutrition and Physical Activity, 9*(1), 1–12.
Heesch, K. C., Giles-Corti, B., & Turrell, G. (2015). Cycling for transport and recreation: Associations with the socio-economic, natural and built environment. *Health & Place, 36*, 152–161.
Heinen, E., van Wee, B., & Maat, K. (2009). Commuting by bicycle: An overview of the literature. *Transport Reviews, 30*, 59.
Hess, P. M., Moudon, A. V., Snyder, M. C., & Stanilov, K. (1999). Site design and pedestrian travel. *Transportation Research Record, 1674*, 9–19. https://doi.org/10.3141/1674-02
Hillnhutter, H. (2022). Stimulating urban walking environments–Can we measure the effect? *Urban Analytics and City Science, 49*(1), 275–289.
Kim, P., & Dumitrescu, E. (2016). *Share the road: Investment in walking and cycling road infrastructure*. UNEP FIA Foundation.
Ma, L., Dill, J., & Mohr, C. (2014). The objective versus the perceived environment: What matters for bicycling? *Transportation, 41*(6), 1135–1152.
Mertens, L. (2016). *Cycling for transport: The role of the physical environment.*
Munn, Z., et al. (2018). Systematic review or scoping review? Guidance for authors when choosing between a systematic or scoping review approach. *BMC Medical Research Methodology, 18*(1), 1–7.
Ohmatsu, S., Nakano, H., Tominaga, T., Terakawa, Y., Murata, T., & Morioka, S. (2014). Activation of the serotonergic system by pedaling exercise changes anterior cingulate cortex activity and improves negative emotion. *Behavioural Brain Research, 270*, 112–117.
Oja, P., Titze, S., Bauman, A., de Geus, B., Krenn, P., Reger-Nash, B., & Kohlberger, T. (2011). Health benefits of cycling: A systematic review. *Scandinavian Journal of Medicine & Science in Sports, 21*(4), 496–509.
Pavlick, D., Faghri, A., DeLucia, S., Gayen, S., Pavlick, D., Faghri, A., DeLucia, S., & Gayen, S. (2020). Human health and the transportation infrastructure. *Journal of Human Resource and Sustainability Studies, 8*(3), 219–248.
Pucher, J., & Buehler, R. (2006). Why Canadians cycle more than Americans: A comparative analysis of bicycling trends and policies. *Transport Policy, 13*(3), 265–279.
Pucher, J., & Buehler, R. (2017). Cycling towards a more sustainable transport future. *Transport Reviews, 37*, 689.
Rabl, A., & de Nazelle, A. (2012). Benefits of shift from car to active transport. *Transport Policy, 19*(1), 121–131.
Rojas López, M. C., & Wong, Y. D. (2017). Attitudes towards active mobility in Singapore: A qualitative study. *Case Studies on Transport Policy, 5*(4), 662–670.
Sager, B. A. (2002). *Is the constitution of a greenway trail network associated with cycling commuter use? – ProQuest*. Retrieved March 10, 2022.
Temeljotov-Salaj, A., & Lindkvist, C. (2021). Urban facility management. *Facilities, 39*(7/8), 525–537.
Trondheim kommune. (2017). *Kommunedelplan: Energi Og Klima 2017–2030*, 48.

van Dyck, D., Cerin, E., Conway, T. L., de Bourdeaudhuij, I., Owen, N., Kerr, J., Cardon, G., Frank, L. D., Saelens, B. E., & Sallis, J. F. (2013). Perceived neighborhood environmental attributes associated with adults' leisure-time physical activity: Findings from Belgium, Australia and the USA. *Health & Place, 19*(1), 59–68.

Victoria Transport Institute – Online TDM Encyclopedia. (2016). Retrieved December 6, 2021, from https://www.vtpi.org/tdm/

Vukmirovic, M., & Gavrilovic, S. (2020). Placemaking as an approach of sustainable urban facilities management. *Facilities, 38*(11/12), 801–818.

Waitt, G., & Stanes, E. (2022). Reactivating commuter cycling: COVID-19 pandemic disruption to everyday transport choices in Sydney, Australia. *Journal of Transport Geography, 98*, 103270.

Warburton, D. E. R., Nicol, C. W., & Bredin, S. S. D. (2006). Health benefits of physical activity: The evidence. *CMAJ, 174*(6), 801–809.

Winters, M., Teschke, K., Grant, M., Setton, E. M., & Brauer, M. (2010). How far out of the way will we travel?: Built environment influences on route selection for bicycle and car travel. *Transportation Research Record, 2190*, 1.

Wu, J., Yang, M., Sun, S., & Zhao, J. (2018). Modeling travel mode choices in connection to metro stations by mixed logit models: A case study in Nanjing, China. *Promet – Traffic & Transportation, 30*(5), 549–561.

Zhang, X. (2016). *Perceived importance and objective measures of built environment walkability of a university campus* (Master thesis). Wuhan University.

Zhang, X., & Mu, L. (2020). Incorporating online survey and social media data into a GIS analysis for measuring walkability. *Global Perspectives on Health Geography*, 133–155.

Chapter 2
Social Sustainability in a Gender-Biased Occupation

Radhlinah Aulin and Vera Rytter

Abstract This paper discussed construction industry's ongoing battle with gender equality. Growing awareness of social sustainability responsibility has led to the mounting importance of human resources in the organisations. One perspective involves the focus on gender equality. Today women's participation in the construction industry is still low except in times of acute labour shortages. This disproportionate gender distribution leads to the construction industry losing competent and skilled resources. The study aim is to identify challenges faced by women employed in the construction industry, factors that make them stay (pull factors) and factors that make them leave (push factors). To achieve this aim, a quantitative study was adopted. A web-based questionnaire was sent out via social media. The web-based questionnaire was open for 3 months, and a total of 124 women had responded. In general, the majority of results from the study are in agreement with the literature. The strong pull factors identified are as follows: interesting and challenging work and good work relations. The push factors are as follows: not the dream job, unsuitable job, poor working conditions/environment and offensive behaviour. The output of this study is the practical strategies on how to attract and retain women in the construction industry: improvement of image through marketing; exposure about the industry through education, female role models; eliminating macho culture, improvement of the working conditions and facilities; and flexible working times to accommodate those with family. With these recommendations, the industry can achieve better gender equality, and this, in turn, may transform it by encompassing a wider set of ethical considerations, including sustainable construction.

Keywords Social sustainability · Gender biased · Women · Construction · Retaining

---

R. Aulin (✉)
Division of Construction Management, Lund University, Lund, Sweden
e-mail: radhlinah.aulin@construction.lth.se

V. Rytter
Division Civil SE South, NCC Sverige AB, Örebro, Sweden
e-mail: vera.rytter@ncc.se

G. Lindahl, S. C. Gottlieb (eds.), *SDGs in Construction Economics and Organization*, Springer Proceedings in Business and Economics,
https://doi.org/10.1007/978-3-031-25498-7_2

2.1 Introduction

Since the Brundtland Report in 1987, there has been an increasing awareness that the construction industry must support the sustainable development vision by including social considerations throughout the entire construction project life cycle. However, existing sustainability studies in construction are largely related to the environmental and economic aspects. Social sustainability has not always received as much attention as more focus is diverted towards the economic and environmental sustainability. It was only recently that a more conscious effort is given to the social dimension of sustainability (Rasouli & Kumarasuriyar, 2016). Many researchers claimed that the social sustainability dimension is a concept in chaos (Vallance et al., 2011; Murphy, 2012). Most definitions base social sustainability on people's needs leading to the well-being of society: diversity, integration, justice, gender equality, inclusivity, participation, quality of life and democratic governance (Sourani & Sohail, 2005; Peters & Allison, 2011; Almahmoud & Doloi, 2015; Rasouli & Kumarasuriyar, 2016).

Growing awareness of sustainability and corporate social responsibility has led to the mounting importance of human resources in most organisations (Amrutha & Geetha, 2020). One perspective involves the focus on gender equality. Today, there are occupations that are defined by gender such as women dominating the education sectors, health care and social work while men dominate a more physically demanding branch such as construction, firefighting and mining. In Sweden, below 10% of the construction workers are women. There exists a wide body of research on women's participation in construction industries around the world with a strong case made for equality and inclusivity. The continuing underrepresentation of women in the construction industry has been debated for decades that lead to political and economic concerns. However, in South Asian countries, women play an important role which consists of performing unskilled tasks for low pay (Wells, 2001). Although women make up at least 20–40% of India's construction workers, they are less recognised than male workers, receive lower pay, often exposed to safety hazards and subjected to sexual harassment. They are integrated into the building workforce at the bottom end of the industry, as unskilled workers or head-load carriers. Access to training is denied to them (Rahul, 2014).

To encourage women's participation in the construction industry, the Swedish government has decided that by 2030, recruitment of new employees must include at least 25% women. This matches the UN development goals: 5 (gender equality), 8 (decent working conditions and economic growth) and 10 (reduce inequalities). Another positive note is that there has been an increase in women employment in the big construction companies. In the short term, the industry is tapping the local resource and filling the skills gap of labour shortage without outsourcing from outside the country (Ginige et al., 2007). Despite the number of recent recruitment initiatives, the industry has failed to make significant progress in recruiting and retaining more women.

2.1.1 Aim

To date, there is limited research examining the factors that enhance or impede gender-based recruitment and their retention in the industry including the construction sector. To identify strategies to tackle this persistent issue, it is important to have knowledge of the reasons why women choose to work in construction and better understand how to improve gender balances. Although this has been researched, there is a need for deeper insights into how to design strategies that effectively promote the inclusion of women in the construction industry. Such insights may benefit the construction industry to change the cultural stereotypes resulting in the reduction of the labour shortage and competence gap.

The study aim is to identify challenges faced by the women employed in construction industry, factors that make them stay (pull factors) and factors that make them leave (push factors). To achieve the aim, a quantitative study (with free text options) was adopted. A web-based questionnaire was sent out via social media platform 'Teknikkvinnor' (Technology Women). The group 'Teknikkvinnor' is a platform and meeting places for women in technology with a focus on career development and personal development. The web-based questionnaire was open for 3 months, and a total of 124 women responded.

2.2 Gender-Biased Occupation

Numerous research studies confirm an accurate picture of the current and historical challenges faced by women in a male-dominated working environment or vice versa. Today, many professions are still gender biased largely due to the career choices made during the education level. For example, 34% of men work in technology and production compared to 11% of women; on the contrary, 24% women work in health care, compared to 8% of men, according to the Swedish National statistic. The bias is apparent both horizontally—simplified as women working with 'people' and men with 'things' (Su et al., 2009)—and vertically, with men dominating superior positions across sectors (Blackburn et al., 2014). This is true even in relatively gender-equal countries, including Sweden. In 2014, The European Commission reports that the gender bias in the labour market leads to recruitment problems for employers, perpetuating the undervaluation of female-dominated work or vice versa and limiting individuals' career opportunities. The Swedish National statistic reports that among the occupations most dominated by women (more than 85%) are pre-school teaching, nursing and health care. On the other hand, more than 90% of carpenters, plumbers, electrician, technicians, machine operators and truck drivers are men. The Act on Equal Rights for Women and Men in the Labour Market was substituted in 1992 for the Gender Equality Act. Today, even though women can choose to work in any occupation, the gender-biased occupations still characterise the labour market. The act has not changed the trend of women entering the workforce where they remain in certain occupational sectors such as education, health and service sectors, notably banking, insurance and the retail trade.

2.2.1 *Women in Construction Industry*

The definition for male-dominated occupation in an industry is where women's participation is below 25% (Bigelow et al., 2015). Today women's participation in the construction industry is still low except in times of acute labour shortages when women have received the opportunity to enter the industry (Clarke et al., 2015). In construction-related occupations, approximately 98% of all employees are estimated to be men (Bigelow et al., 2015). The number of women studying STEM (science, technology, engineering and mathematics) subjects has increased over recent years (Botcherby & Buckner, 2012): but this has been slow to translate into improved employment participation, a picture reflected across the European Union (EU). In Sweden, the same scenario is evident where only 10% of women who studied within the construction programme work in the construction industry, and this figure has been the same for the last 10 years (Sveriges byggindustrier, 2017). The distribution of women studying construction-related subjects is as follows: (i) college level, around 11% women; (ii) bachelor level, around 25% women; and (iii) at the master's level, around 50% women. In the labour market, 11% of the employees in the construction industry are women, and 23% of those work in a large company. A more equal or even higher distribution of women can be observed in the architect's office, specialist competencies and administrators within the construction sector.

2.3 Challenges for Women in Construction

2.3.1 *Industry Image*

According to Ginige et al. (2007), an image, the mental picture, decides attitudes and behaviour of people, and it is built through a combination of both information gained from the environment and relevant past experiences. The construction industry is a good example of an industry-wide problem with 'image', which makes both men and women reluctant or uninterested in the industry. The predominant image of construction is that it is a male-dominated industry requiring physical strength and a good tolerance for outdoor conditions; inclement weather and bad language are cited as having a negative impact to women entering the industry. Other image-related barriers include the dominant male workforce; exclusive networks; informal recruitment; discriminatory sexist behaviour; bad language and sexist jokes; long hours culture; competitive and adversarial ingrained culture characterised by masculinity; conflict and crisis; challenging, dangerous and hostile environment; facilities; training; career progression; and the present level of their participation (Dainty et al., 2004; Amartunga et al., 2006; Aulin & Jingmond, 2011; Clarke et al., 2015). An interesting study conducted by Gale (1994) revealed that both sexes share a common image of the construction industry except for some image factors such as job security, equal opportunities in the industry and consideration of environmental issues. An unavoidable image which is synonymous with the industry is its poor working environment (including both the physical and psychosocial) which makes it a less attractive choice for a career. Besides accidents at the workplace, reports about construction failure, poor quality, effect on the environment, time and cost overrun are some examples of the negative image of the industry.

2.3.2 *Education*

Today everyone is entitled to education in whichever field they choose. However, all educations are far from equal. There are educations that still are gender dominated. It is evident that educational stratification leads to occupational segregation (Fielden et al., 2000). According to the Swedish National Agency for Education, in 2019, the number of women who follow the vocational construction programme (construction, electrical, plumbing,) courses at college level is 7%, while 55% of women follow the Natural Science programme (enabling them to further their studies at the university). One reason for the low interest is society still favours the white-collar employment to blue-collar or the socially acceptable employment when making career choices. At the university level, the education is diverse with various specialisations. The programmes that are most women dominated are 'Health care, Elementary Education and the Domestic spheres' (HEED, pedagogic and teaching, social science, law, and business administration). The technical and data programmes are male dominated.

Tellhed et al. (2016) explained the reason why men's low interest in HEEDs is not due to low self-esteem or how they perceive their ability in the field but due to participation and social affiliation. Gender bias is not only limited to nursing texts but found in the language and image of the nurse perpetuated within the classroom and clinical practice areas (Meadus & Twomey, 2011). In a study by Rytter (2020), majority of students who chose civil engineering programme based it on information from family and friends (45%), from education roadshow (21%), from the career guidance teachers during college (12%) and the rest from brochures and the Internet. Majority of these students made the choices based on their interest, good job opportunity and diverse education programme.

2.3.3 *Career Opportunities*

The potential to develop career opportunities in the construction industry is not transparent both at the operative and professional levels. This includes, which qualifications or merits are required, to be successful. This is especially true for women who enter nontraditional occupations; their choices are based on strong influence from their family, friends and teachers (Agapiou, 2002). The decision of selecting a career in the construction industry should be taken well in advance at schools' level particularly by those who intended to be a construction professional in the future. Lack of sufficient information can also mean that women who choose the construction industry often initially do not have a complete picture/knowledge of what the activity entails. However, when they understand what career opportunities there are and what the industry does, there is a risk that the industry does not meet their expectations. This leads to a reduction in commitment and a preference to move on to other professional roles outside the construction sector (Amartunga et al., 2006). Additionally, there is the conflict between work and family obligations. A study by Lingard and Lin (2004) showed that women in construction adopt an 'either or' approach to career and family. Women who choose to have a family develop lower expectations of the work experience, while women who expect to balance both family and career in the construction industry may experience significant difficulties.

2.3.4 Recruitment and Retaining

The recruitment process is a major challenge for many women who want to enter the construction industry as they feel its terms of employment are aimed at attracting men and not women (Adogbo et al., 2015). Equally important, an integral element in recruitment, is retainment. Retention of workers within an occupation is a measure of their satisfaction and 'fit' in the workplace. When being recruited, it is important to understand not only how they can gain access to that field but also whether or not they would remain there and under what conditions. Workers who quit may not like the job; they may be ill-prepared for or ill-informed about the job, or they may find that the workplace is very unwelcoming or alienating to them. In the construction industry, the barriers resulting in the lower retention of women at the workplace as identified by Green (1992) have changed little over time; they are still relevant today, such as inequities, the assignment of tasks on the jobsite, training and skills acquisition, promotion opportunities, sexual and racial harassment, deployment practices of contractors, the unwillingness of co-workers to teach them and the limitations on career advancement within unions and companies.

Therefore, it is important to untap the women workforce and the best job seekers that match the present and future needs of the industry. An important part of the industry strategy is employer branding—to attract and retain high-quality and talented employees (Sivertzen et al., 2013; Kucherov & Zavyalova, 2012). Recruitment of women in the construction workforce has been identified as a potential solution to overcome the skills gap while enhancing equal opportunities for women within the industry and to bring diversity to construction. Women workforce can provide able skilled labour to fill the gap (Ginige et al., 2007). Chhabra and Sharma (2014) suggested common organisational attributes such as compensation, career prospects and growth, job profile, brand name, corporate culture, employee empowerment, training and development, supportive and encouraging colleagues, innovative employer/novel work practices, humanitarian organisation giving back to society, job security, recognition/appreciation, good supervisor and worker relationship, customer-oriented organisation and acceptance and belonging. These emphasise employers' attractiveness as the envisioned benefits that a potential employee sees in working for a specific organisation are image and branch equity (Berthon et al., 2005).

2.3.5 *Culture and Work Environment*

There is a substantial body of literature emphasising that women who choose to work in the construction industry are aware that it is male dominated with strong macho culture, with stereotype challenges such as discrimination and harassment (Amartunga et al., 2006; Adogbo et al., 2015; Naismith et al., 2017). Women holding a higher position may encounter mistrust from male colleagues and distrust from the operative workers (Aulin & Jingmond, 2011). Another challenge is balancing between family and career; this is not only exceptional to construction industry. However, as common knowledge, the construction industry is project-based with long hours, moving from project to project and having intensive deadlines. This may be a hinderance to women who would like to be successful in both her career and her family. This may result in them prioritising their family and leaving the industry. When working in construction projects, working conditions and working environment can be challenging. Furthermore, the whole workplace culture poses problems to women on site. For example, temporary sanitary facilities are usually unisex, often without privacy and generally not well maintained. On top of proving their technical skills, women workers need to have the ability to fit into the accepted behaviour at the workplace which can be even more problematic.

2.4 Results and Analysis

The survey conducted managed to gather 124 responses. Majority of respondents are below 26 years old portraying that they are newcomers to the industry with less than 5 years working experience. Only 2% are in the oldest age range between 30 and 34 years old. The questions were divided into seven themes: image, education, career opportunities, recruitment process, culture and work environment.

The first question posed was the industry's image the respondent had anticipated upon being employed in the construction sector. Is it of no surprise that 80% of respondents answered male dominated and macho culture, followed by opportunity for career development (46%) and require strong mental strength (30%). Only 26% of respondents claimed that construction is synonym with tough working condition: dirty, dangerous, stressful and bad weather. In the open text, respondents state that the construction industry is exciting and challenging, secures future, stimulates project and benefits the society.

In this study, majority of respondents (73%) chose to follow the engineering programme (with focus on construction) because of interest, while 45% do it because of the good job opportunities. Only 18% made the choice because of family or friends. In the following question, 73% of the respondents agreed that their employment was based on their competencies. The questions posed during recruitment interviews were based on standard formulas. None of the respondents felt gender discriminated during the interview process. Nevertheless, 10% felt that they were being employed because of their gender and the company is filling their quotas to portray gender equality in the organisation. Interestingly, almost 10% of the respondents were asked if they have children or intend to have children in the future. Another interesting question posed by the interviewers is how the recruitment process could be improved to attract more women into the branch. The size of a company does make a difference. Bigger companies usually have gender equality policy in place and work actively to comply with it.

Responses regarding culture and work environment show a disturbing picture that unhealthy working climate still exist: 83% of the respondents were not respected by their male colleagues/management, 62% experienced poor working environment, 41% suffered sexual harassment, 29%% faced discrimination and 20% had difficulty balancing career and family. Only 6% experienced any physically demanding work, and this is reflected by the role of the respondents which is not as operatives. In the free text section, few respondents claimed that prejudices existed due to gender, were ignored by colleagues, belittled and even needed to prove their competence. The newcomers were often addressed condescendingly and unappreciated at the workplace. Many experienced offensive jokes, sexist comments or rude words at the workplace including presence of sexist objects such as press mugs, posters with sexist motives and Viagra in restrooms. One question (free text answer) posed to the respondents asked if they have benefited by being a woman at the workplace. The responses were that they were respected, accepted and received assistance when needed from everyone including from the operatives' level and the management level. They agreed that being a woman can have its advantages in a male-dominated occupation. For example, it creates a better working climate and a calmer work environment; they stand out in the company (because there are so few of them) and have better career opportunity.

One third of the respondents have had experience working on construction sites, and half of them are still working on sites. An interesting question posed to those that are still active is what factors attract them to stay (pull factors) in the industry and to those that left (push factors) is what made them leave. The response for this question is illustrated in Fig. 2.1 presenting the pull factors and the push factors.

Only 12% were concerned about gender equality at the workplace (Fig. 2.1). As each construction project is never the same, it is no surprise that 70% chose to stay due to the interesting work assignments (Fig. 2.1). Good relationship with male colleagues/management, enjoying challenges and being respected for their competence were positive factors that respondents graded high for reasons to stay on. As for the push factor (Fig. 2.2), the dominant reason is that it is not their dream jobs (46%) and 32% respondents find the job unsuitable and poor working conditions/environment. Most of the respondents do not work full time on construction site. Some are

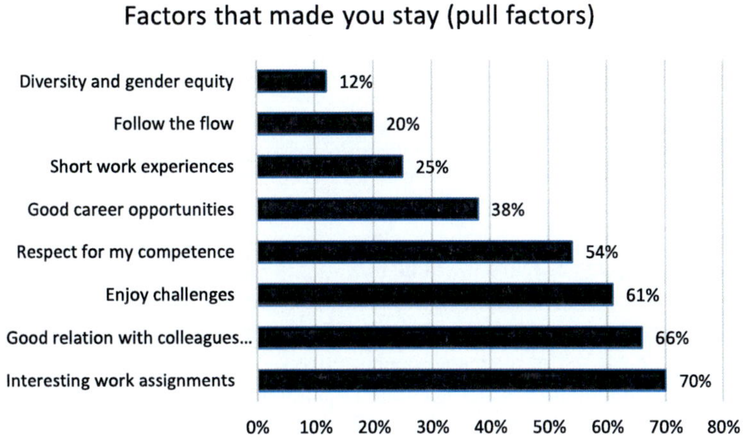

Fig. 2.1 Factors that attract you to stay in the industry (pull factors)

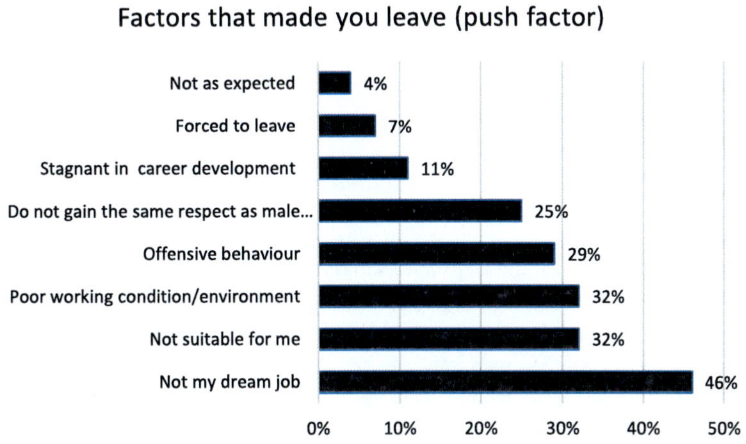

Fig. 2.2 Factors that made you leave (push factors)

consultants or project managers and spend a couple of days a week out on construction sites. Others make site visits and inspections or attend meetings and visit construction sites from time to time. Some respondents have answered that they stay because they simply thrive and find it fun. One respondent wrote:

> Love being part of a team where we build something physical that others can use. And that no days are the same. New problems always arise that we have to solve. Can be stressful for some but others like me are driven by it.

In summary, the survey results agreed with the literature about the industry image. The respondents are aware of the image, tough working conditions and the macho culture that exist before and after entering the industry (Wells, 2001; Ginige et al., 2007; Adogbo et al., 2015). One positive result from the study/survey showed that there exists opportunity for career development in the industry which contradicts Wells (2001) study that says otherwise. Role model could be one potential strategy to attract more women in the industry as mentioned by Bigelow et al., (2015). Regarding education, the results showed that the respondents are missing a female role model or female mentor when choosing programmes at college/universities. Respondents agreed that a mentor programme during work could help the newcomers to have better information about career development and career path. Amartunga et al. (2006) highlighted that having no clear information on career development can make women employees leave the industry.

Besides having role models, respondents claimed that networking for women in this industry can be seen as a support system for the newcomers. The survey results showed that almost 74% of the respondents were employed based on their competencies. Equally important to state that at least 10% felt that they were recruited because of their gender where companies needed to fill their quotas and abide with the gender equality policy. All respondents are in agreement that the recruitment process and the advertisement were gender neutral which contradicted the study by Amartunga et al. (2006). Regarding the question on building a family and having children, majority of respondents are in agreement that companies need to be more open and flexible on this matter. On the question about working environment, majority of respondents claimed that they need to work harder to gain the same respect as their male colleagues, encounter sexual harassment, face sexist jargons and offensive attitude. One-third of the respondents had experienced some form of discrimination as stated in the literature. An interesting result from the survey is that some respondents felt that being a women can be an advantage. They were being accepted, received more help and gained more respect. Another advantage was that they are easily remembered during meetings as there are only a handful of women on construction projects. It is easier to forward their ideas and thoughts. Majority of respondents stated that even the male colleagues agreed that presence of women on projects/offices improved their attitude, they were nicer and they avoided use of sexist remarks. From the survey results, the pull factors, contributing to respondents staying on project sites are: interesting work description, not one project is alike, daily challenges, problem solving, good relation with colleagues/management, receive mutual respect from colleagues/management, good career development and diversity. While the push factors that made some respondents leave are; it was not their dream job, it didn't suit them, poor working conditions, poor working environment, offensive behaviour and not receiving the same respect as their male colleagues. These are in agreement with the literature.

2.5 Recommended Practical Strategies

The survey results match three UN development goals: 5 Gender equality, 8 Decent working conditions and economic growth and 10 Reduce inequalities. To enhance the results, a recommended practical strategy is presented in Table 2.1.

Table 2.1 Practical strategies to attract women to stay in the construction industry

Factors	Recommendations (ownership and responsible for implementation)[a]
Image	*Marketing* Marketing the social factors in the construction industry and what the industry contributes to society to create well-being for people[b] Highlighting gender equality and how construction companies work to achieve it[b, c] Highlighting women who are active in the industry[b, c] Highlight the more positive aspects of the industry[b, c]
Education	*Primary education and below* Gender neutral information about construction through education materials[d] *College* Educate teachers about what and how the industry contributes to society[d] Provide more visits/lectures by women from the construction industry[d] Generate greater cooperation between schools and construction companies[b, d] Organise education campaigns by construction companies[b, d] *Technical college and universities* Inform students in construction-related educations about the reality of working life[b, d] Provide more visits/lectures by women from the construction industry[b, d] Organise network meetings between students and women active in the construction industry[b, d]
Female role model	Increase the opportunity for mentorship for women who are new to the industry. Mentors can provide information and materials to help increase the ability to advance in the career[b] Create and highlight role models that women can relate to[b] Illuminate the women who work in the industry today[b] Construction companies to spread knowledge about diversity and inclusivity[b] Employ more newly graduated women[b] Ensuring equal pay[b] Construction companies to invest in skills development so that the women in the industry today are motivated to stay[b] Female networking[b, d]
Culture and work environment	Eliminate macho culture and discrimination[b–d] Tougher measures against discriminatory behaviour and the use of sexist jargons and enforcement of these measures[b–d] Management to play an active role as a driving force towards gender equality and anchors it among employees[b] Educate employees on gender equality, diversity and inclusivity[b]
Decent working conditions	*Workplace design* Nicer and cleaner office environments[b] Separate changing rooms and sanitary facilities for women and men[b] *PPE* Adapt tools to women's physiques and bodies[b] Workwear design to suit for women[b] *Working time* Introduce more flexible working hours[b] Clear implementation of parental leave[b] Construction companies to investigate the potential for introducing flexible work[b]

[a]Ownership and responsible for implementation
[b]Construction companies, unions, The Swedish Construction Federation
[c]Media (digital and non-digital)
[d]Schools, college, universities

2.6 Conclusions

Results from the survey clearly showed that the construction industry still faces challenges to promote recruitment of women and retaining this workforce. A continued promotion of the notion of a male-dominated industry must cease, in order to encourage greater inclusion of women. The notion that the industry is gender biased that differentiates between men and women will continue unless changes are made in everything from behaviour and attitude to names and occupational terminologies. The changes are to focus on the pull factors on why women want to stay and to improve it, while mitigating the push factors. This paper presents practical strategy on how to attract women to stay in the construction industry which are follows: improve image through marketing, increase exposure about the industry as early as during the primary school, provide female role models at work and even at college/ universities, rigorously implement and enforce policy and regulation regarding discrimination and offensive behaviour (verbal and graphic), improve working conditions and facilities and introduce flexible working hours to accommodate those with family. With these recommendations, the industry can achieve better gender equality that will transform the industry encompassing a wider set of ethical considerations, including sustainable construction, secure employment and better work environment. Theses practical recommendations will allow the construction companies, the unions, The Swedish Construction Federation, schools, college and universities and media to fulfil the three UN development goals: 5 (gender equality), 8 (decent working conditions and economic growth) and 10 (reduce inequalities) to achieve social sustainability in construction.

References

Adogbo, K. J., Ibrahim, A. D., & Ibrahim, Y. M. (2015). Development of a framework for attracting and retaining women in construction practice. *Journal of Construction in Developing Countries, 20*(1), 99–115.

Agapiou, A. (2002). Perceptions of gender roles and attitude towards work among male and female operatives in the Scottish construction industry. *Construction Management and Economic, 20*(8), 697–705.

Almahmoud, E., & Doloi, H. K. (2015). Assessment of social sustainability in construction projects using social network analysis. *Facilities, 33*(3/4), 152–176.

Amartunga, D., Haigh, R., Lee, A., Shanmugam, G., & Elvitigala, G. (2006). *Construction industry and women: A review of the barriers*. Research Institute for the Built and Human Environment, The University of Salford.

Amrutha, V. N., & Geetha, S. N. (2020). A systematic review on green human resource management: Implications for social sustainability. *Journal of Cleaner Production, 247*, 119131.

Aulin, R., & Jingmond, M. (2011). Issues confronting women participation in the construction industry. In J. Mwakali & H. M. Alinaitwe (Eds.), *Advances in engineering and technology – contribution of scientific research in development*, 31st Jan–1st Feb, Uganda, pp. 312–318.

Berthon, P., Ewing, M., & Hah, L. L. (2005). Captivating company: Dimensions of attractiveness in employer branding. *International Journal of Advertising, 24*, 151–172.

Bigelow, B., Bilbo, D., Mathew, M., Ritter, L., & Elliot, J. (2015). Identifying the most effective factors in attracting female undergraduate students to construction management. *International Journal of Construction Education and Research, 11*(3), 179–195.

Blackburn, R. M., Jarman, J., & Racko, G. (2014). Occupational segregation: Its vertical and horizontal dimensions. In G. Razzu (Ed.), *Gender inequality in the labour market in the UK* (pp. 101–121). Oxford University Press.

Botcherby, S., & Buckner, L. (2012). Women in science, technology, engineering and mathematics: From classroom to boardroom. *UK Statistics 2012*. WISE. http://www.wisecampaign.org.uk/files/useruploads/files/wise_stats_document_final.pdf. Available at: Accessed: 2022-01-05

Chhabra, N. L., & Sharma, S. (2014). Employer branding strategy for improving employer attractiveness. *International Journal of Organizational Analysis, 22*, 48–60.

Clarke, L., Michielsens, E., Snijders, S., Wall, C., Dainty, A., Bagilhole, B., & Barnard, S. (2015). *'No more softly softly': Review of women in the construction workforce*. University of Westminster.

Dainty, A. R. J., Bagilhole, B. M., Ansari, K. H., & Jackson, J. (2004). Creating equality in the construction industry: An agenda for change for women and ethnic minorities. *Journal of Construction Research, 5*(1), 75–86.

Fielden, S. L., Davidson, M. J., Gale, A. W., & Davey, C. L. (2000). Women in construction: The untapped resource. *Construction Management and Economics, 18*, 113–121.

Gale, A. W. (1994). Women in non-traditional occupations: The construction industry. *Women in Management Review, 9*(2), 3–14.

Ginige, K. N., Amaratunga, R. D. G., & Haigh, R. (2007). Improving construction industry image to enhance women representation in the industry workforce. In D. Boyd (Ed.), *Proceedings 23rd annual ARCOM conference*, 3–5 September, Belfast, UK, Association of Researchers in Construction Management, pp. 377–385.

Green, C. E. (1992). *Beyond recruitment: Lessons from the workplace about retention of women in the skilled trades* (Ms Thesis). Department of Urban Studies and Planning, MIT.

Kucherov, D., & Zavyalova, E. (2012). HRD practices and talent management in the companies with the employer brand. *European Journal of training and Development, 36*, 86–104.

Lingard, H., & Lin, J. (2004). Career, family and work environment determinants of organizational commitment among women in the Australian construction industry. *Construction Management and Economics, 22*, 409–420.

Meadus, R. J., & Twomey, J. C. (2011). Men student nurses: The nursing education experience. *Nursing Forum, 46*(4), 269–279.

Murphy, K. (2012). The social pillar of sustainable development: A literature review and framework for policy analysis. *Sustainability: Science, Practice, & Policy, 8*(1).

Naismith, N., Robertson, S., & Tookey, J. (2017). Identifying barriers to retaining female professionals in engineering and construction organisations. *EPiC Series in Education Science, 1*, 227–234.

Peters, J., & Allison, M. (2011). *Equality and diversity: Good practice for the construction sector*. Equality and Human Rights Commission.

Rahul. (2014, October 14). *Participation of the female workers in the construction industry in India: A review*. Available at SSRN: https://ssrn.com/abstract=2509916

Rasouli, A. H., & Kumarasuriyar, A. (2016). The social dimension of sustainability: Towards some definitions and analysis. *Journal of Social Science for Policy Implications, 4*(2), 23–34.

Rytter, V. (2020). *Kvinnor i byggbranschen – En studie om hur byggbranschen kan göras mer attraktiv för kvinnor*. LTH Faculty of Engineering, Lund University.

Sivertzen, A. M., Nilsen, E. R., & Olafsen, A. H. (2013). Employer branding: Employer attractiveness and the use of social media. *The Journal of Product and Brand Management, 22*, 473–483.

Sourani, A. I., & Sohail, M. (2005). Realising social objectives of sustainable construction through procurement strategies. In K. Sullivan & D. T. Kashiwagi (Eds.), *Proceeding of the CIB W92/T23/W107 international symposium on procurement systems: The impact of cultural differences and systems on construction performance*, 8–10 February, Las Vegas, Nevada.

Su, R., Rounds, J., & Armstrong, P. I. (2009). Men and things, women and people: A meta-analysis of sex differences in interests. *Psychological Bulletin, 135*(6), 859–884. https://doi.org/10.1037/a0017364

Sveriges Byggindustrier. (2017). *FAKTA om kvinnor och män i byggbranschen*. Sveriges Byggindustrier.

Tellhed, U., Bäckström, M., & Björklund, F. (2016). Will I fit in and do well? The importance of social belongingness and self-efficacy for explaining gender differences in interest in STEM and HEED majors. *Sex Roles, 77*, 86–96.

Vallance, S., Perkins, H. C., & Dixon, J. E. (2011). What is social sustainability? A clarification of concepts. *Geoforum, 42*(3), 342–348.

Wells, J. (2001). *The construction industry in the twenty-first century: Its image, employment prospects and skill requirements*. ILO.

Chapter 3
Addressing Minority Discrimination in a Master's Education Program for Construction Management

Martine Buser and Dimosthenis Kifokeris

Abstract Increasing minorities' participation in high-tier and managerial positions in the construction sector can compensate for the shortage of skilled workers faced by the industry even in Sweden. However, relevant initiatives seem to not have achieved substantial results yet. This is also evident in construction management education, which then creates implications for the industry. In this paper, we attest to shortcomings in tackling the aforementioned issues, as well as present possible solutions. Theoretically, we adopt diversity management and critical diversity theory and then conduct a literature review followed by an empirical focus on a master's education program for construction management in a Swedish university. Our findings show that while methods and policies may exist, they are generally implemented inefficiently. Even more alarmingly, there can be a "diversity washing" through relevant low-budgeted programs, which may serve more as an extraction for underperforming managers rather than serious initiatives. As such, university-proposed solutions may fail, as the organizational structure does not support them, and the responsibility of implementation lies primarily with the teachers. We therefore propose broader initiatives with a strong reflection in praxis – such as following up on students' behavior in the classroom and examining not only the way foreigners can be integrated but also the way the majority is blindly maintaining and reproducing its privilege. Those could allow construction management education to contribute toward a diverse and equitable development of the Swedish construction sector. In that vein, this paper aims to contribute to SDGs 4, 5, 8, and 10.

Keywords Minorities · Diversity · Construction industry · Construction management education · Sweden

M. Buser (✉) · D. Kifokeris
Department of Architecture and Civil Engineering, Chalmers University of Technology, Gothenburg, Sweden
e-mail: buser@chalmers.se; dimkif@chalmers.se

G. Lindahl, S. C. Gottlieb (eds.), *SDGs in Construction Economics and Organization*, Springer Proceedings in Business and Economics,
https://doi.org/10.1007/978-3-031-25498-7_3

3.1 Introduction

Clarke and Gribling (2008) had made the prediction that in many industrialized countries, the construction industry would face a serious shortage of skilled workers (esp. managers) – a prediction that, evidently, still holds true (Haakestad & Friberg, 2020; Jurisic et al., 2021). Increasing minorities' participation – who have so far been underrepresented in the sector's managerial positions – can compensate for this (Choi et al., 2022). Diversity management within companies is addressing this by proposing a systematic plan and commitment to recruiting and retaining employees with diverse backgrounds and abilities (Köllen, 2021).

However, despite numerous relevant initiatives in the Swedish construction industry (e.g., for increasing women's representation in management), the rate of minorities' participation (particularly in the trades and on-site construction management) has not grown significantly over time; women constitute 9% of the total workforce, although only 2% of the workforce involved in craft trades, and 5.25% among site managers. The highest share of women employed in the Swedish construction industry is found in administrative tasks (Byggföretagen, 2020; SCB, 2018). Similarly, whereas foreigners are well represented among blue-collar workers (Thörnqvist & Bernhardsson, 2015), their participation as managers has been identified to be limited (Byrne et al., 2005). Moreover, when these minorities are reaching managerial positions, they may still face stigmatization and discrimination (Conway et al., 2018). So, we once again take a closer look at minorities' discrimination in the construction industry, by specifically examining the ways a university is trying to deal with these issues and prepare students to work as managers in the male-dominated and integrally homogenous Swedish construction sector. Focusing on a master's (MSc) education program in construction management, this paper's purpose is to reflect on the relevant issues, as well as to present possible solutions addressing such issues toward the diverse and equitable development of the Swedish construction sector. As such, this study also aims to contribute to UN Sustainable Development Goals 4 (Quality Education), 5 (Gender Equality), 8 (Decent Work and Economic Growth), and 10 (Reduced Inequalities) (Immler & Sakkers, 2022). It should be noted that due to the paper's delimited scope, we simplify the minority categorization and conventionally use a binary understanding of gender; however, we are aware that such traditional notions are increasingly being challenged in scholarship and social representation (Kosciesza, 2022).

This introduction is followed by the research method, a literature review on diversity management, critical diversity theory, and minority discrimination in construction-related industry and higher education (also in Sweden), the empirical focus on the aforementioned MSc education program, a discussion offering critical insights, and the conclusions.

3.2 Research Method

3.2.1 Literature Review

The literature search was conducted using the concept-centric framework of systematic review augmented by units of analysis (Webster & Watson, 2002). As such, the review was gauged to be completed when no new concepts relevant to the search terms could be found (Webster & Watson, 2002). The main concepts were "minority discrimination in construction," "higher education," "construction management education," "Sweden," and "diversity." The units of analysis emerged as the review was conducted in several iterations – thus facilitating its revision. These units of analysis included, indicatively, "gender" and "intercultural group work." By applying exclusion criteria (Dundar & Fleeman, 2017) in each of the review iterations, the references were eventually reduced to the ones featured in this study.

3.2.2 Empirical Part

For the empirical part of this study, a multi-paradigmatic approach (Lewis & Grimes, 1999) was used, combining elements of autoethnography, action research, and content analysis. Autoethnography is utilized to gain insights and qualitative results through self-observation and self-reflection – also implemented in the construction academia and industry (Grosse, 2019). It is hereby deployed to draw on the authors' own experiences (teaching in the MSc program for, respectively, more than 9 and 3 years). Moreover, the authors utilize action research in the context of higher education (Gibbs et al., 2017) to reflect the "fiduciary responsibilities" (Pecorino et al., 2008) they have toward the students, namely, acting with the intent of benefitting the students' learning, not harming them, and ultimately leading to knowledge coproduction with them (Gibbs et al., 2017). By engaging in autoethnography and action research, we are both subjects and actors in the research process itself, apart from being just observers (Alfaro-Tanco et al., 2021). This can be considered a limitation, as it introduced our own biases in the analysis, but can also be understood as a conditional possibility, because we were able to experience the effects of the researched phenomena ourselves. Finally, content analysis (Donald, 2022) was deployed for reviewing 120 individual assignments of first-year MSc students that self-assessed their performance and collaboration in group work. The students' citations mobilized for this paper are excerpts from these assessments.

3.2.3 *Synthesis*

The synthesis of the insights gained from the literature review and the empirical results follows the abductive reasoning of qualitative research – where observations and explanations of phenomena are developed by working iteratively between theory and data (Bell et al., 2019). Through abduction, critical reflections can be developed gradually (Bell et al., 2019).

3.3 Literature Review

There has been increased pressure for intensifying the participation of minorities and marginalized groups (esp. women) in companies' management – including boards of directors (Choi et al., 2022). Research has shown that two main reasons justify this intensification. First, women are assumed to bring different values and attitudes in the workplace and therefore should improve company performance and profitability (Nielsen & Huse, 2010). Second, there is a shortage of qualified candidates, as leadership becomes more important in the face of globalization, fierce competition, and shorter lifecycles of building concepts – creating pressure on company competences and recruitment processes (Norberg & Johansson, 2021). Diversity management can be mobilized for solving this issue (Köllen, 2021).

Diversity management practices in the workplace are developed, formalized, and implemented by organizations (Yang & Konrad, 2011). This is mainly dealt with by Human Resources (HR) management and covers recruitment, reward, performance, appraisal, employee development, and individual competences in delivering competitive advantage through leadership and teamwork (Yang & Konrad, 2011; Köllen, 2021) – also for higher education professionals (Rani & Kumar, 2021). By having a planned strategy, the minority integration should be facilitated (Köllen, 2021). But to be actually efficient, diversity management "should allow employees to bring their entire set of identities to work rather than requiring employees to suppress important identities in order to assimilate to the dominant organizational culture and use the entire sum of their demographic and cultural knowledge to bear on organizational problems" (Yang & Konrad, 2011). Diversity management usually builds on traditional definitions of diversity (e.g., regarding race, ethnicity, or gender) (Howarth & Andreouli, 2016). The authors are aware that individuals differ in numerous ways, and factors such as migration history, class, material base, sexual orientation, disability, culture, and religion should also be considered. However, university projects and initiatives are still mostly dealing with traditional definitions; as such, the authors revert to those for the purposes of the current study.

As a response to a too-rosy picture of diversity management as an economic solution – downplaying the real minority issues in the workplace – authors have also gathered under the banner of critical diversity theory (Zanoni et al., 2010). Their main critique is that using a strong business rhetoric focusing on individual contribution, diversity management literature fails to address issues related to social inequalities, discrimination, and exclusion, while downplaying power issues and resistance (Zanoni et al., 2010).

In addition to the aforesaid, minorities are defined in many studies against a white, well-shaped, heteronormative, and ambitious male (which itself may not fit even with the general male population), and equality initiatives often imply that minorities should be treated akin to this "ideal" man (Barnard et al., 2010). This understanding assumes that this "ideal" model is the one to follow, and minorities should be assimilated into the dominant group – regardless of differences between (and within) minority and majority groups. However, ignoring those differences might create a false dichotomy and hamper a contextualized definition of a minority or marginalized group. For example, it is naïve to treat women and men as homogeneous groups; it is likely that some women actually enjoy working in a male-dominated sector such as construction and equally likely that some men in the industry find this patriarchal culture problematic (Özbilgin & Tatli, 2011). Regardless, gender segregation is very well documented (also in construction), and many studies are listing the challenges and dominant male culture of the respective industries (also in Sweden, see Styhre, 2011; Johansson et al., 2021). Bridges et al. (2020) identify a relationship between male gender identity, the work of skilled trades, and the body, insisting that the traditional binary conception deems the feminine body as inferior – especially in construction, where work is often associated with physical performance. Existing informal recruitment and hiring practices also reproduce existing inequalities (Bridges et al., 2020), even extending to managerial positions (Choi et al., 2022).

The participation of ethnic minorities in construction management is, however, less documented than gender segregation. Besides, most of the few studies addressing this issue are focusing on the managerial challenges linked to culturally diverse managers on the building site and not on the way the respective minorities may or may not access such managerial positions (Gale & Davidson, 2006; Dainty et al., 2007; Loosemore et al., 2012). However, according to Clarke and Gribling (2008), the adherence to traditional practices such as old-style apprenticeships, craft-based skill structures, an itinerant workforce, and intensive deployment of labor can explain the lack of managerial diversity in the construction industry.

In Sweden, researchers have documented ethnic discrimination, where immigrants face higher rates of residential segregation, unemployment, and criminality. For example, Andersson et al. (2010) have demonstrated that the three successive waves of state anti-segregation policies have failed to deliver the expected results. In education, the Swedish system produces a considerable gap between nonimmigrant and immigrant students' achievement results and completion rates, which is above average when compared to other Western nations (Lundahl & Lindblad, 2018). As it is difficult for minority students to improve their personal position, they may engage in various protective mechanisms – e.g., social and academic disengagement – in response to negative stereotypes and previous experiences with discrimination (Verkuyten et al., 2019). This can even mean that when arriving at the university, minority students may be already conditioned to a long experience of discrimination (Buser & Koch, 2014).

To moderate minority discrimination in the classroom, Verkuyten et al. (2019) have proposed a mixed classroom composition, close student-teacher relation, and a multiculturality-conscious education. Regarding interculturalism and multiculturalism, Jansen (2004), De Vita (2005), Downey et al. (2006), Spencer-Oatey and Dauber (2017), and Poort et al. (2018) have all shown that variability-conscious, intercultural group work in project-based courses within higher education can help diminish minority discrimination and help construct a flexible professional identity that can function in multiple different contexts. Going beyond group work in the classroom, Leask (2009) has noted that a rewarding and motivating integration and interaction among majority and minority students on the curriculum and campus levels is key – something seemingly aligned with high-level Swedish national policies for, overlappingly, internationalization and diminishing of racial discrimination (Swedish Government Inquiries, 2018). On gender, Leicht-Scholten et al. (2009) have argued for institutionalization of integration measures in research and teaching (especially in STEM), while Lindberg et al. (2011) showed that among both teachers and students in Swedish higher education, the distribution of high-tier career paths between the genders is imbalanced – despite the almost equal percentage of women obtaining the relevant degrees. Mellén and Angervall (2021) and Peterson and Jordansson (2021) have been critical in that policy, enrollment, recruitment, and organizational measures tackling gender discrimination in Swedish academia have been relatively "shallow," as they follow a neoliberal market logic and favor an integrationist rather than a transformative translation of gender mainstreaming. Regarding disability, studies focusing either on neurodiversity (Knott & Taylor, 2014; Casement et al., 2017) or physical disabilities (Corrêa et al., 2021) have shown that while higher education staff is reasonably well informed about existing cases, there is a wide discrepancy on the measures and assistive technology implemented to support disabled colleagues or students. The advent of a hybrid higher education with a stronger (but not exclusive) digital element has the potential to improve accessibility; however, ableist dynamics can still remain pervasive (Fernandez, 2021). Finally – and calling back to Buser and Koch (2014) – disabled students may have already been conditioned to be marginalized before arriving at the university, due to factors like lack of parental support and class and economic insecurity (Taneja-Johansson, 2021).

As such, the literature shows that the perpetual reproduction of minority discrimination in the construction sector (including construction management) can also stem from discriminations still existing in academic institutions and the classroom itself, where the future professionals are being shaped. Consequently, tackling such discrimination in higher education can help ameliorate the situation in the professional field. Moreover, the studied efforts on diversity management, critical diversity theory, and discrimination due to ethnic background, multiculturality, gender, and disability, while not exhausting all possible types of discrimination, do show that the responsibility of tackling those issues should not fall solely on teachers but rather emanate from the full program curricula and university policies at large.

3.4 Empirical Part

After years of nondiscrimination policy (i.e., directing no initiatives to support minority representation in the institution, as those had been claimed to reenforce and legitimize existing discriminations), the Swedish university in the current study has made a 180° turn and is expected to spend 300 MSEK (ca. 30 million €) in 2019–2029 to strengthen the representation of minorities in its faculties. The employees are targeted first, as the institution aims at transforming the academic culture, system, and procedures including recruitment. However, such a mobilization is also reaching education. As such, fair treatment of diversity (in particular gender equality) is to be integrated into curricula to improve quality and increase the students' relevant knowledge by adding specific learning goals related to the course topics. Lecturers are to be given different types of support to be able to review and potentially make changes to their courses, in terms of both content and design and in case they have a particular interest in doing so. The focus is on the various ways that traditional professional engineering roles and values are perpetuated, create artifacts, and affect society. In that vein, teachers are expected to assess their course content and education practices and incorporate a higher reflexivity toward the consequences of their subject matter for equality and inclusion (Grzelec, 2021).

The appreciation and efficiency of these measures can primarily be regarded upon the traditional division of gender – and even then, there are shortcomings, as instead of making the education attractive beyond heteronormative male students, relevant projects mostly attempt a quite shallow and not integral "queerification" of engineering education. Statistics show that in the studied MSc program, the representation of female students is not an issue, as they represent on average 43% of the classroom population (see Table 3.1), but there is still a gender division in group work and tasks. Moreover, female students have been reported to express a lack of confidence in their competences, doubts, and insecurity, much more than male students.

However, going beyond gender is even more difficult, as the students' ethnicity, background, or religion cannot be identified according to the Discrimination Act (2008, amended in 2014). As such, we are facing a problem that we cannot properly describe and measure, and the results of the proposed solutions are, accordingly, hard to assess. Nonetheless, to aid such results and assessment, a toolbox aiming at

Table 3.1 Statistics of acceptance of female students in the MSc program

	2017	2018	2019	2020	2021
Women/total number of students	27/65	20/65	39/75	35/74	33/75
Women in %	41%	30%	52%	47%	44%

educational settings is proposed to the teachers. In the department accommodating the MSc program in construction management, this toolbox includes classroom initiatives for guided integration through work in mixed groups, introduction to cultural relativity, diversity in the education team, and lectures and exercises in conflict negotiation. Alarmingly, those do not always bear fruit. There has been documentation of mixed group integration being perceived as "forced" – with students preferring to revert to a Swedish/non-Swedish division, and given the chance, choosing a homogenous grouping. There is an apparent mistrust toward "foreigners" (a "we–them" mentality), and emerging conflicts are taken at the level of nationality instead of differences between individuals. "It is difficult to work with the foreigners" is often quoted in the students' assessment of groups that have met difficulty. On the other hand, international students may complain about local students' lack of motivation and commitment, as "they don't care about the grades." Some local students do express a clear hostility to the presence of international ones as written by a Swedish student: "The foreigners should not think that they will have a free ride in our education and steal our jobs afterwards," in a statement that can be characterized as racist. The attribution of identity as a minority is not only exercised by the dominant groups (Swedish and/or male) but is also mobilized by local students to account for problematic situations. Students who face collaboration issues may attribute the situation to the minority they feel they belong. For example, in the case of a five-person group consisting of a subgroup of two Swedish female students and a subgroup of three second-generation male immigrants, the teachers were contacted separately by each subgroup for complaints against the other, correspondingly, for being victims of sexism or racist discrimination. The insights also underline the experiences of second-generation immigrants with displayed language and cultural affinities who, although schooled in Sweden before entering the university, often prefer to work together, rather than with Swedish students having no apparent immigration background.

Such perceptions, related to either preconceived conditioning and/or lived experiences, are held not only by Swedish-educated students but also by newcomers belonging to national, ethnic, and racial minorities (as those are realized in the context of current MSc program). However, whereas the local students have the possibility to verbalize and act upon what they perceive as segregation, the international students may not share the same opportunity.

A foreign student arriving in Sweden at the start of the program's first course stated: "I become more compromised when working with new people, esp. when they have unfamiliar cultural backgrounds to me. For several times, I spent time figuring out some suggestions on modifying project process and results, but eventually I gave them up, because I do not want to be hated for giving too much pressure on my team members. But when I worked with my country classmates before, I usually stick to my ideas unless there are more convincing ways to do." Another international student arriving at the start of an 8-week course has shared: "Every meeting was an exciting time and a new attempt to communicate and learn more Swedish language for communication. […] The most exciting aspect of our meetings for me was always practicing speaking Swedish … then I was able to learn critical thinking, how to objectively question myself, and ask questions were confused amidst my colleagues, esp. those moments when they unconsciously switch to talk in Swedish on a fast mode and I am not able to follow the conversation. [...] Our team meetings were practically the only times I had during the week to bond with people, so I think I made judicious use of that time. [...] I had a view to doing more, adding more elements to the slides also, but it was not generally agreed. Thence, I obliged to my colleagues' preference and built on it and it all eventually worked well." Language is a discrimination factor for international students who feel excluded whenever the interactions are taking place in Swedish.

It should be noted that there are shades in those perceptions, and the students are often conscious about them – as described in another personal note by a student of an explicitly stated mixed Swedish and French background: "The first day we met, I detected directly distinctive ways of communications to each other. One member of our group couldn't speak Swedish at all and had some difficulties to talk in English fluently. The other two members seemed to know each other from the past which created some kind of close bound between them. This made an invisible communication wall between them, and it didn't help when they, sometimes, only talked in Swedish. Personally, being half French, half Swedish, I'm very used to these kinds of gaps of communication. What I began to do is to always respond in English when they tried to talk to me in Swedish. Therefore, without having to force the two Swedish members, I gave them a social pressure to continue the conversation in English, which I think the third non-Swedish-speaking member enjoyed. It usually worked. The two Swedish members were very used to work in one way, by taking the assignment task more as a direction giver for the project, whereas the non-Swedish was used to follow the exact words by the letter of the task."

Situations where more than one minority attributes are in place (e.g., gender and ethnicity) can be even more complicated. In such a case, an uneven number of groups had to pair to review the results of their project work, and a group of two foreign female students was left aside – even after the teacher openly asked bigger groups to divide and work with one or both of them. Eventually, one of the two students said loudly: "If they feel uneasy about working with us, they should not be forced to do so." Eventually, one of the mixed groups invited them to join.

Such situations and established perceptions among the students seem to also have their counterparts in the teachers. As shown in Table 3.2, more than half of the courses are homogenously taught by Swedish teachers, while mixed teaching teams in the rest of the courses include almost exclusively Europeans. The students, both local and international, recognize a difference in treatment according to the use of language. Swedish teachers tend to build their course on Swedish case studies

Table 3.2 Gender and ethnic distribution of teachers in the MSc program

MSc program	Swedish only	Mixed (European)	Male only	Mostly male (>75%)	Gender-mixed	Mostly female (>75%)
9 courses	5	4	2	4	1	2

referring to Swedish texts, slides, and vocabulary, which favors local students and demotivates the international ones – especially when they are told that they "can find the translation by using their mobile phones!" Moreover, the courses' teaching staff still consists of primarily men. This solidified teacher representation of mainly Swedish male colleagues makes it hard to propose minority-aware role models and convincing examples of a diverse gender or cultural background. This can create a self-feeding cycle, especially between Swedish students and Swedish teachers – regardless of honest tries by some teachers to tackle this. However, it has to be said that during course evaluations, the students have not been able to find examples of gender discrimination attributed to the teaching staff.

3.5 Discussion

The insights gained by the literature and the empirical investigation show that integrating awareness and preparing soon-to-be professionals in construction management for adequately responding to gender – and other minority-based challenges need to be first tackled during their education – in the level of the classroom, syllabi, curricula, and even university policies.

The local students seem to be able to identify and act upon discriminating behaviors, at least when they feel they are victims of such (micro)aggressions. However, at least part of them does not mobilize this knowledge to analyze and solve discrimination issues when they are part of such interactions with international students. The local students are embedded in the institution culture and routines, which gives them a clear advantage when interacting with the teaching staff or planning and executing project work. But this is not the case of the new international students who do not feel legitimized to act upon such treatments.

So, the efficiency of the actually proposed and implemented methods is lacking. Even if learning goals on ethics and discrimination are added to the curriculum, there is no follow-up, reflection, or assessment of these goals in term of the students' classroom behavior and practices. Whereas awareness is being created around minority discrimination (e.g., with teachers trained to use gender-neutral vocabulary), microaggressions are still part of the daily life for the international students. Such microaggressions can be defined as minor and delicate instances of marginalization, conveying negative messages toward minorities and gradually building up a negative attitude (Ogunyemi et al., 2020). Underlining a foreigner's "good" language mastering or commenting on one's achievements by referring to their country of origin is microaggressively perpetuating stereotypes, even if expressed as compliments; such perpetuations are worsened when ignoring or dismissing an idea, question, or student's presence (Hinton Jr. et al., 2020). Even more alarming, there can be instances of "diversity washing" through relevant low-budgeted programs, which may serve more as an extraction for underperforming managers rather than serious initiatives. In this context, university-proposed solutions are bound to fail, as the organizational structure does not seriously support them.

Lastly, other minorities (e.g., LGBTQI+ people) should also be benefitted from nondiscrimination policies, practices, and measures. Tackling such issues has still a long way to go even in the most "aware" higher education institutions, and continuous work is needed.

3.6 Conclusions

In engineering (and specifically construction management) higher education and its reflection into the student's future professional career, the notions of minority integration, and developing consciousness on diversity presupposed a movement from outside the minorities themselves – with the minority members having to move into the majority groups. This approach focuses on the problem met by the minorities and renders the behavior of the majorities invisible. This can embed another type of segregation in the classroom (even for, e.g., second-generation immigrants), where disengagement appears to be a coping strategy.

The lacking efficiency of measures against minority discrimination in education is evident, while awareness is being created, discrimination from students and microaggressions from teachers are still part of the international students' daily life. Not following up on the development of the students' awareness and surface-level "diversity washing" through relevant low-budgeted programs may trivialize the importance of these topics. Pushing the responsibility of implementing some of the relevant tools on the levels of the university classroom and the teachers is evidently a shortcoming – even if such a responsibility is claimed to be on an institutional level, with policies implemented in syllabi and curricula. Therefore, broader initiatives with a strong reflection in praxis should be implemented, such as following up on students' classroom behavior or examining not only the way foreigners can be integrated but also the way the majority is blindly maintaining and reproducing its privilege.

References

Alfaro-Tanco, J. A., Avella, L., Moscoso, P., & Näslund, D. (2021). An evaluation framework for the dual contribution of action research: Opportunities and challenges in the field of operations management. *International Journal of Qualitative Methods, 20*, 1–16.

Andersson, R., Bråmå, Å., & Holmqvist, E. (2010). Counteracting segregation: Swedish policies and experiences. *Housing Studies, 25*(2), 237–256.

Barnard, S., Powell, A., Baginhole, B., & Dainty, A. (2010). Researching UK women professionals in SET: A critical review of current approaches. *International Journal of Gender, Science and Technology, 2*(3), 361–381.

Bell, E., Bryman, A., & Harley, B. (2019). *Business research methods* (5th ed.). Oxford University Press.

Bridges, D., Wulff, E., Bamberry, L., Krivokapic-Skoko, B., & Jenkins, S. (2020). Negotiating gender in the male-dominated skilled trades: A systematic literature review. *Construction Management and Economics, 38*(10), 894–916.

Buser, M., & Koch, C. (2014). Tales of the suburbs? The social sustainability agenda in Sweden through literary accounts. *Sustainability, 6*(2), 913–934.

Byggföretagen. (20 december 2020). Förvärvsarbetande i byggindustrin. https://byggforetagen.se/statistik/forvarvsarbetande-i-byggindustrin/

Byrne, J., Clarke, L., & Meer, M. V. D. (2005). Gender and ethnic minority exclusion from skilled occupations in construction: A Western European comparison. *Construction Management and Economics, 23*(10), 1025–1034.

Casement, S., de los Pinos, C. C., & Forrester-Jones, R. (2017). Experiences of university life for students with Asperger's Syndrome: A comparative study between Spain and England. *International Journal of Inclusive Education, 21*(1), 73–89.

Choi, J. O., Shane, J. S., & Chih, Y. Y. (2022). Diversity and inclusion in the engineering-construction industry. *Journal of Management in Engineering, 38*(2), 02021002.

Clarke, L., & Gribling, M. (2008). Obstacles to diversity in construction: The example of Heathrow Terminal 5. *Construction Management and Economics, 26*(10), 65.

Conway, Z., Wade, F., & Smith, S. D. (2018). A participant observation study of gender dynamics on construction sites. In C. Gorse & C. J. Neilson (Eds.), *Proceedings of the 34th annual ARCOM conference* (pp. 291–300). Belfast: Association of Researchers in Construction Management.

Corrêa, A. Z. A. H., Masuchi, M. H., Baeta, N. C. D. C. C., Takiuchi, L., & Bianco, B. (2021). Disability inclusion in higher education: Knowledge and perceptions of the academic community. *Disability and rehabilitation: Assistive Technology, 16*(7), 735–740.

Dainty, A., Moore, D., & Murray, M. (2007). *Communication in construction: Theory and practice*. Routledge.

De Vita, G. (2005). Fostering intercultural learning through multicultural group work. In J. Carroll & J. Ryan (Eds.), *Teaching international students : Improving learning for all* (pp. 75–83). Routledge.

Donald, W. E. (2022). Content analysis of metadata, titles, and abstracts (CAMTA): Application of the method to business and management research. *Management Research Review, 45*(1), 47–64.

Downey, G. L., Lucena, J. L., Moskal, B. M., Parkhurst, R., Bigley, T., Hays, C., Jesiek, B. K., Kelly, L., Miller, J., Ruff, S., Lehr, J. L., & Nichols-Belo, A. (2006). The globally competent engineer: Working effectively with people who define problems differently. *Journal of Engineering Education, 95*(2), 107–122.

Dundar, Y., & Fleeman, N. (2017). Applying inclusion and exclusion criteria. In A. Boland, G. Cherry, & R. Dickson (Eds.), *Doing a systematic review: A student's guide* (2nd ed.). SAGE.

Fernandez, S. (2021). Making space in higher education: Disability, digital technology, and the inclusive prospect of digital collaborative making. *International Journal of Inclusive Education, 25*(12), 1375–1390.

Gale, A. W., & Davidson, M. J. (Eds.). (2006). *Managing diversity and equality in construction: Initiatives and practice*. Routledge.

Gibbs, P., Cartney, P., Wilkinson, K., Parkinson, J., Cunningham, S., James-Reynolds, C., Zoubir, T., Brown, V., Barter, P., Sumner, P., MacDonald, A., Dayananda, A., & Pitt, A. (2017). Literature review on the use of action research in higher education. *Educational Action Research, 25*(1), 3–22.

Grosse, H. (2019). An insider's point of view: Autoethnography in the construction industry. *Construction Management and Economics, 37*(9), 481–498.

Grzelec, A. (2021). Comprehensive and practical method for integration of Equality, Diversity and Inclusion (EDI) into engineering education. In *Proc. 8:e Utvecklingskonferensen för Sveriges ingenjörsutbildningar*. Karlstadt: Karlstadt Universitet.

Haakestad, H., & Friberg, J. H. (2020). Deskilling revisited: Labour migration, neo-Taylorism and the degradation of craft work in the Norwegian construction industry. *Economic and Industrial Democracy, 41*(3), 630–651.

Hinton, A. O., Jr., Vue, Z., Termini, C. M., Taylor, B. L., Shuler, H. D., & McReynolds, M. R. (2020). Mentoring minority trainees: Minorities in academia face specific challenges that mentors should address to instill confidence. *EMBO Reports, 21*(10), e51269.

Howarth, C., & Andreouli, E. (2016). "Nobody wants to be an outsider": From diversity management to diversity engagement. *Political Psychology, 37*(3), 327–340.

Immler, N. L., & Sakkers, H. (2022). The UN-sustainable development goals going local: Learning from localising human rights. *The International Journal of Human Rights, 26*(2), 262–284.

Jansen, D. E. (2004). Developing the intercultural competence of engineering students: A proposal for the method and contents of a seminar. *World Transactions on Engineering and Technology Education, 3*(1), 24–28.

Johansson, H., Stensönes, W., & Sundquist, V. (2021). *En jämställd byggprocess för en attraktiv byggbransch*. SBUF.

Jurisic, B. B., Galic, M., & Marenjak, S. (2021). Review of the construction labour demand and shortages in the EU. *Buildings, 11*(1), 17.

Knott, F., & Taylor, A. (2014). Life at university with Asperger syndrome: A comparison of student and staff perspectives. *International Journal of Inclusive Education, 18*(4), 411–426.

Köllen, T. (2021). Diversity management: A critical review and agenda for the future. *Journal of Management Inquiry, 30*(3), 259–272.

Kosciesza, A. J. (2022). Intersectional gender measurement: Proposing a new metric for gender identity and gender experience. *Feminist Media Studies*. https://doi.org/10.1080/14680777.2021.2018008

Leask, B. (2009). Using formal and informal curricula to improve interactions between home and international students. *Journal of Studies in International Education, 13*(2), 205–221.

Leicht-Scholten, C., Weheliye, A. J., & Wolffram, A. (2009). Institutionalisation of gender and diversity management in engineering education. *European Journal of Engineering Education, 34*(5), 447–454.

Lewis, M. W., & Grimes, A. J. (1999). Metatriangulation: Building theory from multiple paradigms. *Academy of Management Review, 24*(4), 672–690.

Lindberg, L., Riis, U., & Silaner, C. (2011). Gender equality in Swedish higher education: Patterns and shifts. *Scandinavian Journal of Educational Research, 55*(2), 165–179.

Loosemore, M., Raftery, J., Reilly, C., & Higgon, D. (2012). *Risk management in projects*. Routledge.

Lundahl, L., & Lindblad, M. (2018). Immigrant student achievement and education policy in Sweden. In *Immigrant student achievement and education policy* (pp. 69–85). Springer.

Mellén, J., & Angervall, P. (2021). Gender and choice: Differentiating options in Swedish upper secondary STEM programmes. *Journal of Education Policy, 36*(3), 417–435.

Nielsen, S., & Huse, M. (2010). Women directors' contribution to board decision-making and strategic involvement: The role of equality perception. *European Management Review, 7*, 16–29.

Norberg, C., & Johansson, M. (2021). "Women and "ideal" women": The representation of women in the construction industry. *Gender Studies, 38*, 1–24.

Ogunyemi, D., Clare, C., Astudillo, Y. M., Marseille, M., Manu, E., & Kim, S. (2020). Microaggressions in the learning environment: A systematic review. *Journal of Diversity in Higher Education, 13*(2), 97–119.

Özbilgin, M., & Tatli, A. (2011). Mapping out the field of equality and diversity: Rise of individualism and voluntarism. *Human Relations, 64*(9), 1229–1253.

På tal om kvinnor och män, SCB. (2018). https://www.scb.se/hitta-statistik/statistik-efter-amne/levnadsforhallanden/jamstalldhet/jamstalldhetsstatistik/pong/publikationer/pa-tal-om-kvinnor-och-man.-lathund-om-jamstalldhet-2018/

Pecorino, P. A., Kincaid, S., & Gironda, B. (2008). Research and experimentation in teaching effectiveness: The ethical review process and the IRB. *International Journal for the Scholarship of Teaching and Learning, 2*(1), 1–11.

Peterson, H., & Jordansson, B. (2021). Gender mainstreaming in Swedish academia: Translating policy into practice. *Journal of Gender Studies, 31*(1), 87–100.

Poort, I., Jansen, E., & Hoffman, A. (2018). Intercultural group work in higher education: Costs and benefits from an expectancy-value theory perspective. *International Journal of Education Research, 93*, 218–231.

Rani, N., & Kumar, A. (2021). Impact of diversity management practices on performance of teachers in higher education sector. *Journal of Commerce and Trade, 16*(1), 36–41.

Spencer-Oatey, H., & Dauber, D. (2017). The gains and pains of mixed national group work at university. *Journal of Multilingual and Multicultural Development, 38*(3), 219–236.

Styhre, A. (2011). The overworked site manager: Gendered ideologies in the construction industry. *Construction Management and Economics, 29*(9), 943–955.

Swedish Government Inquiries. (2018). *Internationalisation of Swedish higher education institutions. SOU 2018:78.* Swedish Government Official Reports.

Taneja-Johansson, S. (2021). Facilitators and barriers along pathways to higher education in Sweden: A disability lens. *International Journal of Inclusive Education*, 1–15. https://doi.org/10.1080/13603116.2021.1941320

Thörnqvist, C., & Bernhardsson, S. (2015). Their own stories – How Polish construction workers posted to Sweden experience their job situation, or resistance versus life projects. *Transfer, 21*(1), 23–36.

Verkuyten, M., Thijs, J., & Gharaei, N. (2019). Discrimination and academic (dis) engagement of ethnic-racial minority students: A social identity threat perspective. *Social Psychology of Education, 22*(2), 267–290.

Webster, J., & Watson, R. T. (2002). Analyzing the past to prepare for the future: Writing a literature review. *MIS Quarterly, 26*(2), xiii–xxiii.

Yang, Y., & Konrad, A. M. (2011). Understanding diversity management practices: Implications of institutional theory and resource-based theory. *Group & Organization Management, 36*(1), 6–38.

Zanoni, P., Janssens, M., Benschop, Y., & Nkomo, S. (2010). Unpacking diversity, grasping inequality: Rethinking difference through critical perspectives. *Organization, 17*(1), 9–29.

Chapter 4
Pursuing Organisational Change in Construction Education by Implementing Change Projects

Nicolaj Frederiksen, Henrik Kryger Madsen, Hanne Tønder Ludvigsen, Mikkel Bruun Primdahl, and Marianne Forman

Abstract The university colleges in Denmark are progressively confronted with new educational policies, where adaptation presupposes that the individual educational department facilitates organisational change. In this study, we describe and analyse how the Department of Construction at University College Lillebaelt (UCL) has strived to adapt to new educational policies by implementing change projects. Specifically, we examine the implementation of the two change projects entitled 'Bridge building' and 'Capacity development' and explain how aspects of organisational life (i.e. structure, identity and routines) develop and alter as a result of the change processes. The study empirically draws on interviews and workshops conducted between 2018 and 2021. In the analysis, we show that the department, in the Bridge building project, implements changes in the study programme on Architectural Technology and Construction Management to adapt to educational policies demanding a strengthened quality of education. In addition, we show that the department, in the Capacity development project, implements changes internally in the department to adapt to educational policies demanding an increased involvement in research and development activities. In the discussion and conclusions, we explain that the top management of UCL redirects the practical efforts required to adapt to new educational policies into the educational departments. We also demonstrate that the department is experiencing a pressure to implement change projects in order to adapt to new educational policies and thereby ensure quality education.

N. Frederiksen (✉) · M. B. Primdahl · M. Forman
Department of the Built Environment, Aalborg University, Copenhagen, Denmark
e-mail: nif@build.aau.dk; mbpr@build.aau.dk; mfor@build.aau.dk

H. K. Madsen · H. T. Ludvigsen
Department of Construction, University College Lillebaelt, Odense, Denmark
e-mail: hkma@ucl.dk; htl@ucl.dk

G. Lindahl, S. C. Gottlieb (eds.), *SDGs in Construction Economics and Organization*, Springer Proceedings in Business and Economics,
https://doi.org/10.1007/978-3-031-25498-7_4

Keywords Change project · Construction education · Educational policy · Organisational change · University college

4.1 Introduction

Development in the field of education has been prompted as an important source of fostering new skills and entrepreneurship that can address societal challenges (George et al., 2016). An implication of this is that the field of education is progressively subjected to new educational policies demanding the individual educational institution to act by facilitating organisational change. Adaptation to new educational policies has also been highlighted as critical to meeting the fourth United Nations Sustainable Development Goal on "quality education" (cf. Boeren, 2019). In the study, we demonstrate how a university college department adapts to new educational policies by implementing change projects to ensure quality education.

Since 2013, the university colleges in Denmark have received funding from the Finance Act to conduct and advance their involvement in practical and application-oriented research and development activities (R&D). Later in 2014, the revised Act on University Colleges of Higher Education was enacted, which made it compulsory for university colleges to conduct R&D (Uddannelses- og Forskningsministeriet, 2014). The main purpose of the legal requirement is to strengthen the knowledge base of the study programmes offered by the university colleges and to ensure that the programmes contribute to production of practical as well as scientific knowledge (Danmarks Evalueringsinstitut, 2017).

A direct consequence of the legal requirement is that the responsibilities of the university colleges and the related job descriptions for the teaching staff have undergone significant changes. Prior to 2014, the university colleges in Denmark were mainly obliged to conduct teaching activities, and the staff affiliated with the educational departments primarily served as 'teachers' who provided lectures and supervision. Following the enactment of the revised Act on University Colleges of Higher Education in 2014, the staff were, in addition to the conventional teaching activities, also legally obliged to conduct R&D. The legal requirement on R&D thereby positioned the educational departments in a difficult situation with two strategic options for ensuring compliance with the legal requirement. The first involved a major staff turnover, where the existing teaching staff had to be replaced with new employees who possess competencies on R&D. The second involved a profound upskill of the teaching staff to cultivate competencies on R&D.

In the study, we describe and analyse how the Department of Construction at University College Lillebaelt (UCL), through a bottom-up approach, adapts to new educational policies by implementing change projects to upskill staff. The department chose the bottom-up approach in the belief that internal change is a responsibility of the individual educational department and furthermore provides an opportunity to build on existing conditions in the department. This is a rather different approach compared to other educational departments at UCL where internal changes primarily are pursued by making changes in the staff. The overall purpose of the change projects is to cultivate competencies on R&D internally in the department. A change project can be considered as a temporary organised effort of work aiming to change aspects of organisational life in accordance with a particular rationale (Alvesson & Sveningsson, 2008). Specifically, we examine the two change projects entitled 'Bridge building' and 'Capacity development'. In brief, Bridge building aims to improve the study programme offered by the department by strengthening the students' analytical competencies so that they are qualified to study for a master's degree. Analogously, Capacity development aims to (re)organise the department and foster academic environments where the staff can exchange experiences and jointly develop competencies on R&D that can strengthen the knowledge base in the teaching activities. By mobilising an analytical framework based on the literature on organisational change, we analyse empirical material from the two change projects to obtain an understanding of the internal changes pursued by the department.

4.2 Organisational Change: From Prescriptive Models to a Normal Condition

Change is a prerequisite for contemporary organisations to be able to develop and alter internal structures, identities and routines (Battilana et al., 2010; Weick & Quinn, 1999) thereby addressing and adapting to changing external conditions and pressures (Fernie et al., 2006; George et al., 2016; Gottlieb & Frederiksen, 2020). Because of this centrality, organisational change has been an enduring examined phenomenon in the fields of management and organisation research (Pettigrew et al., 2001; Suddaby & Foster, 2017). Studies on organisational change have traditionally distinguished between episodic changes and continuous changes (cf. Weick & Quinn, 1999). Episodic changes are those that occur when an organisation deliberately initiates activities to transform from its present state to a desired future state (Stouten et al., 2018). Analogously, continuous changes are those that are perpetual, developmental and cumulative (Pettigrew et al., 2001).

According to Haridimos Tsoukas and Robert Chia (2002), organisational change, whether episodic or continuous, is distinctly pervasive, indivisible and a rather normal condition of organisational life. Studies on organisational change should therefore treat processes of change as a normal condition for organisations instead of exceptional ventures accomplished through prescriptive models such as Michael Beer and colleagues' (1990) model to effective change or John Kotter's (1995) model for transforming an organisation. Specifically, studies should examine how aspects of organisational life such as structures, identities and routines become temporary instantiations of perpetual change processes (cf. Langley et al., 2013; Tsoukas & Chia, 2002). In this perspective, the analytical focus thereby shifts from understanding the change process itself (as per Beer et al., 1990; Kotter, 1995) to understand how aspects of organisational life are developed and altered as a result of change processes.

4.2.1 Studying the Impact of Change Processes on Organisational Life

A critical starting point when studying organisational change is to construct clarity regarding the notion of change (cf. Quattrone & Hopper, 2001; Suddaby & Foster, 2017). In order to grasp the impact of change processes on organisational life, studies should moreover apply a research design that allows examination across multiple contexts and levels of analysis (Alvesson & Sveningsson, 2008; Langley et al., 2013; Pettigrew et al., 2001). In this regard, the study sets out to explore how the change projects trigger structural, identity and routine changes in the examined department. These three analytical categories are core aspects of organisational life and are fundamental in understanding organisational change processes (Becker et al., 2005; Gioia et al., 2013; Hannan & Freeman, 1984). We therefore perceive the notion of change as incremental or radical developments in organisational structures, identities or routines that lead to more or less permanent changes in organisational life. Moreover, we join the group of researchers who perceive organisational structures, identities and routines as dynamic processes of organisational life (e.g. Howard-Greenville et al., 2016; Langley et al., 2013; Tsoukas & Chia, 2002) rather than stable entities of an organisation. Each of the analytical categories is briefly described in the following.

An organisational structure encompasses a series of contingencies and regularities characterising the organisation such as its authority, distribution of tasks, patterns of interaction, resource allocation, rules and strategy (Chia, 1997; Ranson et al., 1980). Changes in one or more of these contingencies and regularities usually affect the structure at large and thus exhibit a structural change (Král & Králová, 2016). Changes in the organisational structure often occur with ambitions to address and adapt to external conditions and pressures (Hannan & Freeman, 1984). An organisational identity refers to an organisation's self-perception. The identity of an organisation is said to be a result of its members' shared perceptions of features that are central to the organisation and differentiate it from other comparable organisations (Albert & Whetten, 1985). Change in an organisational identity can take place over relatively shorter durations of time and is assumed to be fundamental for organisational change (Gioia et al., 2013). An organisational routine can be understood as a regular and predictable behavioural pattern of an organisation (Nelson & Winter, 1982). Routines provide an organisation with capabilities to perform and accomplish diverse types of work (Howard-Greenville et al., 2016). A common assumption on routines is that organisations change their everyday activities when they make changes in their routines (cf. Becker et al., 2005).

4.3 Case Description and Research Design

UCL is a university college that offers more than 40 academy and professional bachelor's programmes within the three educational areas: (1) business science and technology, (2) education and social sciences and (3) health sciences. Each of the three areas has a number of research centres that are responsible for initiating and supporting R&D carried out by the staff affiliated with the educational departments. As of 2022, UCL has 11,300 students. Approximately 550 of these are enrolled in the professional bachelor's programme on Architectural Technology and Construction Management.

The Department of Construction, which is under scrutiny in this study, belongs to the educational area of business science and technology and corresponds with the research centre of trade. The department consists of 25 persons who manage the study programme on Architectural Technology and Construction Management and conduct R&D. According to the executive order in force, the purpose of the professional bachelor's programme on Architectural Technology and Construction Management is to qualify the students to be able to independently plan, lead and handle technical and administrative work in the design and execution of building and construction tasks.

Since 2014, when the university colleges became legally obliged to conduct R&D, the department has strived to strengthen its involvement in R&D in several ways. For instance, a significant number of the staff have undergone an upskill from adjuncts to lecturers. A so-called knowledge centre was also established in the department in an effort to create a central unit that could manage external projects and function as the formal link between the department and the industry. The knowledge centre, however, was closed by UCL with the argument that there should be a greater focus on cultivating research activities. Consequently, two PhD projects were commenced to strengthen the research activities in the department. Commencement of PhD projects is generally a widespread and internally recognised strategy for obtaining research competencies in the educational departments at UCL. Unfortunately, only one of the PhD projects was completed, and the candidate subsequently applied for a job outside UCL. Also, the department has unsuccessfully attempted to recruit persons holding a PhD degree or having relevant experience on R&D.

In 2018, the department revised its internal strategy and started searching for assistance outside UCL. A result of this quest was the establishment of a formal collaboration with Aalborg University. Accordingly, the department and Aalborg University have in recent years collaborated throughout a series of change projects with the aim to develop and support the department in its current development and involvement in R&D. Two of the change projects are Bridge building and Capacity development. These are the subject of analysis in this study.

4.3.1 Methods and Collection of Empirical Material

The research presented draws on a qualitative research paradigm based on principles from action research (McNiff & Whitehead, 2002) where the authors have both functioned as project facilitators and sequentially met to discuss and reflect on the experiences. This was to obtain insights into the actual change work performed (Alvesson & Sveningsson, 2008) and to advance practical as well as scientific knowledge of the change projects (Van de Ven, 2018). We collected empirical material in the examined change projects between 2018 and 2021 using interviews and workshops. The collection of empirical material is elaborated in the following.

4.3.1.1 Interviews

In summer 2021, we interviewed six informants from different organisational levels at UCL. Three of the informants were heads of departments, while the other three were lecturers in the Department of Construction. Each of the interviews was audio-recorded by agreement with the informants and selected passages of the audio-recordings were subsequently transcribed verbatim. The interview conversations had an average duration of 49 minutes. All interviews were open-ended and pragmatic by nature (cf. Lamont & Swidler, 2014) and covered the following five aspects: (1) UCL's current situation, (2) understandings and requirements on R&D, (3) the daily work with R&D, (4) relevant competencies when conducting R&D and (5) the relationship between R&D and teaching activities.

4.3.1.2 Workshops

A total of six workshops were held throughout the study period. Three of the workshops dealt with conceptualising the collaboration between the department and Aalborg University to meet the set aspirations. Two workshops involved group exercises with the staff in the department to get insight into their daily work and the perceived core competencies and values characterising the department. The last workshop was held with the staff in the department who have been involved in R&D to obtain an understanding of the experienced challenges and needs. Each of the workshops had a duration of 6–8 hours and notes were prepared accordingly.

4.3.1.3 Internal Meetings

Internal meetings have been held between the department and Aalborg University (i.e., the authors) about every 3 weeks. At the meetings, the ongoing collaboration and the progression of the change projects and future efforts were discussed. Each of the meetings had a duration of 1–2 hours.

4.4 Analysis

In this section, we describe and analyse the two change projects. We specifically focus on the efforts made in the change projects to facilitate structural, identity and routine changes in the department and the change work performed in this pursuit. Insights from the workshops form the empirical basis in the first part of the analysis on Bridge building, while the interview conversations are mainly used in the second part of the analysis on Capacity development.

4.4.1 Bridge Building

The Bridge building project was initiated in September 2018 and completed ultimo January 2019. The project originated from a report prepared by the Danish Ministry of Higher Education and Science in which it was proposed that all bachelor's programmes in Denmark incorporated a philosophy course. The stipulated aim of the philosophy course was to develop the individual student as a professional and to promote curiosity, critical thinking and judgment (Uddannelses- og Forskningsministeriet, 2018).

Although most of the professional bachelor's programmes at UCL at that time already touched upon scientific theories and methods, the Department of Construction found it necessary to integrate philosophy more profoundly in the study programme on Architectural Technology and Construction Management. In this effort, the department entered a collaboration with Aalborg University on the project entitled Bridge building. The purpose of the project was twofold. The first was to strengthen the students' analytical competencies so that they were prepared to study for a master's degree. The second was to reinforce the general education in the study programme on Architectural Technology and Construction Management (Forman & Gottlieb, 2019).

4.4.1.1 Implementation of Bridge Building

A starting point for the project was to compare and understand the differences between the 3.5-year professional bachelor's programme on Architectural Technology and Construction Management (offered by UCL) and the 2-year master's programme on Construction Management and Informatics (offered by Aalborg University). The master's programme has traditionally attracted many graduates with a background from the study programme on Architectural Technology and Construction Management. The comparison was made to obtain an understanding of the different didactically approaches, understandings and values characterising the two study programmes. The comparison was conducted by examining the curricula of the study programmes and through workshops.

As part of the comparison, it was found that the bachelor's programme didactically drew on principles where the students were introduced to and had to master the practical use of a wide range of tools and profession-oriented work functions. Analogously, the master's programme was based on principles from problem-based learning where the students explored practical and theoretical problems in the construction industry and mobilised scientific theories and methods to understand and potentially solve the identified problems. This difference has previously been highlighted to be a significant challenge for graduated bachelors who are admitted for the master's programme as they experience the curriculum and the lectures to be somewhat abstract (Forman & Gottlieb, 2019). An outcome of the comparison therefore was the ambition to integrate scientific theories and methods into the bachelor's programme while maintaining the practical focus on the use of tools and profession-oriented work functions. This is elaborated by a project participant:

> Among other things, we wanted to change the focus from solely understanding how to use a technology such as a drone to cultivate a more profound understanding of a drone as a technological device that produces data.

In practice, a series of initiatives were incorporated in the last two semesters of the bachelor's programme (i.e., the sixth and seventh semesters) to promote and strengthen the students' use of scientific theories and methods. In relation to the sixth semester, which involves an internship, the students were introduced to new learning targets, which presupposed the application of scientific theories and methods to reflect on their own practice and the experiences from the internship. These reflections were to be documented in an individual internship report that had to be prepared to pass the semester. Prior to the internship, the students were accordingly taught scientific theories and methodologies such as cultural analysis, discourse analysis and ethnographic and qualitative methodologies.

In relation to the seventh semester where the students prepare their bachelor projects, the students were encouraged to further develop the insights obtained through the internships (i.e., dilemmas, paradoxes and contradictions) to formulate bachelor projects that were relevant to practice as well as grounded in scientific theories and methods. To support this, a mandatory 10 ECTS introduction course to the bachelor project was added. The introduction course entailed lectures in academic writing, formulation of project description, methodological considerations as well as theoretical approaches to structure and analyse empirical material. The introduction course was handled by the department's staff who were expected to supervise the students on their bachelor projects.

The ambition to integrate scientific theories and methods more profoundly in the study programme derived several discussions in department regarding which core competencies graduates should acquire in the study programme. This is emphasised by a Lecturer:

> There have been several discussions in the department about which core competencies our graduates should acquire throughout the study. I think such discussions are important as they push us to debate and reflect on how we can improve the study programme and maintain our relevance.

4.4.1.2 Change Work Performed and the Effects of Bridge Building

In this first part of the analysis, we showed how the Department of Construction through the Bridge building project performed change work in order to anchor scientific theory and methods in the study programme. Structurally, the department made changes in the curriculum and introduced new learning targets to create a balance between a practice-oriented focus and a scientific focus. In relation to identity, the department developed shared perceptions and started debating what core competencies graduates should acquire in the study programme. As opposed to previously, these perceptions and debates concerned not only how to prepare students for a job in the industry but also about preparing the students to study for a master's degree. In line with the increased focus on scientific theories and methods in the study programme, the department's staff also acquired new routines accordingly. For instance, where the staff traditionally provided lectures about the practical use of tools and profession-oriented work functions, they now teach the application of scientific theories and methods and how to understand practical insights through a scientific approach.

4.4.2 *Capacity Development*

The capacity development project was initiated in June 2021 and is currently planned to run until ultimo 2022. The objective of the project has been to (re)organise the Department of Construction and upskill the staff with competencies so that they can better conduct R&D. R&D forms an increasing part of the activities in the department, which has simultaneously created a pressure to develop capacity that supports the staff's involvement in R&D. This is emphasised by a Head of Department:

> With the legal requirement on R&D and the new job descriptions, many of our current employees are not qualified to possess their own jobs, which means that they have to undergo a significant capacity development.

Specifically, the stated ambition of the department has been, through the Capacity development project, to strengthen its involvement in R&D by developing capacity within (1) research management, (2) formation of academic environments and (3) the teaching activities. These are elaborated in the following.

4.4.2.1 Preliminary Results from the Implementation of Capacity Development

Historically, the research centres and the educational departments at UCL have been essentially detached from one another. Specifically, employees have been either responsible for conducting R&D in the research centres or responsible for carrying out teaching activities in the study programmes. UCL has structurally attempted to bridge the research centres and the study programmes by moving employees from the research centres into the educational departments. This is elaborated by a Head of Department:

> Back in the days, employees were either affiliated the research centres or the educational departments. We still have some specialists affiliated the research centres such as docents and statisticians but most of our employees are now hired directly to the educational departments, which means that they must conduct teaching activities as well as R&D. This decision was made in the belief that teaching activities and R&D can enrich one another and should go hand in hand.

As explained in the quote, the motive for bridging the research centres and the educational departments has been to strengthen both areas. However, the department was unaffected by this effort because the research centre within trade (where the department belongs) did not have any employees with relevant R&D competencies within construction. A Head of Department elaborates:

> The research centre within trade is very small and don't have any employees with relevant research experience in construction, which makes it really difficult to support R&D in the Department of Construction.

Consequently, the most significant outcome of the structural change was that the department, in addition to teaching activities, has also been obliged to conduct R&D, which historically has otherwise been the responsibility of the research centre. With ambitions to strengthen the involvement in R&D and increase the general understanding of research management, the department invited researchers from Aalborg University to seminars to share experiences on R&D and approaches to structure academic environments.

An outcome of the seminars was the idea that the department should develop its internal structure containing the four units: (1) building design, (2) building statics and HVAC, (3) communication and (4) construction planning and management, into academic environments. The rationale behind this idea was to foster academic environments with specialised knowledge that could collectively develop the R&D competencies and translate new knowledge generated through R&D into the teaching activities. However, the top management at UCL has also expressed an expectation that the department strengthens its engagement in dissemination activities such as article publication. This is explicated by a Lecturer:

> The department units (i.e., academic environments) are spaces where we can foster academic environments as well as distribute tasks and develop R&D competencies related to the specific academic environment. However, the top management at UCL still expects us to acquire advanced writing skills so that we can publish articles based on our involvement in R&D.

In an attempt to develop advanced writing skills, the department has on its own initiative been establishing a so-called writing workshop. According to several of the informants, the writing workshop is relevant because dissemination in written format is a new discipline in the department that requires extensive training to master. Although the establishment of the writing workshop generally has been appreciated by the staff in the department, it has also given rise to debate in the top management of UCL. The debate has been concerning whether persons without a PhD degree have the capacity to produce and publish scientific articles. The following presents the different views of two heads of departments on this issue:

> You don't necessarily have to hold a PhD degree to produce scientific articles as long as you work in accordance with the recognised scientific approaches and methods or collaborate with experienced researchers.

> There's no one in the Department of Construction with a PhD degree and I therefore think it is very unlikely that the department can produce research. Academia is a closed club that you can only be part of if you hold a PhD degree. The department needs to commence PhD projects to comply with the legal requirements on R&D.

Several of the informants have been expressing an interest in writing scientific articles but at the same time point out that the articles must benefit the teaching activities in the study programme. Moreover, the informants have been emphasising that the link between R&D and teaching activities is fragmented and that future initiatives should aim at merging R&D and teaching activities better. This is elaborated by a Lecturer:

> We're expected to produce scientific articles. However, it's not valuable for us (i.e., the staff) if the only purpose is to demonstrate that we can do it and it doesn't benefit our teaching activities. In the future, we should work to create coherency between R&D and teaching activities and identify what tools and approaches we can apply to ensure a harmony.

4.4.2.2 Change Work Performed and the Preliminary Effects of Capacity Development

In this second part of the analysis, we showed how the Department of Construction through the Capacity development project performed change work to (re)organise the department and upskill the staff with competencies so that they can better conduct R&D. Structurally, the department has been embracing the idea that the existing four units in the department must be developed into distinct academic environments, which means that each employee in the department will belong to an academic environment. A writing workshop has also been established across the specialised units wherein the staff are supposed to meet and collectively advance skills on scientific writing and dissemination. In relation to identity, the department has with the new internal structure based on academic environments attempted to foster internal awareness that employees with shared professional interest are to increase specialisation within their domains and link activities in the environments with the teaching activities. In addition, the academic environments are expected to collectively develop routines that ensure integration of knowledge gained through R&D in the teaching activities.

Table 4.1 Organisational changes pursued in the change projects

Project:	Bridge building	Capacity development
Project period:	September 1, 2018, to January 31, 2019	June 1, 2021, to December 31, 2022 (planned)
Structural changes:	Changes in the structure of the study programme (new curriculum and new learning targets)	Establishment of academic environments and writing workshop
Identity changes:	Awareness that students must both be prepared for a job in the industry and for studying a master's degree	Awareness that the staff must increase specialisation and link knowledge obtained through R&D in the teaching activities
Routine changes:	Development of new routines in the teaching of scientific theories and methods	Development of new routines that bridge R&D and teaching activities

4.5 Discussion

In the analysis, we have described how the Department of Construction has strived to facilitate structural, identity and routine changes by implementing the two change projects Bridge building and Capacity development. In this section, we first present a table summarising the main findings from the analysis. Subsequently, we reflect on the department's aspirations to develop and alter aspects of organisational life through change projects to adapt to new educational policies.

The structural, identity and routine changes pursued in the Bridge building and Capacity development projects are shown in Table 4.1.

4.5.1 Pursuing Organisational Change in Construction Education Through Change Projects

It is well known that modern societal conditions are under rapid change (e.g. Alvesson & Sveningsson, 2008). A consequence hereof is that organisations, including educational institutions as demonstrated in this study, are progressively confronted with demands and expectations to adapt to changing societal conditions to ensure quality education and consolidate their relevance in society (Boeren, 2019; George et al., 2016). A recognised approach to adapting to changing societal conditions is through the implementation of change projects aimed at changing aspects of organisational life (Alvesson & Sveningsson, 2008).

In the analysis, we have provided insights into the efforts made by the Department of Construction at UCL to develop and alter aspects of organisational life by implementing the two change projects Bridge building and Capacity development. Both change projects were initiated to adapt to educational policies that demand an increased quality of education and an increased involvement of university colleges in R&D. In addition, both change projects involved structural, identity and routine changes in the department (as illustrated in Table 4.1). The change projects were bottom-up initiatives formulated, initiated and implemented by the department. As illustrated in the analysis, UCL is obliged to integrate new educational policies into the study programmes offered by UCL. The practical efforts required to adapt to new educational policies are, however, redirected to the educational departments, which simultaneously experience a pressure from the top management of UCL to facilitate internal changes accordingly. In line with this, the requirements of the top management of UCL in relation to changes in the educational departments are ambiguous. This is most significantly exemplified in the second part of the analysis, where two heads of departments have different views on whether the Department of Construction can produce research and thereby join ‘the academic club’ without having employees with PhD degrees.

4.6 Conclusions

In the study, we have described and analysed how the Department of Construction at University College Lillebaelt has strived to adapt to educational policies by implementing change projects. Our interest was to obtain an understanding of the internal changes pursued by the department and the change work performed in this regard. In the study, we have shown that the department, through a bottom-up approach, implements change projects to adapt to new educational policies demanding a strengthened quality of education and an increased involvement of university colleges in R&D. We have also shown that the department successfully has made changes in its internal structure, identity and routines. An implication hereof is that the department at large has attained a greater focus on scientific theories and methods as well as the need to foster academic environments and specialisation. Although it is a long-term and comprehensive effort, the bottom-up approach is considered a fruitful approach to facilitate change in the department in the sense that structures, identities and routines are developed and altered with respect to existing conditions in the department.

Finally, we have throughout the study demonstrated how a university college department strives to consolidate its relevance by adapting to the needs of society (expressed through educational policies). The study thus contributes with practical insights into how the fourth United Nations Sustainable Development Goal on quality education is embraced by an educational department through the implementation of change projects aimed at facilitating internal change accordingly.

References

Albert, S., & Whetten, D. A. (1985). Organizational identity. In L. L. Cummings & M. M. Staw (Eds.), *Research in organizational behaviour* (pp. 263–295). JAI.

Alvesson, M., & Sveningsson, S. (2008). *Changing organizational culture: Cultural change work in progress*. Routledge.

Battilana, J., Gilmartin, M., Sengul, M., Pache, A.-C., & Alexander, J. A. (2010). Leadership competencies for implementing planned organizational change. *The Leader Quarterly, 21*(3), 422–438.

Becker, M. C., Lazaric, N., Nelson, R. R., & Winter, S. G. (2005). Applying organizational routines in understanding organizational change. *Industrial and Corporate Change, 14*(5), 775–791.

Beer, M., Eisenstat, R. A., & Spector, B. (1990). Why change programs don't produce change. *Harvard Business Review, 68*(6), 158–166.

Boeren, E. (2019). Understanding sustainable development goal (SDG) 4 on "quality education" from micro, meso and macro perspectives. *International Review of Education, 65*(2), 277–294.

Chia, R. (1997). Essai: Thirty years on: From organizational structures to the organization of thought. *Organization Studies, 18*(4), 685–707.

Danmarks Evalueringsinstitut. (2017). *Forskning og udvikling på professionshøjskolerne som vej til uddannelseskvalitet*. Danmarks Evalueringsinstitut.

Fernie, S., Leiringer, R., & Thorpe, T. (2006). Change in construction: A critical perspective. *Building Research & Information, 34*(2), 91–103.

Forman, M., & Gottlieb, S. C. (2019). *Brobygning mellem bygningskonstruktøruddannelsen og kandidat i ledelse og informatik i byggeriet*. Statens Byggeforskningsinstitut.

George, G., Howard-Greenville, J., Joshi, A., & Tihanyi, L. (2016). Understanding and tackling societal grand challenges through management research. *Academy of Management Journal, 59*(6), 1880–1895.

Gioia, D. A., Patvardhan, S. D., Hamilton, A. L., & Corley, K. G. (2013). Organizational identity formation and change. *Academy of Management Annals, 7*(1), 123–192.

Gottlieb, S. C., & Frederiksen, N. (2020). Deregulation as socio-spatial transformation: Dimensions and consequences of shifting governmentalities in the Danish construction industry. *Environment and Planning C: Politics and Space, 38*(3), 484–502.

Hannan, M. T., & Freeman, J. (1984). Structural inertia and organizational change. *American Sociological Review, 49*(2), 149–164.

Howard-Greenville, J., Rerup, C., Langley, A., & Tsoukas, H. (2016). Introduction: Advancing a process perspective on routines by zooming out and zooming in. In J. Howard-Greenville, C. Rerup, A. Langley, & H. Tsoukas (Eds.), *Organizational routines: How they are created, maintained, and changed*. Oxford University Press.

Kotter, J. P. (1995). Leading change: Why transformation efforts fail. *Harvard Business Review, 2*(1), 1–10.

Král, P., & Králová, V. (2016). Approaches to changing organizational structure: The effect of drivers and communication. *Journal of Business Research, 69*(11), 5169–5174.

Lamont, M., & Swidler, A. (2014). Methodological pluralism and the possibilities and limits of interviewing. *Qualitative Sociology, 37*(2), 153–171.

Langley, A., Smallman, C., Tsoukas, H., & Van de Ven, A. (2013). Process studies of change in organization and management: Unveiling temporality, activity, and flow. *Academy of Management Journal, 56*(1), 1–13.

McNiff, J., & Whitehead, J. (2002). *Action research: Principles and practices*. Routledge Falmer.

Nelson, P. R., & Winter, S. (1982). *An evolutionary theory of economic change*. Harvard University Press.

Pettigrew, A. M., Woodman, R. W., & Cameron, K. S. (2001). Studying organizational change and development: Challenges for future research. *Academy of Management, 44*(4), 697–713.

Quattrone, P., & Hopper, T. (2001). What does organizational change mean? Speculations on a taken for granted category. *Management Accounting Research, 12*(4), 403–435.

Ranson, S., Hinings, B., & Greenwood, R. (1980). The structuring of organizational structures. *Administrative Science Quarterly, 25*(1), 1–17.

Stouten, J., Rousseau, D. M., & De Cremer, D. (2018). Successful organizational change: Integrating the management practice and scholarly literatures. *Academy of Management, 12*(2), 752–788.

Suddaby, R., & Foster, W. M. (2017). History and organizational change. *Journal of Management, 43*(1), 19–38.

Tsoukas, H., & Chia, R. (2002). On organizational becoming: Rethinking organizational change. *Organization Science, 13*(5), 567–582.

Uddannelses- og Forskningsministeriet. (2014). *Bekendtgørelse af lov om professionshøjskoler for videregående uddannelser. LBK nr 936 af 25/08/2014.* Uddannelses- og Forskningsministeriet.

Uddannelses- og Forskningsministeriet. (2018). *Inspirationsgruppen om nyt filosofikum: Afrapportering til uddannelses- og forskningsministeren.* Uddannelses- og Forskningsministeriet.

Van de Ven, A. H. (2018). Academic-practitioner engaged scholarship. *Information and Organization, 28*(1), 37–43.

Weick, K. E., & Quinn, R. E. (1999). Organizational change and development. *Annual Review of Psychology, 50*(1), 361–386.

Chapter 5
BIM for Construction Education in Nigeria

Theophilus Olowa, Emlyn Witt, and Irene Lill

Abstract Introducing advanced information technologies (AITs) such as a BIM-enabled learning environment (BLE) into architectural, engineering, construction, and facility management (AEC/FM) education might be an indirect but effective means of eradicating poverty and ensuring upward socioeconomic mobility for people all over the world, as the UN rightly recognizes in their mandate. In this study, we focus on a data collection tool to investigate the current state of the art of BIM for construction education among academics in an emerging economy (Nigeria). In the same framework, we strive to understand their perspectives, the benefits of BIM for construction education, and its perceived limitations. This preliminary study will assist us to later identify and compare structural features in these contexts to those in Europe and then assess the outcomes and repercussions of the proposed BLE in a developing economy like Nigeria. Using qualitative survey research design and analyzing with the aid of NVivo, nine themes emerged from the preliminary findings, viz., advantages, beyond present area, BIM in teaching, BIM in BIM-enabled learning, constraints, general views about BIM in Nigeria, stakeholder attitudes, and teaching method. These suggest that the designed data collection tool is sufficient in eliciting information from the population that is associated with BIM-enabled learning among academics in emerging economies which would help in meeting the fourth goal (Quality Education) of the UN's Sustainable Development Goal (SDG) plan.

Keywords Building Information Modeling (BIM) · BIM-enabled learning · BIM education · Virtual learning environment · AEC-FM

T. Olowa (✉) · E. Witt · I. Lill
Department of Civil Engineering and Architecture, Tallinn University of Technology, Tallinn, Estonia
e-mail: theophilus.olowa@taltech.ee; emlyn.witt@taltech.ee; irene.lill@taltech.ee

G. Lindahl, S. C. Gottlieb (eds.), *SDGs in Construction Economics and Organization*, Springer Proceedings in Business and Economics,
https://doi.org/10.1007/978-3-031-25498-7_5

5.1 Introduction

As part of its aim as a member of the Global Education Coalition, the United Nations (United Nations, n.d.) is assisting countries "...in mobilizing resources and implementing innovative and context-appropriate solutions to provide education..." globally. This, they stated, is in line with the fourth goal of their 2030 agenda for Sustainable Development Goals (SDGs). While 2030 is only a few years away, the agenda has been negatively disrupted by COVID-19 as most schools were shut down for an extended period. Digitalization is not only disrupting the construction industry but also revolutionizing the way that knowledge in architectural, engineering, construction, and facility management (AEC/FM) disciplines is conceived, developed, delivered, and reconstructed between academics and students. This offers an opportunity to prevent (or at least limit) the negative impact that may arise in the future from similar disasters. Building Information Modeling (BIM) is pivotal in this revolution. Digitalization has permeated human existence to the extent that it defines generations, based on how influenced and digitally exposed people are to it. For example, we have Gen X, Y, and Z that are all defined according to peoples' affinity to digitalization. The change in knowledge creation, construction, reconstruction, and the desire to contribute to sustainable development by seeking equitable solutions and universal access to education is also attracting researchers' attention to consider an effective means of how these new bodies of knowledge are stored, communicated, shared, transmitted, and evaluated. In response to these challenges, the development of a BIM-enabled learning environment (BLE) is an initiative in its developmental stage to augment competency-based teaching and learning in a common but location independent, open, accessible, self-paced, and collaborative environment by leveraging advanced information technologies.

The structures of the proposed BLE have been delineated into three domains (BIM, collaborative, and VLE functions) and interpreted using Adaptive Structuration Theory (AST) in our earlier work (Olowa et al., 2022). However, the identification, isolation, formulation, and validation of the existing different structural features and spirit under these domains are based on data collected from only three European countries, viz., Italy, Finland, and Estonia. The geographical context from which the structural features and spirit are formulated imposes a limitation of whether a globally relevant set of structural features and spirit have been captured. Therefore, the aim of this study is to find out if there are other context-specific structural features and spirit that are missing in the validated three domains especially from the perspective of developing economies.

In this study, we focus on the data collection tool by using the Will, Skill, Tool (WST) model among academics for exploring the general state of the art of BIM for construction education in developing economies. We try to understand their views on stakeholders' attitudes, advantages of BIM for construction education, and perceived constraints in the same context. Afterward, we intend to delineate and compare the identified structural features in this context to the European context and eventually use Adaptive Structuration Theory (AST) to evaluate the outcomes and consequences of the proposed BLE in an emerging economy like Nigeria.

The structure of this paper is as follows: the next paragraph briefly examines the relevant literature followed by the methodology and results sections, respectively. Conclusions are drawn at the end of the paper.

5.1.1 Literature Review

BIM education for both graduates and construction professionals has not progressed nor advanced as quickly as in the industry deployment of BIM (Casey, 2008; Forgues & Becerik-Gerber, 2013), and this presents a challenge for the education of architectural, engineering, construction, and facility management (AEC/FM) students (Ambrose, 2012). Firstly, graduates and construction professionals must be BIM natives to meet the immediate industry needs and challenges if they are to successfully carry out their professional tasks especially as global industry players. Many higher education institutions (HEIs) and faculty or practitioner researchers around the world are already standing up to this challenge, whether in bits and pieces or in major chunks, by altering existing curriculums at topical level within a course or carrying out a total overhaul and redesign of academic programs (Gerber et al., 2015; Olowa et al., 2019). Secondly, education tends to also miss out on new opportunities arising from BIM which have the potential to enhance AEC/FM education and benefit the industry in return (Olowa et al., 2020). These include opportunities for improving the integration of curricula within and between construction-related disciplines using real project data and creating learning environments that more closely correspond to industry realities.

Previous studies in advanced economies have adduced different reasons for the slow reflection of BIM in AEC/FM education generally. For example, Puolitaival and Forsythe (2016), Gier (2008), Klotz et al. (2009), Wong et al. (2011), Becerik-Gerber et al. (2011), Clevenger et al. (2012), Codinhoto et al. (2013), Forsythe et al. (2013), and Underwood and Ayoade (2015) argue that unavailability of appropriate teaching and learning resources for BIM; difficulty in finding the balance between theory and practice, technology and process, and traditional and emerging Construction Project Management (CPM) methods; and facilitating staff's professional development as the major challenges facing BIM in construction education. Despite these challenges, some of the authors (e.g., Puolitaival & Forsythe, 2016) opined that theoretical educational resources are available, easy, and straightforward to locate. Notwithstanding the availability of these educational resources, they maintained that having actual building models is a challenge in terms of preparing and optimizing usage of the model for high-quality educational purposes. These challenges and benefits are largely from studies carried out in the developed economies.

In the emerging economies context which Nigeria represents, there is a growing corpus of study that attempts to probe into the challenges/barriers and benefits of BIM in the sense of the construction industry in general (e.g., Amuda-Yusuf et al., 2017) and in quantity surveying profession specifically (e.g., Babatunde et al., 2018;

Babatunde & Ekundayo, 2019). Babatunde and Ekundayo (2019) identified 17 major barriers to the incorporation of BIM into quantity surveying undergraduate curriculums as follows:

- The lack of IT infrastructure or poor internet connectivity
- Intensive resource requirement of executing BIM protocols
- The inability of government to lead/direct BIM implementation
- High cost associated with staff/lecturer training or retraining
- Unavailability of staff with BIM expertise to take BIM courses
- The frequent need for BIM software upgrading
- The nonexistence of accreditation standards and requirements to guide the implementation of BIM
- Inadequate/erratic power supply
- The lack of collaboration with industry expert
- BIM is problematic for people with weak general IT skills
- Resistance to change
- Difficulty in introducing BIM in an already well-established curriculum
- Need for industry involvement, i.e., the need to engage expert industry practitioners in the development and delivery of a BIM curriculum
- The lack of university management support
- The lack of ICT literacy of staff or technical expertise
- Problem with integrating different areas of the curriculum to realize the multidisciplinary aspect of BIM
- BIM demands new teaching methods
- The lack of BIM-specific materials and textbooks, as well as other educational resources for students

Furthermore, the study also identified 12 BIM drivers and 14 BIM benefits of incorporating BIM into undergraduate quantity surveying curriculums using quantitative approach.

While previous research on BIM in HEIs identified barriers, drivers, and benefits based on a mono-discipline quantitative data from students and faculties (Babatunde et al., 2018; Babatunde & Ekundayo, 2019), this study explores the general state of the art of BIM education by considering it from a multidisciplinary perspective among the AEC-FM faculties using a qualitative approach to have a deep insight into the contextual advantage, stakeholders' attitudes, and constraints of BIM-enabled education in Nigerian HEIs.

5.2 Methodology

This study is a preliminary phase of an exploratory sequential mixed methods (Creswell, 2013) approach to explore the level of BIM for construction education among academics in higher education institutions (HEIs) in Nigeria with a focus on the development of a data collection tool. The anticipated solution being proposed toward promoting dispersed and self-paced learning in a BIM-enabled environment would be principally beneficial to educators especially those in HEIs, i.e., academics. As such, academics were recruited to participate in this preliminary study.

5.2.1 *Site*

Interviews were carried out among professors in the faculty of environmental sciences at the university of Ilorin (UNILORIN) using purposive sampling technique. This was done to overcome part of the limitations of an earlier study carried out within the European Union by finding additional evidence from a different social context and making a comparison. Although the University of Ilorin itself was founded in 1976, the faculty of environmental sciences was not established until 2013 starting with five departments, namely, architecture, estate management, quantity surveying, surveying and geo-informatics, and urban and regional planning departments. The number of departments is still what it used to be as at the time of this study. The staff structure is such that most faculty members are either experienced professionals from the industry or other academic institutions with very few inexperienced junior academics. There are only two departments that currently run master's programs at present, namely, departments of architecture and urban and regional planning. This institution was chosen for this preliminary study because of its reputation as a stable (i.e., not given to unexpected closures due to strikes, etc.) academic institution in the country.

5.2.2 *Theoretical Model*

The Will, Skill, Tool (WST) framework is used as a general framework to give meaning to the experiences of academics participating in this study. The WST model has been extensively used in both quantitative and qualitative studies as either an indicator or predictor of teachers' readiness to integrate technology into their teaching (Agyei & Voogt, 2011). Sasota et al. (2021) argue that WST model is versatile in explaining the integration of technology in teaching practice by academics. There are three important components in the model, namely will, skill, and tool. Will is described as the desire to use technology, skill as the ability to acquire the requisite knowledge and use the desired technology, while tool is the availability and accessibility of the technology. The will and skill are internal factors considered to be within the control of academics to some extent. The tool, however, is regarded as external factor because the technology (i.e., hardware and software) is expected to be provided by the employer. The model showing the connection between these attributes required for integrating technology in teaching practice is shown in Fig. 5.1.

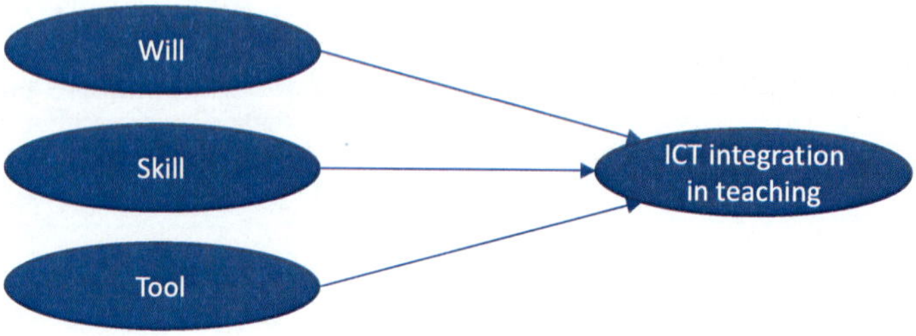

Fig. 5.1 WST model of ICT integration in teaching. (Based on Velazquez, 2007)

5.2.3 *Instrument*

A semi-structured interview was used as a tool to explore the academics' experiences and teaching practices in construction education because it allows the respondents to describe their experiences in a broad way (Krahn & Putnam, 2003) yet channels the conversation toward issues of research interest. This tool's flexibility and openness allowed customizing questions to the specific circumstances and discursive meanings of participants (Hildebrand & Markovic, 2007). The main questions asked were as follows:

1. What are your general views about BIM in Nigeria, stakeholder attitudes, advantages, constraints?
2. Please describe your teaching practice with respect to subject(s) and target audience.
3. Do you currently use BIM in delivering your teaching/training?

If yes:

   (a) How do you use BIM in the delivery? (e.g., for visualizations, project data, communication, etc.)

if no:

   (b) Could you use BIM to help deliver your teaching/training and for what? (e.g., for visualizations, project data, communication, etc.)

4. Beyond your present area(s) of teaching/training, how do you think BIM could be used in BIM-enabled learning?

5.2.4 *Analysis*

The analysis of the interview was carried out using thematic analysis. It seemed logical to use thematic analysis, as its qualitative nature provides a method particularly suitable for understanding subjective meanings and the process of social construction of individual narratives surrounding a particular theme. This was achieved through content coding by using NVivo (QSR International Pty Ltd, 2020) for the substantive contents of the interview.

5.3 Results

This section provides the preliminary findings of the study.

5.3.1 Demographics

Table 5.1 shows the breakdown of the participants' demographics.

Table 5.1 shows that there were ten faculty members interviewed from the University of Ilorin with only one of them being female while the rest were males. Breakdown of their cadres reveal that there was an associate professor, three technologists, and six lecturers (who were either in Lecturer I or II positions) that participated in this study. They represented three departments from the institution, namely, architecture, urban and regional planning, and estate management with five, one, and four participants from the departments, respectively. With respect to their highest educational qualifications, there was only one PhD degree holder, seven master's degree holders, and two first degree holders.

Table 5.1 Demographics of participants

Characteristic	Number of interviewees	Percentage (%)
Gender		
Male	9	90
Female	1	10
Academic title		
Associate professor	1	10
Technologist/instructor	3	30
Lecturer (I and II)	6	60
Length of service in present university		
5–10 years	10	100
Discipline		
Architecture	5	50
Urban and regional planning	1	10
Estate management	4	40
Highest qualification		
PhD	1	10
Masters	7	70
Bachelors	2	20

5.3.2 *Themes Emerging from the Interviews Based on WST Model*

Table 5.2 shows the themes generated through content analysis, the number of coded references generated in each, and the number of participants from whose transcripts the codes were extracted.

5.3.3 *Description of Themes Relating to Will*

5.3.3.1 Academics' Perception on Stakeholder Attitudes

In describing the attitudes of the stakeholders, the participants attempted to distinguish among the various stakeholders. They opined that there are some stakeholders that they believed are interested in BIM and its application, whereas there are others that are averse to and skeptical about its use. Some examples of those interested in BIM and its application as suggested by participants include the public sector and some private organizations, which are large. The use of BIM by academics and among other stakeholders is viewed as almost nonexistent.

5.3.3.2 General Views of Academics About BIM

Participants viewed BIM in Nigeria as a welcome development which allows for more realistic designs and quicker design development, thereby leading to overall time saving in construction. They also opined that the advent of BIM would see a lot of job loss especially those that relate to draughts men's and job men. They also feared that BIM gives an unrestrained avenue for nonprofessionals to promote quackery in different AEC disciplines.

Table 5.2 Emergent themes from the interviews

Themes	Number of references	Number of interviewees
Will		
Stakeholder attitudes	5	3
General views about BIM	10	6
BIM in teaching	6	5
Skill		
Advantages	6	4
Teaching method	10	6
Tool		
Beyond present area	3	3
Constraints	18	8
Total	57	9[a]

[a]While the total was ten, one of the interviewees' transcripts had no relevant data for this study

5.3.3.3 BIM in Teaching

In response to how the interviewees reflect BIM in their teaching, the majority revealed that it does not reflect in their teaching at all while a few indicated that they only carry out BIM awareness education.

5.3.4 Description of Themes Relating to Skill

5.3.4.1 Advantages

The advantages ascribed to the use of BIM by the interviewees include: speed of design accuracy of design and calculation parameters; usefulness in prefabrication; saves time and gives good results; ability to design more complex structures; and ability to simulate and plan construction activities.

5.3.4.2 Teaching Method

Many of the participants said that they use "hybrid" methods in their class teaching. By this, they give notes or handouts to students and sometimes project on the wall. In addition to this traditional method of teaching, some include a method they referred to as "learning by doing" which involves giving briefs to students for development. Assessments are carried out formatively by tests, writing of term papers, and presenting in seminars. Summative assessments are carried out by writing semester exams.

5.3.5 Description of Themes Relating to Tool

5.3.5.1 Constraints

The major constraints to the wide spread of BIM in Nigeria that were expressed by the interviewees include procurement and renewal cost, accreditation bodies not mandating BIM knowledge or skills for students, outdated curriculums, lack of access to technological hardware (e.g., computers, internet, and electricity), lack of BIM education, less emphasis on practical knowledge, inadequate security of lives and properties in the country, staff training and empowerment, and lack of time.

5.3.5.2 Possible Areas Beyond Present Use

It seemed difficult for most of the interviewees to suggest how BIM could be used in other areas beyond their present teaching practice as most of them already revealed that they do not interact with BIM in their practice. However, one area that was suggested is in 3D house printing, where BIM would be the backbone on which the operations take place. The other area was in virtual learning, where BIM would enhance the ability of students to learn remotely given the ongoing experience of the COVID-19 pandemic.

5.4 Conclusions

COVID-19 has underscored the importance of having an effective way to engage learners in distributed learning globally if a meaningful achievement is to be made toward providing sustainable education for all by 2030. Introducing advanced information technologies (AITs) like the BIM-enabled learning environment (BLE) into AEC/FM education might be a veritable means to achieving this. This in turn will serve the 2030 agenda of the United Nation's SDG to eliminate poverty and guarantee upward socioeconomic mobility of peoples around the globe. This is because the BLE will promote location-free, competency-based, common, open, compatible, convenient, and accessible platform for AEC/FM disciplines at graduate level and offer opportunity for both self and continuous professional development toward sustainable human development and growth. To achieve this goal, a fully robust BLE must first be developed and made operational.

This study is a preliminary exploration of academics in AEC/FM disciplines to understand the state of the art of BIM education in Nigeria, their views about stakeholders' attitudes, opportunities, and constraints. Nine themes emerged from the preliminary findings and were grouped using the WST model, viz., advantages, beyond present area, BIM in teaching, BIM in BIM-enabled learning, constraints, general views about BIM in Nigeria, stakeholder attitudes, and teaching method. With these responses and the analyzed results, it is now possible to proceed in eliciting information from a much larger and country-wide population sample. This will enable the identification of structural features (structures and spirit) associated with BIM-enabled learning among academics in emerging economies. This would contribute to meeting the 4th goal (Quality Education) in the UN's SDG plan.

In response to how the interviewees reflect BIM in their teaching, the majority revealed that it does not reflect in their teaching at all while a few indicated that they only carry out BIM awareness education. This condition could be ascribed to several reasons associated with will, skill, and tool. In many cases, it is important to be aware of a technology and what it can do before the concepts of will and skill (or skill and will) can take effect. It also seemed difficult for most of the interviewees to suggest how BIM could be used in other areas beyond their present teaching practice as most of them already revealed that they do not interact with BIM in their practice. These realities among the academics underscore the need for an avant-garde platform such as BLE that will promote teaching and learning of BIM-enabled education in construction economics and organization. This will empower and provide both academics and learners who might not have BIM authoring experience nor BIM teaching and learning resources (e.g., modules) to have access to BIM-enabled education with just their browser and the internet.

The WST model suggests that it is possible to achieve a high level of technological integration into teaching by enhancing the will and skill of the academics and providing an enabling environment for the will and skill to be demonstrated. It is when this happens that the academics in emerging economies may constructively offer suggestions for improving the BLE to ensure its robustness and dependability for global use irrespective of the developmental context.

There are limitations in this study some of which will be addressed in future research. An important one is about the number of participants sampled in this study. In future, the study will consider a greater number of academics from a number of higher education institutions and representative of a more diverse range of AEC/FM disciplines.

Acknowledgments This research was supported by the BIM-enabled Learning Environment for Digital Construction (BENEDICT) project (grant number: 2020-1-EE01-KA203-077993), the Integrating Education with Consumer Behavior relevant to Energy Efficiency and Climate Change at the Universities of Russia, Sri Lanka, and Bangladesh (BECK) project (grant number: 598746-EPP-1-2018-1-LT-EPPKA2- CBHE-JP) and Strengthening University-Enterprise Collaboration for Resilient Communities in Asia (SECRA) project (grant number: 619022-EPP-1-2020-1-SE-EPPKA2-CBHE-JP), all co-funded by the Erasmus+ Programme of the European Union. The European Commission support to produce this publication does not constitute an endorsement of the contents which reflect the views only of the authors, and the Commission cannot be held responsible for any use which may be made of the information contained therein.

References

Agyei, D. D., & Voogt, J. M. (2011). Exploring the potential of the will, skill, tool model in Ghana: Predicting prospective and practicing teachers' use of technology. *Computers and Education, 56*(1), 91–100.

Ambrose, M. A. (2012). Agent provocateur – BIM in the academic design studio. *International Journal of Architectural Computing., 10*, 53.

Amuda-Yusuf, G., Adebiyi, R. T., Olowa, T. O. O., & Oladapo, I. B. (2017). Barriers to building information modelling adoption in Nigeria. *Journal of Research Information in Civil Engineering., 14*(2), 1574–1591.

Babatunde, S. O., & Ekundayo, D. (2019). Barriers to the incorporation of BIM into quantity surveying undergraduate curriculum in the Nigerian universities. *Journal of Engineering, Design and Technology, 17*(3), 629–648.

Babatunde, S. O., Ekundayo, D., Babalola, O., & Jimoh, J. A. (2018). Analysis of the drivers and benefits of BIM incorporation into quantity surveying profession. *Journal of Engineering, Design and Technology, 16*(5), 750–766.

Becerik-Gerber, B., Gerber, D. J., & Ku, K. (2011). The pace of technological innovation in architecture, engineering, and construction education: Integrating recent trends into the curricula. *Electronic Journal of Information Technology in Construction, 16*, 411–432.

Casey, M. J. (2008). Work in progress: How building informational modeling may unify IT in the civil engineering curriculum. In *2008 38th annual frontiers in education conference*, S4J-5-S4J-6.

Clevenger, C., Glick, S., & del Puerto, C. L. (2012). Interoperable learning leveraging building information modeling (BIM) in construction education. *International Journal of Construction Education and Research, 8*(2), 101–118.

Codinhoto, R., Kiviniemi, A., Kemmer, S., & Gravina da Rocha, C. (2013). BIM-FM Implementation. *International Journal of 3-D Information Modeling, 2*(2), 1–15.

Creswell, J. W. (2013). In V. Knight (Ed.), *Qualitative inquiry and research design: Choosing among five approaches* (3rd ed.). SAGE Publications, Inc.

Forgues, D., & Becerik-Gerber, B. (2013). Integrated project delivery and building information modeling: Redefining the relationship between education and practice. *International Journal of Design Education, 6*(2), 47–56.

Forsythe, P., Jupp, J., & Sawhney, A. (2013). Building information modelling in tertiary construction project management education: A programme-wide implementation strategy. *Journal for Education in the Built Environment, 8*(1), 16–34.

Gerber, D. J., Khashe, S., & Smith, I. F. C. (2015). Surveying the evolution of computing in architecture, engineering, and construction education. *Journal of Computing in Civil Engineering, 29*(5), 04014060.

Gier, D. M. (2008). *What impact does using building information modeling have on teaching estimating to construction management students?* (Unknown).

Hildebrand, J., & Markovic, D. (2007). Systemic therapists' experience of powerlessness. *Australian and New Zealand Journal of Family Therapy, 28*(4), 191–199.

Klotz, L., Johnson, P. W., Leopard, T., Johnson, P., Maruszewski, S., Horman, M., & Riley, D. (2009). Campus construction as a research laboratory: Model for intracampus collaboration. *Journal of Professional Issues in Engineering Education & Practice, 135*(4), 122–128.

Krahn, G., & Putnam, M. (2003). Qualitative methods in psychological research. In M. C. Roberts & S. S. Ilardi (Eds.), *Handbook of research methods in clinical psychology* (pp. 176–195). Blackwell Publishers.

Olowa, T. O. O., Witt, E., & Lill, I. (2019). *BIM for construction education: Initial findings from a literature review* (Vol. 2, pp. 305–313).

Olowa, T., Witt, E., & Lill, I. (2020). Conceptualising building information modelling for construction education. *Journal of Civil Engineering and Management, 26*(6), 551–563.

Olowa, T., Witt, E., Morganti, C., Teittinen, T., & Lill, I. (2022). Defining a BIM-enabled learning environment—An adaptive structuration theory perspective. *Buildings, 12*(3), 292.

Puolitaival, T., & Forsythe, P. (2016). Practical challenges of BIM education. *Structural Survey, 34*(4/5), 351–366. https://doi.org/10.1108/SS-12-2015-0053

QSR International Pty Ltd. (2020, March). *NVivo*.

Sasota, R. S., Cristobal, R. R., Sario, I. S., Biyo, J. T., & Magadia, J. C. (2021). Will–skill–tool (WST) model of technology integration in teaching science and mathematics in the Philippines. *Journal of Computers in Education, 8*(3), 443–464.

Underwood, J., & Ayoade, O. (2015, March). Current position and associated challenges of BIM education in UK higher education. *BIM-Education*, 47.

United Nations. (n.d.). Education – United Nations Sustainable Development. *Qualit y education: Why it matters*. Retrieved February 9, 2022, from https://www.un.org/sustainabledevelopment/education/

Velazquez, C. M. (2007). Testing predictive models of technology integration in Mexico and the United States. *Computers in the Schools, 24*(3–4), 153–173.

Wong, A., Wong, K. F., & Nadeem, A. (2011). *Building information modelling for tertiary construction education in Hong Kong*.

Part II
Business Processes and the Circular Economy

Chapter 6
Theoretical Framework of Circular Business Model Innovation for Building Contractors

Lin Kjerulf and Kim Haugbølle

Abstract The construction industry's traditional linear business models of "take, make, and dispose" are being challenged in practice due to the increased market demands and the political agenda that points towards sustainable and circular practices. The majority of existing research within sustainable and circular business models is limited to short-term consumer products, but the research is poorly matching the conditions of buildings as capital-intensive goods with significantly longer life cycles. The data collection builds on an integrative literature review that will synthesize current findings on sustainable and circular business models by assessing the four major business model dimensions of value proposition, value delivery, value creation, and value capture. The aim of this paper is to identify and discuss important shortcomings in the general framework of sustainable business models with regard to application in the construction and real estate sector and more specifically larger contractor firms. The main findings conclude there is a research gap in terms of developing new business models that both capture the distinctive characteristics of the construction industry and innovate building contractors' traditional practices towards the sustainable and circular transition. This paper points out that future circular business models of larger contractor firms will probably include elements of the business models "Orchestrator" and "inclusive value creation" due to the increased need for coordination and early project involvement with multiple stakeholders in the value chain for the co-creation of long-term and valuable partnering agreements as well as participation in new types of procurement.

Keywords Circular economy · Business model · Material flows · Innovation · Contractors

---

L. Kjerulf (✉) · K. Haugbølle
Department of the Built Environment, Aalborg University, Copenhagen, Denmark
e-mail: link@build.aau.dk; khh@build.aau.dk

G. Lindahl, S. C. Gottlieb (eds.), *SDGs in Construction Economics and Organization*, Springer Proceedings in Business and Economics,
https://doi.org/10.1007/978-3-031-25498-7_6

6.1 Introduction

Sustainable development is accelerating in the Danish construction industry, and contractor firms are increasingly experiencing new sustainable market demands, which challenge their existing work processes and business models. In particular, the term circular economy has received significant attention as the new economic paradigm (Kirchherr et al., 2017), and the definition is according to the Ellen MacArthur Foundation (Webster, 2017): "A circular economy is one that is restorative by design, and which aims to keep products, components and materials at their highest utility and value, at all times." The work processes of waste management aiming at closing material loops and moving further up the waste hierarchy are traditionally not considered part of the contractor's role. Circular demolition and sourcing of materials could potentially be an integrated part of future building projects. Thus, the European Commission has launched a circular economy action plan and a recently updated EU Taxonomy report that include technical screening criteria within circular principles for real estate investments (European Commission, 2020). Nationally, the launch of the voluntary sustainability class in the Danish Building Regulations includes requirements within life-cycle analysis, resource consumption at the building site, and gradually increasing CO_2 limits for new buildings above 1000 m^2 will be implemented as mandatory from 2023 with the aim of reducing the CO_2 emissions from 12 to 7.5 kg $CO_2eq/m^2/year$ in 2030 (Ministry of the Interior and Housing, 2021). The new political instruments are rapidly driving circular initiatives in the built environment and enable construction companies to rethink their current business models. In addition, the almost exponential growth in new DGNB certified projects is pushing for a sustainable transformation of Danish construction.

The research objective of this paper is to identify and discuss important shortcomings in the general framework of sustainable business models with regard to application in the construction and real estate sector. As a result, the paper will demonstrate the needed considerations and conditions that are applicable in the construction industry when innovating existing business model structures. The findings are followed by a discussion focusing on larger building contractors' possibility to engage in new types of sustainable business models. This research contributes to insights for developing new business models for the sustainable transformation in construction, which is an important step towards realizing the UN's Sustainable Development Goals (SDGs) of SDG 12 "Responsible consumption and production" and SDG target 17.17 "Encouraging effective partnerships" (United Nations, 2016).

6.2 Research Methodology

The preliminary data collection consists of an integrative literature review on traditional and sustainable business models by mapping and synthesizing the presented common ideas and concepts in the chosen individual papers (Torraco, 2016). Thus, the quality assessment involves a critical review on identifying the most general

Table 6.1 Data collection of the chosen studies

Category/similarities	Studies in total	Knowledge gap
Commonly used business models	3	The traditional business models do not include dimensions for circular economy thinking, e.g., end of life
Business models in construction	1	The business model archetypes for different roles in construction are only considering the present status
Sustainable and circular business models	6	Narrow focus on products and services with the aim of breaking shorter life cycles
Sustainable and circular business models in construction	3	Adapted to solve case-based specific issues
Sustainability strategies and circular economy	7	Broad overview and conceptualizations of current state of the art within circular economy and sustainability
Sustainable transformation and circular economy in construction	8	The studies are missing the link to specific implementation business strategies

trends in commonly used business models as well as the new wave of prevailing sustainable and circular business models. Thus, the five steps of the integrative review include (Russell, 2005) (1) definition of research problem, (2) generation of literature search, (3) evaluation of the data, (4) analysis of data, and (5) interpretation and presentation of findings. The databases included the university library and Google Scholar and resulted in 28 studies to serve as the basis of the analysis, and Table 6.1 demonstrates the categories of the literature. The chosen studies are peer-reviewed and assessed by the methodology and content quality as well as the relevance to the subjects of circular economy, business model innovation, and sustainability in the construction industry.

The examination of the data (Table 6.1) is also focusing on the various knowledge gaps regarding the practical applications for the sustainable transition of the construction and real estate sector. Based on the framework developed by Lüdeke-Freund et al. (2019) that encompass an analysis of 26 current circular business models from literature, the patterns will be evaluated and compared with other studies to point at the distinctions of conceptualizations in existing business model frameworks. The business model patterns are discussed in the following section and divided into the four major dimensions of "value proposition," "value delivery," "value creation," and "value capture." The findings will provide a broad overview of the coupling between current literature within sustainable business models and the real estate sector in order to understand the construction firms' room for repositioning within the existing structures and frameworks. Furthermore, the discussion will reflect on larger contractors' future business model when adapting towards sustainable practices. The perspective of larger contractor firms, that functions as the construction management role, is dominating the assessments, as they represent the role of being responsible for implementing sustainable and circular initiatives in practice and the fact that their role is often overlooked in the academic literature.

6.3 Findings: Sustainable and Circular Business Models

A business model is defined as "the rationale of how an organization creates, delivers and captures value" (Osterwalder & Pigneur, 2010). The traditional research concerning business models has established the development of models and tools, e.g., the well-known business model canvas and value proposition design (Osterwalder & Pigneur, 2010; Osterwalder et al., 2014) as well as the mapping of diverse strategies for business model innovation (Gassmann et al., 2014). Shifting towards sustainable and circular business models, the value creation is to maintain the economic value embedded in products (Rosa et al., 2019) and support the closing of resource flows (Lüdeke-Freund et al., 2019) by long-lasting design, maintenance, repair, reuse, remanufacturing, refurbishing, and recycling (Geissdoerfer et al., 2017). Also, the increased apprehensions of the existing economic and capitalistic structures are probably part of the growing interest in other types of business models (Schaltegger et al., 2016; Porter & Kramer, 2011). Furthermore, the current linear approach has proven problematic in terms of the transformation process towards a more sustainable economic system and its ineffectiveness for handling global challenges (Weigend et al., 2020). Hence, the change into a circular economy should consider an effective economy at all scales and not only an efficient productive economy as in contrast to the linear financial economy, which dominates the current state (Webster, 2021).

The following sections will present the four major dimensions applied by the study of Lüdeke-Freund et al. (2019), and the overall six business model patterns are as follows:

1. Repair and maintenance
2. Reuse and redistribution
3. Refurbishment and remanufacturing
4. Recycling
5. Cascading and repurposing
6. Organic feedstock

The patterns represent different design strategies and striving for either prolonging life cycles of products, reusing for next product life, remanufacturing of products, closing loops in production, or recovering resources (Lüdeke-Freund et al., 2019).

6.3.1 *Value Proposition*

A sustainable value proposition must consider the trade-offs between the ideal product/service performance and the optimized social and environmental effects (Boons & Lüdeke-Freund, 2013). Another literature review points at the fact that the value proposition in a circular business model is the creation of a product/service that includes and intentionally uses a circular strategy to create value (Nussholz, 2017). The dimension of value proposition considers products and services with the aim of prolonging the life cycle of products or providing services related to take-back management, education, maintenance, or waste handling (Lüdeke-Freund et al., 2019). A study from the Netherlands (Van den Brink et al., 2017) aimed at the construction industry points out that the value proposition includes a service provider offering a product that is completely tailored to the clients' needs and delivered in a sustainable manner (Van den Brink et al., 2017). Thus, the study suggests that an external service provider is in charge of finding the most optimal sustainable solutions and consider buildings as customized and unique products. The study argues the need for a third-party service provider, as the delegation of ownership to the contractor or developer is not a feasible option due to the complexity of multiple owners (Van den Brink et al., 2017). Another study from Finland by Ritala et al. (2018) suggests nine different value propositions for sustainable business model innovation as a result of analyzing 500 of the largest global corporations: (1) maximize material and energy efficiency, (2) create value from waste, (3) substitute with renewables, (4) deliver functionality, (5) adopt a stewardship role, (6) encourage sufficiency, (7) repurpose for society, (8) create inclusive value (e.g., suppliers own materials instead of the client), and (9) develop scale-up solutions. In comparison with the study by Lüdeke-Freund et al. (2019), the study by Ritala et al. (2018) is not suggesting the prolonging of life cycle by, e.g., maintaining; however, the value propositions are considering other types of ownership models and services related to material/energy efficiency. A case study of a Danish architectural firm included a circular business model, where the value proposition for testing a mobile concrete recycling plant included (1) circular economy solutions by reducing the carbon footprint and (2) same standards, e.g., price, architectural value, and quality (Nussholz et al., 2019). The case study only focuses on a single material fraction, namely, concrete, and by that facilitates the possibility of setting up clear goals in the value proposition.

6.3.2 *Value Delivery*

Value delivery includes target customers such as quality conscious, cost conscious, green customers, business-to-business (B2B) customers, business-to-customer (B2C) suppliers, B2B suppliers, and customer-to-customer (C2C) suppliers (Lüdeke-Freund et al., 2019). The case study of a Danish architectural firm suggests the focus on customers who are interested in more environmentally friendly solutions, e.g., public housing organizations (Nussholz et al., 2019). When comparing the two frameworks (Lüdeke-Freund et al., 2019; Nussholz et al., 2019), the customer segment in the construction industry is significantly limited compared to other industries, as the customers are few and typically include public institutions or private investors. Furthermore, the value delivery in the study by Lüdeke-Freund et al. (2019) also consists of value delivery processes and suggests connecting suppliers and customers, providing product-based services, and providing used/take-back used/sharing "products/components/materials/waste." The value delivery processes suggested by Lüdeke-Freund et al. (2019) are highly relevant for "stand-alone" products and components with shorter life spans where, e.g., "take-back" services would be difficult to implement for a building that requires predictions related to the building's future usage, function, and owners that can differ from the users or residents. The study by Van den Brink et al. (2017) that focuses on "building" level states that the value delivery includes a service provider that performs all the activities from designing to operating by being responsible for the operational lease solution. Moreover, the service provider is responsible for choosing the products and ensuring that the building meets the performance level (Van den Brink et al., 2017). Thus, the study is especially focusing on connecting the construction phase and the operational phase and thereby inspired by the Energy Service Company (ESCO) models (Van den Brink et al., 2017). However, these types of models are difficult to practice in reality due to the multiple stakeholders involved at different project phases.

6.3.3 Value Creation

Value creation includes "partners and stakeholders" and "value creation processes." Partners and stakeholders suggest suppliers, manufacturers, retailers, service providers, public institutions, collectors of products/components/materials/waste, and others (e.g., researchers) (Lüdeke-Freund et al., 2019). Another study points at the importance of establishing alliances with other stakeholders, generating value for multiple value chain partners, or finding new ways of shaping economic transactions between partners (Nussholz, 2017). According to the study of Van den Brink et al. (2017), the partners will always include one or more of the following: the client, external suppliers and/or financiers. The Danish architectural firm strives to engage in partnerships and joint ventures with a concrete recycling plant and a gravel mining company and form a network to develop a certification standard (Nussholz et al., 2019). When comparing the studies (Lüdeke-Freund et al., 2019; Van den Brink et al., 2017; Nussholz et al., 2019), multidisciplinary partnerships are significantly important in the construction industry, as to find new types of solutions by combining competencies from different technical fields. The value creation processes consist of maintaining/repairing, refurbishing/remanufacturing, recycling, upgrading/upcycling, reselling, taking back, winning back base materials, using used products/components/materials/waste, and designing products/components/materials (Lüdeke-Freund et al., 2019). The processes are focusing on stand-online components or products, and the value creation is therefore difficult to adapt when applying the same logic to "buildings." The case of the built environment involves increased complexity as "buildings" represent long-term capital goods with longer life cycles that include repairs and refurbishments throughout a building's life span. Specifically, buildings consist of numerous standardized stand-alone products; however, when assembled, buildings are unique, complex, long-lasting and transformed units (Pomponi & Moncaster, 2017). Nevertheless, the majority of the academic literature within circular economy in construction focuses on the macro-level of cities and neighborhoods or the micro-level of construction materials, e.g., by life-cycle assessments; thus, there is a lack of focus on the meso-level of buildings (Pomponi & Moncaster, 2017).

6.3.4 Value Capture

Financial models need a transition from "price per unit" to pricing the "job to be done" and value the compliance of needs rather than selling amounts of products (Boons & Lüdeke-Freund, 2013). The value capture dimensions can be developed by capitalizing on additional revenue sources, cost reductions, or nonfinancial benefits related to circular efficiency strategies (Nussholz, 2017). According to the study by Lüdeke-Freund et al. (2019), the value capture's subcategories are "revenues" and "costs." Revenue suggests "additional product revenues, payments per unit of service, payments for functions or results, and price premiums." "Costs" include labor, repair/maintenance, waste handling/processing, manufacturing, resource inputs, transportation/logistics, and supply risks (Lüdeke-Freund et al., 2019). The case study of a Danish architectural firm identifies cost for labor, materials, as well as research and development (Nussholz et al., 2019). The revenue is coming from contractors and public funding for innovation of technology development (Nussholz et al., 2019). The case study represents a special case, as the business model is based on research and external funding that require additional effort for a construction company to initiate. According to another study from the Netherlands that examines three specific case studies from the construction industry, circular buildings must entail a new perspective on ownership of the materials that are only temporarily stored/embedded in a building (Leising et al., 2018). Nevertheless, elements embedded in buildings such as facades or roof are considered as fixtures and therefore an integrated part of the real estate, which challenge the circular economy concept of closing material loops, as the distinction of moveable objects and fixtures often is not considered in the academic literature (Ploeger et al., 2017). The use of more standardized interfaces and Building Information Modeling (BIM) could be applied for marking the components belonging to the supplier and eventually the owner by a continuous documentation process (Ploeger et al., 2017). Other studies point to the importance of providing green taxes in order to change the current economic system and by that generate incentives for driving energy savings (Tsai et al., 2011; De Jesus & Mendonça, 2018; Smol et al., 2020).

The literature on circular and sustainable business models demonstrates either a broad perspective of many studies combined to develop archetypes/patterns (Lüdeke-Freund et al., 2019; Ritala et al., 2018; Bocken et al., 2014) or a theoretic framework (Pomponi & Moncaster, 2017; Nussholz, 2017; Boons & Lüdeke-Freund, 2013; Schaltegger et al., 2016) or individual case studies aimed for solving specific issues (Nussholz et al., 2019; Leising et al., 2018; Van den Brink et al., 2017). Thus, few studies are adapted towards sustainable and circular business models in the construction industry, and the focus on the contractor firms' role appears neglected in the literature.

6.4 Discussion: Future Business Model of Contractors

Based on the previous analysis of the current literature, the discussion will emphasize the narrower focus on construction companies, which have an essential role in implementing sustainable solutions in practice. Thus, the discussion will include an evaluation of contractors' current business model and reflect on the transformation process towards a more sustainable business model. The main theory will discuss relevant models from the framework of Gassmann et al. (2014) work on 55 business models and Ritala et al. (2018) study on sustainable business model adoption among 500 firms that consists of nine business model archetypes.

Larger contractor firms' current business model is best characterized as a combination of "Long Tail" and "Orchestrator." The Long Tail business model is based on focusing on selling small quantities of a wide range of products. Thus, the Long Tail model includes lower profit margins and smaller volume sales of individual products. The model means that companies can sell niche products and therefore gives customers an advantage by having a wide range of options and therefore increasing the chances of finding the product that suits their individual needs (Gassmann et al., 2014). Contractors' business model is typically not to sell "standard houses," but they are often involved in tendering processes with long negotiations based on the chosen selection and award criteria. The purpose is to calculate the expenses and describe the work processes related to the building project and at the same time possess a high level of flexibility aimed at meeting the customers' myriad of individual needs and considerations (Winch & Cha, 2020). In addition, the business model is also "cost-driven" due to the strong price competition in construction (Berg et al., 2019), and practices include increasing the productivity at the building sites.

"Orchestrator" is a business model where the company focuses exclusively on its core competencies, and activities that fall outside these competencies must be delegated to specialized service providers who have the necessary skills to perform the task successfully. As a management player of the value chain, "the Orchestrator" will spend a large part of its time coordinating time and matching individual value-creating activities (Gassmann et al., 2014). Thus, the business model for larger contractors is categorized by offering niche products by evaluating the buildability (Berg et al., 2019) and also managing highly complex projects, which includes finding suitable collaboration partners, e.g., subcontractors and optimizing planning processes.

When observing contractors' future business model towards addressing the sustainability demands, a suggestion could be the "Trash-to-Cash" business model that includes a value proposition based on recycling or reusing old materials/products. Used materials/products are collected and either resold or transformed into new products. The model assumes that the acquisition of resources includes a low or no expense associated with developing new products (Gassmann et al., 2014). However, this is not the case when it comes to used building materials/products, which often result in a more costly process compared to the procurement of virgin materials partly due to inflexible building regulations and the lack of standardization in the area (Nordby, 2019; De Jesus & Mendonça, 2018). While this may change in line with new CO_2 requirements in the Danish Building Regulations from 2023 (Ministry of the Interior and Housing, 2021), where scarce resources of recycled/reused products may undertake a development towards becoming cost competitive due to the increasing demand, it is deemed to be a long and slow change process. Nevertheless, there is a greater chance of the "Trash-to-Cash" core business to be managed by waste treatment plants or material/product manufacturers rather than contractors who are used to waste handling, technical processing, and transforming resources. The process could involve take-back schemes, e.g., as a result of materials potentially being owned by manufacturers or real estate companies (Stephan & Athanassiadis, 2017). Hence, new material manufacturers may arise due to the increasing demand for reused materials that also incorporate documentation of the quality. One example is the Danish company "Old Bricks" (in Danish: Gamle Mursten) that recover and resell bricks from demolition (Nussholz et al., 2019). Thus, contractor firms are an important link when it comes to delivering and receiving used materials/products, but the business model itself is far from their core disciplines.

The business model "Make More Of It" enables companies to offer know-how or other resources to other companies in order to generate additional revenue besides the core revenue (Gassmann et al., 2014). Contractor firms' position in the value chain is in development, as the tendency is that contractors are moving towards becoming consulting contractors by an earlier involvement in construction projects (Berg et al., 2019). The increase in project complexity provides a business opportunity for contactor firms to sell "consultancy" services, e.g., related to early buildability advising with the aim of winning the project in the final tendering process. The "Add-On" business model is about pricing the core offering competitively, but additional "extra" services will raise the final price (Gassmann et al., 2014). Thus, the business model generates the possibility of working with options that meet the customer's specific individual needs. As mentioned earlier, the construction industry is highly dominated by competitive pricing and offering sustainability initiatives will often lead to extra costs. As a result, "Add-On" could play a role in the tendering process, as to offer additional services related to environmental sustainability and thereby influence the building client in a greener direction. Also, the contractor can demonstrate potential solutions or initiatives that they are working with to drive the innovation despite the services being outside the economical boundaries of the project. Both business models could be integrated simultaneous dependent on the specific project and the building clients' willingness to engage in new types of interactions either in the early market dialogues or in the tendering negotiations.

Another suggestion is to handle new sustainability requirements according to the "Orchestrator" business model (Gassmann et al., 2014) and by that retain known work practices. The advantage of the model is a close collaboration with external partners, whose innovative solutions can help strengthen contractor firms' production. Thus, the model prescribes the unnecessary state of optimizing the in-house competencies but rather finds suitable partners who are better equipped to handle the tasks. Contractor firms already hold strong competencies in being a management contractor, so the increased coordinator role as "Orchestrator" is not new but rather an extension of an already existing modus operandi. The role of being the "Orchestrator" will probably change in terms of requiring more time spent on coordination and the establishment of valuable partnering agreements with multiple stakeholders from the value chain. There is no doubt that the sustainable agenda will mean the conclusion of cooperation agreements with innovative partners in the value chain. Moreover, the shift in consumer preferences towards increased sustainability demands (De Jesus & Mendonça, 2018) would probably require upgrading employees' skills to actively contribute to sustainable initiatives and possess qualified knowledge in the dialogue with engineers, architects, and other partners (Brooks & Rich, 2016). However, the level of knowledge might not fall under the category of in-depth technical expertise, thus to a higher degree anticipating the need for long-term partnering agreements for the handling of sustainable initiatives (Aarseth et al., 2017) or participate in new types of procurements (Tang et al., 2019; Häkkinen & Belloni, 2011). One strategy could include hiring in-house sustainability consultants to handle new sustainability requirements. Another strategy is to "decentralize" the knowledge for a high number of employees, e.g., project leaders and site managers instead of having few experts that are centralized in their own team as a support function. In this way, the knowledge on sustainability is more integrated into the organization's portfolio of projects, while the in-depth technical expertise is derived from collaboration with external partners. This strategy has the advantage of not having to rely on few critical resources within the field but the disadvantage of potentially having inadequate in-house knowledge concurrently with increasing sustainability demands.

As mentioned earlier, the study by Ritala et al. (2018) suggests nine archetypes for sustainable business model innovation: (1) maximize material and energy efficiency, (2) create value from waste, (3) substitute with renewables, (4) deliver functionality, (5) adopt a stewardship role, (6) encourage sufficiency, (7) repurpose for society, (8) inclusive value creation, e.g., suppliers own materials instead of the client, and (9) develop scale-up solutions. The business model innovation "create value from waste" is comparable with the business model "Trash-to-Cash" and is concerned with closing the resource loops of materials and products (Ritala et al., 2018). "Deliver functionality" rather than ownership and "inclusive value creation" are examples of business model approaches with alternative types of ownership, e.g., sharing economy. A known example in the construction industry is the ESCO models, where the energy service company is responsible for carrying out the energy services without the client's own capital, and the energy savings will repay the company. Nevertheless, the long lifetimes of buildings complicate the realizability and the profitability of these types of leasing arrangements for the application of circular economy services (Van den Brink et al., 2017). Thus, an advanced circular economy service, e.g., for the facade of a building based on leasing would potentially lead to an arrangement of consortia, e.g., similar to public-private partnership solutions (Van den Brink et al., 2017). Furthermore, the study also concludes that lease solutions with suppliers would probably only make sense if they work together in longer project commitment for multiple projects, e.g., as comakers or chain partners (Van den Brink et al., 2017). "Adopt a stewardship role" is about taking stewardship as a company by demonstrating additional responsibility to address social or environmental issues, whereas "repurpose for society" is about transforming the corporate structure for sustainability by the striving of using the power of markets (Ritala et al., 2018). Both business models are similar to "the Orchestrator" by possessing a leader role in the industry and directing the project organization towards solving social or environmental requirements set by the building client in the procurement. As a large contractor, it is possible to influence the market and the building clients' decision-making by suggesting sustainable initiatives in early market dialogues and in the tender material.

Current frameworks for sustainable business model innovation (Lüdeke-Freund et al., 2019; Gassmann et al., 2014; Ritala et al., 2018; Bocken et al., 2014) are only to a limited extent considering the distinctive characteristics of the construction industry due to the narrow focus on industry with a production chain like "make-to-stock" (Van den Brink et al., 2017). As a result, the future business model of contractors indicates an increased strategy of applying the "Orchestrator" business model (Gassmann et al., 2014), as the transition points in the direction of an additional coordinator role by participating in earlier project involvement with greater complexity due to the interdisciplinary character of sustainable measures. Namely, sustainable measures are also generating the need for "inclusive value creation" (Ritala et al., 2018) to form new types of long-term partnering agreements for the handling of sustainable initiatives or participate in new types of procurements.

6.5 Conclusions

The main findings include that value proposition must consider a different logic than breaking the shorter lifespans of products, as buildings already include long life cycles of 50–100 years. The value delivery in the construction industry differs in terms of customer segments compared to industry, as they typically represent the role as an "ordered or investors" and are therefore cost intensive, few, and repeatable. Value creation processes are focused on stand-online components or products, but buildings consist of numerous standardized stand-alone products that function as unique, complex, long-lasting, and transformed units when assembled. Finally, the value capture must consider buildings' fixtures such as facades or roofs, which is an integrated part of the real estate that challenges the circular economy concept of closing material loops.

The prevalent business model of large contractors is characterized by a combination of offering niche products, "long tail," and having the role as the construction management, "Orchestrator." The adaptation towards a new circular and sustainable business model points out that current frameworks are not designed with the purpose for application in the construction industry because the focus is primarily on products aimed at breaking shorter life cycles. The future circular business models of larger contractor firms will probably include elements of the business models "Orchestrator" and "inclusive value creation" due to the increased need for coordination and early project involvement with multiple stakeholders in the value chain for the co-creation of long-term and valuable partnering agreements as well as participation in new types of procurements.

The research supports the SDG 12 titled "Responsible consumption and production," where the development of new sustainable business models in the construction industry is essential in order to meet target 12.2 of "sustainable management and use of natural resources" and target 12.5 of "substantially reduce waste generation' (United Nations, 2016). Furthermore, it can be argued that new sustainable business models in construction are potentially also in line with supporting SDG target 17.17 "encouraging effective partnerships" (United Nations, 2016), because of the potential alternative stakeholder formations when dealing with sustainable measures.

The findings indicate the need for developing new business models that both capture the distinctive characteristics of the construction industry and innovate building contractors' traditional practices towards the sustainable and circular transition. In conclusion, future work will investigate the case study of a Danish contractor firm and map its current business model and the related challenges it faces in the shift towards a new sustainable and circular business model.

References

Aarseth, W., Ahola, T., Aaltonen, K., Økland, A., & Andersen, B. (2017). Project sustainability strategies: A systematic literature review. *International Journal of Project Management, 35*(6), 1071–1083.

Berg, B., Thuesen, C., Ernstsen, S. K., & Jensen, P. A. (2019). Constructing archetypes: Mapping business models in the construction value chain. In *35th conference and annual general meeting*. ARCOM.

Bocken, N. M., Short, S. W., Rana, P., & Evans, S. (2014). A literature and practice review to develop sustainable business model archetypes. *Journal of Cleaner Production, 65*, 42–56.

Boons, F., & Lüdeke-Freund, F. (2013). Business models for sustainable innovation: State-of-the-art and steps towards a research agenda. *Journal of Cleaner Production, 45*, 9–19.

Brooks, A., & Rich, H. (2016). Sustainable construction and socio-technical transitions in London's mega-projects. *The Geographical Journal, 182*(4), 395–405.

De Jesus, A., & Mendonça, S. (2018). Lost in transition? Drivers and barriers in the eco-innovation road to the circular economy. *Ecological Economics, 145*, 75–89.

European Commission. (2020). *Taxonomy: Final report of the technical expert group on sustainable finance*. Available at: https://ec.europa.eu/info/sites/default/files/business_economy_euro/banking_and_finance/documents/200309-sustainable-finance-teg-final-report-taxonomy_en.pdf

Gassmann, O., Frankenberger, K., & Csik, M. (2014). *The business model navigator: 55 models that will revolutionise your business*. Pearson.

Geissdoerfer, M., Savaget, P., Bocken, N. M., & Hultink, E. J. (2017). The circular economy–A new sustainability paradigm? *Journal of Cleaner Production, 143*, 757–768.

Häkkinen, T., & Belloni, K. (2011). Barriers and drivers for sustainable building. *Building Research & Information, 39*(3), 239–255.

Kirchherr, J., Reike, D., & Hekkert, M. (2017). Conceptualizing the circular economy: An analysis of 114 definitions. *Resources, Conservation and Recycling, 127*, 221–232.

Leising, E., Quist, J., & Bocken, N. (2018). Circular economy in the building sector: Three cases and a collaboration tool. *Journal of Cleaner Production, 176*, 976–989.

Lüdeke-Freund, F., Gold, S., & Bocken, N. M. (2019). A review and typology of circular economy business model patterns. *Journal of Industrial Ecology, 23*(1), 36–61.

Ministry of the Interior and Housing. (2021). *National Strategy for Sustainable Construction*. Available at: https://im.dk/Media/637602217765946554/National_Strategy_for_Sustainable_Construktion.pdf

Nordby, A. S. (2019). Barriers and opportunities to reuse of building materials in the Norwegian construction sector. *IOP Conference Series: Earth and Environmental Science, 225*(1), 012061, 1–012061, 4.

Nussholz, J. L. (2017). Circular business models: Defining a concept and framing an emerging research field. *Sustainability, 9*(10), 1810.

Nussholz, J. L., Rasmussen, F. N., & Milios, L. (2019). Circular building materials: Carbon saving potential and the role of business model innovation and public policy. *Resources, Conservation and Recycling, 141*, 308–316.

Osterwalder, A., & Pigneur, Y. (2010). *Business model generation: A handbook for visionaries, game changers, and challengers*. Wiley.

Osterwalder, A., Pigneur, Y., Bernarda, G., & Smith, A. (2014). *Value proposition design: How to create products and services customers want*. Wiley.

Ploeger, H. D., Prins, M., Straub, A., & van den Brink, R. (2017). *Circular economy and real estate: Alternatives for operational lease* (pp. 164–176). IRC2019: Shaping Tomorrow's Built Environment.

Pomponi, F., & Moncaster, A. (2017). Circular economy for the built environment: A research framework. *Journal of Cleaner Production, 143*, 710–718.

Porter, M., & Kramer, M. (2011). The big idea – Creating shared value. *Harvard Business Review, 89*(1/2), 62–77.

Ritala, P., Huotari, P., Bocken, N., Albareda, L., & Puumalainen, K. (2018). Sustainable business model adoption among S&P 500 firms: A longitudinal content analysis study. *Journal of Cleaner Production, 170*, 216–226.

Rosa, P., Sassanelli, C., & Terzi, S. (2019). Towards Circular Business Models: A systematic literature review on classification frameworks and archetypes. *Journal of Cleaner Production, 236*, 117696.

Russell, C. L. (2005). An overview of the integrative research review. *Progress in Transplantation, 15*(1), 8–13.

Schaltegger, S., Hansen, E. G., & Lüdeke-Freund, F. (2016). Business models for sustainability: Origins, present research, and future avenues. *Organization & Environment, 29*(1), 3–10.

Smol, M., Marcinek, P., Duda, J., & Szołdrowska, D. (2020). Importance of sustainable mineral resource management in implementing the circular economy (CE) model and the European green deal strategy. *Resources, 9*(5), 55.

Stephan, A., & Athanassiadis, A. (2017). Towards a more circular construction sector: Estimating and spatializing current and future non-structural material replacement flows to maintain urban building stocks. *Resources, Conservation & Recycling, 129*, 248–262.

Tang, Z. W., Ng, S. T., & Skitmore, M. (2019). Influence of procurement systems to the success of sustainable buildings. *Journal of Cleaner Production, 218*, 1007–1030.

Torraco, R. J. (2016). Writing integrative literature reviews: Using the past and present to explore the future. *Human Resource Development Review, 15*(4), 404–428.

Tsai, W. H., Lin, S. J., Liu, J. Y., Lin, W. R., & Lee, K. C. (2011). Incorporating life cycle assessments into building project decision-making: An energy consumption and CO2 emission perspective. *Energy, 36*(5), 3022–3029.

United Nations. (2016). *Transforming our world: The 2030 agenda for sustainable development*. Available at: https://sustainabledevelopment.un.org/content/documents/21252030%20Agenda%20for%20Sustainable%20Development%20web.pdf

Van den Brink, R., Prins, M., Straub, A., & Ploeger, H. D. (2017). Finding the right incentives: Circular business models for the construction industry. *IRC2019: Shaping Tomorrow's Built Environment*, 189–200.

Webster, K. (2017). *The circular economy a wealth of flows* (2nd ed.). Ellen MacArthur Foundation Publishing.

Webster, K. (2021). A circular economy is about the economy. *Circular Economy and Sustainability, 1*, 1–12.

Weigend, R., Pomponi, F., Webster, K., & D'Amico, B. (2020). The future of the circular economy and the circular economy of the future. *Built Environment Project and Asset Management, 10*(4), 529–546.

Winch, G. M., & Cha, J. (2020). Owner challenges on major projects: The case of UK government. *International Journal of Project Management, 38*(3), 177–187.

Chapter 7
Circular Construction Platforms: A Systematic Literature Review

Christian Thuesen, Claes Seehausen Lindtofte, Sophie Führer, Alexander Olin Barfoed, Thor Biering Rohdin, and Janak Raja

Abstract Construction practices are critical for building our societies, but despite increasing focus on sustainability in the industry, the practices are inherently unsustainable – in absolute terms. The circular economy has been identified as a crucial paradigm shift to keep the construction within absolute boundaries of sustainability, covering waste and resources. The needed transformation comes with added complexity, uncertainty, and a requirement to innovate areas that historically have challenged the industry. This paper outlines preliminary research into the challenges of circular construction and how platform thinking can catalyze the sustainable transformation of construction toward circularity. The paper is based on initial findings from two systematic literature reviews of circularity and platform thinking in construction. The review identifies core circular economy challenges like (1) high variance, low volume, (2) short-term project-based optimization, (3) tough price competition, (4) industry fragmentation, and (5) lacking documentation of material flows and performance. Most of these challenges can be addressed by core features of platform thinking like (1) balancing and economy of scale, (2) long-term strategic thinking, (3) value and cost optimization, (4) value chain integration and coordination, and (5) documentation of platform architectures and performance. Thus, the paper finds platform thinking a promising strategy for enabling circular economy in construction, directly addressing SDG 12, and indirectly SDGs 8, 9, 11, and 13.

C. Thuesen (✉) · A. O. Barfoed · T. B. Rohdin · J. Raja
Department of Management Engineering, Technical University of Denmark, Lyngby, Denmark
e-mail: chth@dtu.dk

C. S. Lindtofte
RGS Nordic, Copenhagen, Denmark
e-mail: csli@rgsnordic.com

S. Führer
Group for Sustainability and Technology, ETH Zürich, Zürich, Switzerland
e-mail: sfuehrer@ethz.ch

G. Lindahl, S. C. Gottlieb (eds.), *SDGs in Construction Economics and Organization*, Springer Proceedings in Business and Economics,
https://doi.org/10.1007/978-3-031-25498-7_7

Keywords Circular economy · Platform thinking · SDG 12 · Construction transformation · Sustainability

7.1 Introduction and Purpose

The construction sector has been key in building our modern societies, and today's increasing population and urbanization create global demand for construction. Employing 18 M people and accounting for 9% of the European GDP (EC, 2016), the construction sector plays an important economic and societal role. However, the global industry's productivity rate has stagnated at a 1% growth rate per year over the past 20 years (Barbosa et al., 2017). Pushed by the covid pandemic, McKinsey foresees a transformation of construction "[...] from a highly complex, fragmented, and project-based industry to a more standardized, consolidated, and integrated one" (Bartlett et al., 2020). The 2018 World Urbanization Prospects report from the United Nations (2019) forecasts that 68% (2018, 55%) of the global population will be living in cities by 2050, with a daily expansion of the global urban area by 200,000 people. Metaphorically speaking, the demand for housing and other infrastructure accumulates every month to "another New York City" (Gates & Gates, 2019). Extrapolated to the demand of virgin raw material and emissions, the built environment accounts for 39% of greenhouse gas (GHG) emissions globally (UNEP, 2019) and consumes 40% of material resources; in the EU, the construction sector produces 30% of the waste (European Commission, 2019). Coupled with the forecasted increase in resource consumption, in absolute measures, current construction practices are unsustainable as these contribute to the exceedance of environmental boundaries and resource supply horizons.

The notion of circular economy (CE) is regarded as a promising and necessary solution to cope with future resource demands by aiming at closed material and energy loops (Geissdoerfer et al., 2017; EMF, 2013). Various circular solutions emerge in the construction industry. López Ruiz and colleagues (2020) identified waste management and the recirculation of recovered materials and their application of secondary building materials as promising solution pathways. The needed transformation comes with added complexity, uncertainty, and a requirement to innovate historically challenging the industry. In this context, several academic and industry reviews identify the importance and challenges of implementing CE in construction (e.g., Styles et al., 2018; Osobajo et al., 2020; Ottosen et al., 2021). Various initiatives have been initiated to address the circularity challenges and support the transformation of construction. Despite their potential to address the negative impacts, current practices and the application of circular strategies in the industry remain limited (González et al., 2021). The inherent complexity in the construction sector as a whole and the application of circular solutions specifically call for further actions. These challenges and opportunities related to value chain integration (Osobajo et al., 2020; Ottosen et al., 2021) and documentation (Styles et al., 2018) are currently not supported by systematic research and development activities. Platform thinking is a concept that has developed significantly over the past decade and has proven as a core strategy for handling the increasing complexities of value production. With inherent innovation capabilities and productivity gains (Jones et al., 2021), platforms provide a promising enabler to tackle various challenges of adopting fully circular construction practices and supporting the industry's productivity, leaving this paper with the opportunity of filling this gap.

Following this, the paper sets out to investigate and answer "how can platform thinking act as an enabler of circular economy in the construction industry?" The research applies a literature review to explore the intersections of the construction industry with both the circular economy and platform thinking. This constitutes the foundation for identifying relevant hypothesis of circular construction platforms (CCP). The paper ends by connecting the challenges of transforming toward a circular construction industry with platform thinking as a catalyst.

7.2 Methodology

This paper reviews existing literature within policymaking, industry, and academia. A systematic literature review extracts state-of-the-art knowledge from previous academic studies combined with central publications from industry and policymakers. The investigated topics concern circular economy, platform thinking, and the construction industry creating cross-sectional fields of studies as illustrated in Fig. 7.1.

The field of studies initially guided definitions of the search strings. The search strings were subsequently modified to include different terms of the same topics, ensuring a more fine-grained identification of relevant articles. Table 7.1 presents the reviewed search strings, including the number of papers reviewed.

A structured assessment of literature was conducted, undergoing four steps. The first step presented all papers generated from the search strings, followed by the

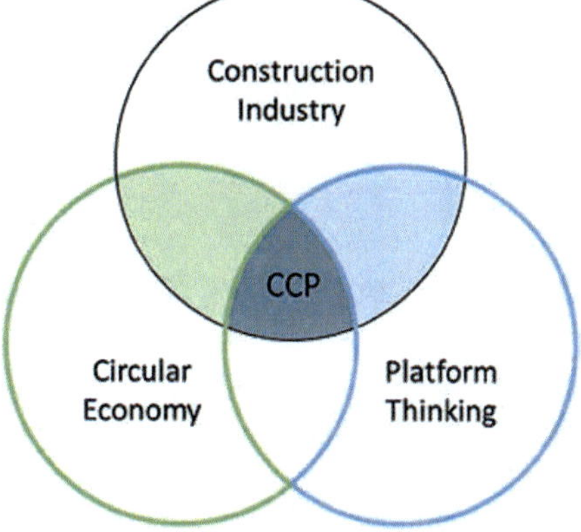

Fig. 7.1 The field of research investigating the integration of (1) the circular economy and (2) platform thinking within the construction industry to derive the intersection for circular construction platforms (CCP)

Table 7.1 Summary of the systematic literature review process and main steps, including the search strings, process, and number of reviewed papers

Search strings	Academic database	Total no. of papers	No. of papers extended	No. of papers screening of titles	No. of papers screening of abstracts
(“Circular economy” and “construction”)	Scopus	1.345	557	100	115
	Web of Science	790	386		
	Dimensions	1.052			
(“Circular economy” or “cradle to cradle”) and (“construction” or “construction industry” or “built environment”)	Scopus	1.489	643	200	
	Web of Science	874			
	Dimensions	584	239		
(“Product platform” and “construction”)	Scopus	86		86	87
	Web of Science	16			
	Dimensions	29			
((“Product platform*” or “mass customization” or “modular”) and (“construction” or “industrial housing” or “built environment”))	Scopus	18.344			
	Web of Science	6.111			
	Dimensions	885			

second step only including English-written papers and open access papers. The most relevant articles were selected based on headings in the third step. The final selection was made by reading the abstracts. Table 7.1 presents the process and number of papers.

7.3 Findings

In the following, we will present the preliminary findings from the literature review concerning circularity and platform thinking, identifying possibilities for developing a framework for circular platforms applicable to the construction industry.

7.3.1 Circularity in Construction the Construction Industry

To attain international goals of keeping resource consumption within certain planetary boundaries, there is a need to reduce the consumption footprint and double the circular material use rate (European Commission, 2020). A key challenge in the construction industry is that all construction and demolition wastes (CDW) derived from buildings' end-of-life stages are currently reused or recycled at a very low rate. Despite a low level of waste disposal sent to landfills, CDW's actual reuse or recycling in Denmark represents less than 36% of materials, where the major part (55%) is recovered and utilized in low-quality applications (MST, 2020). Hossain et al. (2020) likewise address the importance for CE in upgrading the quality of reused and recycled materials and components. An established consensus points toward circular solutions as the tool enabling the construction industry to perform a sustainable transformation (EU, 2020; Osobajo et al., 2020). Two essential topics are presented concerning circular solutions (Osobajo et al., 2020): waste management and resource reuse.

Taking the point of departure today, there are two dimensions describing waste management and resource reuse by either being proactive, enabling circular solutions for future constructions, or applying circular solutions addressing the current building stock. The common demand for both perspectives is data for decision-making. It is crucial to establish reliable CDW data and whereabouts of materials to support decision-makers and policies concerning CDW management and waste reduction in the construction industry (Styles et al., 2018; Hossain et al., 2020). Heinrich and Lang (2019) also address the importance of generating data for decision-making and suggest a forecast model of secondary raw materials acting as the foundation for recovery strategies and recovery mechanisms. A bottom-up approach taken by Lanau and Liu (2020) aims to map materials embedded in construction, such as buildings and roads. These approaches face one major challenge in ensuring and documenting the quality of materials, which act as a crucial parameter for reuse or recycling. Life cycle assessment (LCA) is a well-recognized tool for assessing sustainable potentials and consequences; however, the complexity of generating data is an obstacle to fully integrating this method in decision-making and comparisons to conventional solutions (Ipsen et al., 2021).

Styles et al. (2018) present comprehensive research resulting in 11 specific strategies, including proactive and reactive solutions aiming to reduce, reuse, and recycle CDW, such as designing waste and economic instruments creating economic incentives to use recycled materials. Five design strategies are presented by Ipsen et al. (2021), where Design for Disassembly (DfD) has achieved a lot of attention in the literature due to its simple logic of separating elements or materials and applying them in another construction setting. Styles et al. (2018) argue that DfD has the potential of designing out waste. Despite DfD's popularity, this solution is not unambiguous the correct solution. Critics indicate that a comparison based on a LCA considering DfD concrete elements and upcycled concrete elements does not put DfD in favor but rather proposes a combination of approaches. Furthermore, DfD does not directly produce solutions regarding resource scarcity and how to reuse or recycle the resources of today's buildings. It is well-defined that the construction industry requires circular solutions to accommodate the resource scarcity issue. In contrast, especially concrete has been of high priority for research due to the heavy carbon dioxide (CO_2) emissions (Styles et al., 2018; Gebremariam et al., 2020; Frederiksen & Madsen, 2016) and therefore included as an example of a circular resource.

Reusing concrete aggregates in new concrete elements further implies heavy transportation and processing while facing fierce cost competition against virgin materials to achieve commercial success (Styles et al., 2018; Gebremariam et al., 2020). Alternative solutions such as carving out concrete and applying them for other constructions can overcome the costly and CO_2 heavy transportation and processing, according to Frederiksen and Madsen (2016). However, neither of the mentioned solutions has experienced a commercial breakthrough, not caused by technological ability but rather economical and excess complexity of a reverse supply chain. An established consensus in literature addresses circular solutions' lack of competitiveness due to extended initial investment costs as one of the greatest barriers to adoption in construction (Ipsen et al., 2021; Orsini & Marrone, 2019; Hart et al., 2019). Furthermore, Hossain et al. (2020) present challenges of a reverse material flow where topics such as new business models, modified supply chains, and new processes are important. Aspects of missing policies and legislation, such as requirements for the quality of recovery and financial incentives, also act as barriers to integrating circular solutions when competing against regular solutions (Ipsen et al., 2021; Hart et al., 2019). However, a policy and legislation change can also accelerate market demand in the short and long term.

Sanchez and Haas (2018) take another perspective when addressing circular economy in construction. They argue that the lifecycle impacts of adaptive reuse of existing buildings are superior to new buildings. Furthermore, they present that current pre-project planning tools are insufficient for evaluation circularity. The research focuses on a framework for decision-making called Project Definition Rating Index (PDRI), developed by Construction Industry Institute (CII) and applied for several decades. Their research proposes four new parameters to evaluate the potential of existing buildings equally to new buildings. However, the project price is highly case-specific and can fluctuate, potentially even increasing the cost of new buildings (Table 7.2).

7.3.2 *Platform Thinking in the Construction Industry*

Platform thinking has developed significantly over the past decade and proven as a core strategy for handling the increasing complexities of value production, whether considering Meyer and Lehnerd's (1997) seminal work on product platforms or the rise of platform economies based on digital markets, such as with Airbnb (Parker et al., 2016). Platforms are generally described from either a technical or ecosystem perspective. The technical perspective (Baldwin & Woodard, 2009) views a platform as "a set of stable components that support variety and evolvability in a system by constraining the linkages among the other components" (p. 19). The ecosystem perspective focuses on the actors around the platform ecosystem, where Robertson and Ulrich (1998) define platforms as a collection of assets, such as components, processes, knowledge, people, and relationships all shared between several products. Platforms have proven successful strategies (Gawer, 2011) for achieving

Table 7.2 Summary of the challenges to the adoption of circularity in the construction industry

Challenges	Description	Author, year
1. High variance, low volume	Current construction practices focus on the realization of unique projects. This challenges the adoption of a circular economy since the material from demolitions is case-specific, resulting in high variance and low volume. This is detrimental to the recirculation rate and enforces low-quality application of secondary materials	Osobajo et al. (2020), Hossain et al. (2020), and MST (2020)
2. Short-term project-based optimizations	The traditional industry performance is challenged due to short-term success criteria with minimal long-term organizational learning. This hinders effective learning processes for systematically developing circular solutions that can be leveraged across projects	Styles et al. (2018)
3. Price competition	Cost represents one of the greatest drivers in current construction practices. Circular solutions experience excess and fluctuating costs compared to traditional solutions, e.g., initial investments. This impacts the demand because of lacking possibility of exploiting economy of scale to achieve cost reductions	Sanchez and Haas (2018), Styles et al. (2018), Gebremariam et al. (2020), and Frederiksen and Madsen (2016)
4. Industry fragmentation	The industry is highly fragmented and complex and lacks facilitation of continuity of assets, such as products or services, processes, or teams/people, enabling innovation and economy of repetition. Circular practices introduce new companies, processes, and solutions and thus add additional complexity and fragmentation	Hossain et al. (2020)
5. Lacking documentation of material flows and performance	Existing construction materials used for realizing unique projects are highly regulated and standardized. This challenges circular materials adoption as these lack data and documentation of "production" and formal quality levels	Styles et al. (2018), Ipsen et al. (2021), Heinrich and Lang (2019), and Hossain et al. (2020)

long-term strategic benefits in the automotive, aerospace, and defense industries. A key advantage of platforms is meeting market demands without requiring excessive resources (Robertson & Ulrich, 1998). By achieving high commonality and product adjustability, platforms can exploit economy of scale and still deliver variance in the form of differentiated products. A prerequisite for doing this is the ability to understand the architecture of platforms. Several frameworks exist to analyze and optimize a platform's ability to generate value. This includes Modular Function Deployment (Erixon, 1998), Product Family Master Plans (Harlou, 2006), and Product Variant Master (Hvam et al., 2008). Most of them include different perspectives, coordinating the customer's needs, product design, production processes, and resources.

Research concerning platform thinking in construction is limited (Thuesen & Hvam, 2011; Jones et al., 2021). Traditionally, the sector struggles to resolve the conflict between standardization to minimize cost and the variations in customer demands (Gibb, 2001), making platform thinking a valid candidate to consider when solving this challenge. Thuesen and Hvam (2011) present how a platform building on standardization, strategic partnerships, and continuous platform optimization enables a housing company to improve efficiency and solve the conflict between standardization and variation. By enabling the repetition of assets from project to project, the company's heightened experience, increased efficiency, and reduced costs while meeting the demanded value of the market. Furthermore, Thuesen and Hvam (2011) emphasize that the platform's success depends on initiating and maintaining long-term strategic partnerships between platform participants enabling continuous and structured innovation across the value chain.

Other case studies, including Wörösch et al. (2013) and Kudsk (2013a, b), document how product platforms can supply low-cost housing with a high level of customization. Platform strategies can be implemented from two levels: bottom-up through standardization of components/parts and top-down from building typologies. Wörösch et al. (2013) further suggest that standards and platforms promote the usage of drawings, photos, and prototypes in working descriptions rather than text, which improves communication and productivity. Other research has focused on less practical issues and more on the idea of using platforms in the AEC industry: Jansson et al. (2014) apply a redefinition of the definition of platforms by Robertson and Ulrich (1998), to adapt to the project-based and engineer to order context (ETO).

Today, construction's value and supply chain follow an institutionalized division of labor organized in short-term projects as the primary mode of production; this leads to a fragmented industry along three dimensions (Jones et al., 2021: 2). "Vertically, where different companies deliver different phases of a project (Alashwal & Fong, 2015), Horizontally, when different actors deliver complementary products and services (provided by specialists) at, approximately, the same stage of a process (Fellows & Liu, 2012); and longitudinally, where continuity of teams is disrupted by reassignment at the end of a project, taking any tacit, accumulated knowledge with them (Fergusson & Teicholz, 1996)." The fragmentation challenges the performance of the industry (McKinsey, 2020), driving specialization around institutionalized roles and archetypical business models (Berg et al., 2021) rather than organizing the value chain toward "building better buildings."

The organizational importance of platform thinking is also realized in Jones' et al. (2021) investigation of digital product platforms in UK construction firms. The well-defined product platform facilitates the development of capabilities and allows the integration of a fragmented industry. The horizontal integration captures specialized capabilities within the digital product platform and facilitates repetition across projects. Looking at the vertical integration, an essential finding is how the coordination between the design and manufacturing phase can be achieved by contracting instead of acquisitions while still managing to capture the tacit knowledge generated and exploiting the innovative bottom-up solutions (Table 7.3).

Table 7.3 Summary of enablers caused by platforms

Enablers	Description	Author, year
1. Balancing customization and economy of scale/ repetition	Platforms enable customization with near mass production efficiency as it shares the theoretical underpinning with mass customization	Gibb (2001), Bonev et al. (2015), and Kudsk (2013a, b)
2. Long-term strategic thinking	Platforms enable long-term strategic development as it systematically looks for similarity across customers and avoids sub-optimization in projects	Thuesen and Hvam (2011) and Kudsk (2013a, b)
3. Value and cost optimization	Platforms enable systematic productivity development by separating the value production and cost reduction inspired by lean thinking	Robertson and Ulrich (1998) and Thuesen and Hvam (2011)
4. Value chain integration and coordination	Platforms enable integration and coordination of value chains through standardization, repetition, strategic partnerships, and structured innovation	Jones et al. (2021) and Thuesen and Hvam (2011)
5. Documentation of platform architectures and performance	Platforms enable transparency and comparability-based standardization and repetition among strategic partners. This is a central prerequisite to generating data from and assessing performance and consequences across products and processes	Thuesen and Hvam (2011), Harlou (2006), and Kudsk (2013a, b)

7.4 Discussion

7.4.1 Hypotheses of Circular Construction Platforms

Based on the preliminary results from the literature review, it should be apparent that platform thinking can address some of the core challenges of circular construction. Table 7.4 juxtaposes the identified challenges of CE with the enablers of platform thinking. We will in the following discuss how this translates into several hypotheses for circular construction platforms.

H1: Platforms enable variance of circular solutions and secondary materials The platform's ability to handle customizations while leveraging similarity across projects can address the current challenges of high variability and low volume of secondary materials. However, this will also challenge the current understanding of platform thinking that traditionally targets "end users" with variability. The concept of "end users" is problematic in a circular economy because there potentially is no end to using materials. Thus, platforms need to handle not only the variability toward construction but also the process of deconstructing buildings.

H2: Platforms enable the development of circular solutions that are relevant through time The current practices favor one-of-a-kind solutions that are contextualized to specific projects. While this customization strategy can potentially ensure the reuse of a small volume of secondary materials, the developed solutions would be subject to suboptimization that hinders long-term value delivery. On the

Table 7.4 Contrasting the challenges posed to the adoption of CE in construction with the enabling potential of platform thinking

Circularity challenges		Platform enablers
High variance, low volume	1	Balancing customization and economy of scale/repetition
Short-term project-based optimization	2	Long-term strategic thinking
Price competition	3	Value and cost optimization
Industry fragmentation	4	Value chain integration and coordination
Lacking documentation of material flows and performance	5	Documentation of platform architectures and performance
…		…

other hand, platforms enable long-term optimization that supports the basic principles of the circular economy.

H3: Platforms enable productivity development of circular solutions Current circular solutions are not competitive in the market, focusing on short-term costs at the expenses at long-term costs and value. The long-term perspective of platforms creates different incentive structures that separate and optimize value and cost, targeting the required productivity development for making secondary materials competitive. This can also be realized through increasing taxes and enforcing the role of politics and regulation.

H4: Platforms enable organizational specialization toward complex circular solutions The current fragmentation of the industry leads to lost knowledge from project to project, inefficient decision-making due to lack of knowledge and communication, and know-how remaining at the individual level (Jones et al., 2021). Platforms enable the pursuit of value-adding repetitions driving specialization in increasing complex circular solutions for certain markets and customers.

H5: Platforms enable detailed documentation of circular practices and solutions The current industry practices rely on general standards for documenting processes and products that are abstract, not always verifiable and comparable. This makes continuous improvement of the performance of circular solutions difficult, hampering wider diffusion in the industry. Platforms create shared standards for organizing products, processes, and organizations that are followed across projects. This makes the necessary infrastructure for monitoring the development of circular solutions.

7.4.2 Limitations and Further Research

Where the current construction value chain is thus not fit for purpose, handling the increasing complexity and uncertainty of the circular economy, the above-identified hypothesis documents the potential for introducing platform thinking as a catalyst for transforming construction toward circularity.

Today, there are only limited examples of leveraging platforms for sustainability – and even less so within the area of circularity. Minunno et al. (2018) and Mignacca et al. (2020) represent noteworthy exceptions targeting prefabrication and energy infrastructure, respectively. Thus, there is both industrial and academic potential for developing platforms that systematically improve the competitiveness of construction in general and of circular construction solutions in particular.

Platforms offer an organizational and technical order that allows specialization within a certain market and types of projects. Platforms create organizational learning and innovation infrastructures that allow for specialization and optimization across the supply and value chain. Moreover, it obstructs the development and implementation of circular practices requiring system-based long-term thinking, detailed understanding of the materials used, their assembly, and performance throughout the life cycle. This should be subject to further research and innovation in the industry.

While platforms represent a promising catalyst, they should not be seen as a silver bullet for handling all challenges of circular construction. Further research is also needed into the documentation of materials (Ottosen et al., 2021) and the broader shaping of markets through industry standardization and regulation (Ipsen et al., 2021; Hart et al., 2019).

7.5 Conclusion

CE takes two overall perspectives on the construction industry. Firstly, design solutions can enable future buildings to adopt circularity, and secondly, the potential of today's CDW as substitutes for primary resources. Combining the two perspectives into a reverse supply chain delivers processed or directly recycled CDW to manufacturers that act as suppliers of material or components for the construction industry. However, the literature indicates several barriers to withholding CE's integration in the construction industry. Platforms connecting relevant parties have proven effective in driving innovation and cost reduction and operating complex settings as established in construction. Thus, platform thinking holds the promise of overcoming the barriers identified to complete a sustainable transformation toward circularity.

This is important for the realization of the SDGs. For the global community to attain the international goal SDG 12 of keeping the resource consumption within certain planetary boundaries, there is a need to reduce the consumption footprint and double the circular material use rate (EU, 2020). The provision of sustainable cities and communities (SDG 11) is directly tied to the countries' ability to reduce their energy consumption and carbon emissions through circular construction practices. The application of circular construction platforms can directly address SDGs 12 and 11 and indirectly enable reduced climate change (SDG 13) and contribute to SDGs 8 and 9.

Acknowledgments The authors want to thank several persons and organizations for supporting the project. We want to thank Realdania for financially supporting the project and the companies, Næste, RGS Nordic, GxN & Enemæke & Pedersen, for verifying the preliminary findings.

References

Alashwal, A. M., & Fong, P. S.-W. (2015). Empirical study to determine fragmentation of construction projects. *Journal of Construction Engineering and Management, 141*(7), 04015016.

Baldwin, C. Y., & Woodard, J. J. (2009). The architecture of platforms: A unified view. In A. Gawer (Ed.), Platforms, Markets and Innovation (pp. 19–44). Edward Elgar: Cheltenham, UK and Northampton, MA, USA

Barbosa, F., Woetzel, J., Mischke, J., Ribeirinho, M. J., Sridhar, M., Parsons, M., Bertram, N., & Brown, S. (2017). *Reinventing construction: A route to higher productivity*. McKinsey Global Institute.

Bartlett, K., Blanco, J., Fitzgerald, B., et al. (2020). *Rise of the platform era: The next chapter in construction technology*. McKinsey & Company.

Berg, J. B., Thuesen, C., Ernstsen, S. K., & Jensen, P. A. (2021). Reconfiguring the construction value chain: Analyzing key sources of friction in the business model archetypes of AEC companies in strategic partnerships. *Construction Management and Economics, 39*, 1–16.

Bonev, M., Wörösch, M., & Hvam, L. (2015). Utilizing platforms in industrialized construction: A case study of a precast manufacturer. *Construction Innovation, 15*(1), 84–106.

Ellen McArthur Foundation [EMF]. (2013). *Towards the circular economy. Economic and business rational for an accelerated transition.*

Erixon, G. (1998). *Modular function deployment.*

European Commission [EC]. (2016). *The European construction sector: A global partner.*

European Commission. (2019). Trend paper. EU construction sector: In transition towards a circular economy. European Construction Sector Observatory.

European Commission. (2020). A new circular economy action plan. European Topic Centre on Waste and Materials. European Union.

Fellows, R., & Liu, A. M. M. (2012). Managing organizational interfaces in engineering construction projects: Addressing fragmentation and boundary issues across multiple inter-faces. *Construction Management and Economics, 30*(8), 653–671.

Fergusson, K. J., & Teicholz, P. M. (1996). Achieving industrial facility quality: Integration is key. *Journal of Management in Engineering, 12*(1), 49.

Frederiksen, L. K., & Madsen, U. S. (2016). *Idékatalog over nye designstrategier for genanvendelse* (p. 112). Et InnoBYGprojekt. ISBN: 9788778303899.

Gates, B., & Gates, M. (2019). We didn't see this coming – Nine surprises that have inspired us to act. *GatesNotes*. Available at: https://www.gatesnotes.com/2019-Annual-Letter#ALChapter3. 29/01/2022.

Gawer, A. (2011). Platforms, markets and innovation. Edward Elgar: Cheltenham, UK and Northampton, MA, USA.

Gebremariam, T., et al. (2020). Innovative technologies for recycling End of Life concrete waste in the built environment. *Resources, Conservation and Recycling, 163*, 104911. ISSN: 18790658. https://doi.org/10.1016/j.resconrec.2020.104911

Geissdoerfer, M., Savaget, P., Bocken, N., & Hultink, E. (2017). The Circular Economy – A new sustainability paradigm? *Journal of Cleaner Production, 143*, 757–768.

Gibb, A. G. (2001). Standardization and preassembly distinguishing myth from reality using case study research. *Construction Management & Economics, 19*(3), 307–315.

González, A., Sendra, C., Herena, A., Rosquillas, M., & Vaz, D. (2021). Methodology to assess the circularity in building construction and refurbishment activities. *Resources, Conservation & Recycling Advances, 12*, 12.

Harlou, U. (2006). *General rights developing product families based on architectures: Contribution to a theory of product families*. Technical University of Denmark.

Hart, J., Adams, K., Giesekam, J., Tingley, D. D., & Pomponi, F. (2019). Barriers and drivers in a circular economy: The case of the built environment. *Procedia CIRP, 80*, 619–624. https://doi.org/10.1016/j.procir.2018.12.015

Heinrich, M. A., & Lang, W. (2019). Capture and control of material flows and stocks in urban residential buildings. *IOP Conference Series: Earth and Environmental Science, 225*, 1. ISSN: 17551315. https://doi.org/10.1088/1755-1315/225/1/012001

Hossain, M. U., Ng, S. T., AntwiAfari, P., & Amor, B. (2020). Circular economy and the construction industry: Existing trends, challenges and prospective framework for sustainable construction. *Renewable and Sustainable Energy Reviews, 130*, 109948. https://doi.org/10.1016/j.rser.2020.109948

Hvam, L., Mortensen, N. H., & Riis, J. (2008). *Product customization*. Springer Science & Business Media.

Ipsen, K. L., et al. (2021). How lack of knowledge and tools hinders the eco-design of buildings a systematic review. *Urban Science*. https://doi.org/10.3390/urbansci5010020

Jansson, G., Johnsson, H., & Engström, D. (2014). Platform use in systems building. *Construction Management and Economics, 32*(1), 70–82.

Jones, K., Mosca, L., Whyte, J., Davis, A., & Glass, J. (2021). Addressing specialization and fragmentation: Product platform development in construction consultancy firms. *Construction Management and Economics., 40*(11-12), 918–933.

Kudsk, A., Hvam, L., Thuesen, C., O'Brien Grønvold, M., & Olsen, M. H. (2013a). Modularization in the construction industry using a top-down approach. *Open Construction & Building Technology Journal, 7*, 88–98. https://doi.org/10.2174/1874836801307010088

Kudsk, A., O'Brien Grønvold, M., Olsen, M. H., Hvam, L., & Thuesen, C. (2013b). Stepwise modularization in the construction industry using a bottom-up approach. *Open Construction & Building Technology Journal, 7*, 99–107. https://doi.org/10.2174/1874836801307010099

Lanau, M., & Liu, G. (2020). Developing an urban resource cadaster for circular economy: A case of Odense, Denmark. *Environmental Science and Technology, 54*(7), 4675–4685. ISSN: 15205851. https://doi.org/10.1021/acs.est.9b07749

López Ruiz, L. A., Ramón, X. R., & Domingo, S. G. (2020). The circular economy in the construction and demolition waste sector–A review and an integrative model approach. Journal of Cleaner Production, 248, 119238.

McKinsey. (2020, June). *The next normal in construction: How disruption is reshaping the world's largest ecosystem*.

Meyer, M. H., & Lehnerd, A. P. (1997). *The power of product platforms*. Simon and Schuster.

Mignacca, B., Locatelli, G., & Velenturf, A. (2020). Modularisation as enabler of circular economy in energy infrastructure. *Energy Policy, 139*, 111371. https://doi.org/10.1016/j.enpol.2020.111371

Miljø- og førevareministeriet [MST]. (2020). *Affaldsstatestik 2019*. ISBN: 978-87-7038-249-6.

Minunno, R., O'Grady, T., Morrison, G. M., Gruner, R. L., & Colling, M. (2018). Strategies for applying the circular economy to prefabricated buildings. *Buildings, 8*(9), 125.

Orsini, F., & Marrone, P. (2019). Approaches for a low-carbon production of building materials: A review. *Journal of Cleaner Production, 241*, 118380. ISSN: 09596526. https://doi.org/10.1016/j.jclepro.2019.118380

Osobajo, O. A., Oke, A., Omotayo, T., & Obi, L. I. (2020). *A systematic review of circular economy research in the construction industry Smart and Sustainable Built Environment., 11*(1), 39–64.

Ottosen, L. M., et al. (2021). Implementation stage for circular economy in the Danish building and construction sector. *Detritus., 17*, 2021. https://doi.org/10.31025/2611-4135/2021.15110

Parker, G., Van Alstyne, M. W., & Jiang, X. (2016). *Platform ecosystems: How developers invert the firm*. Boston University Questrom School of Business Research Paper (2861574). Chicago.

Robertson, D., & Ulrich, K. (1998). Planning for product platforms. *Sloan Management Review, 39*(4), 19.

Sanchez, B., & Haas, C. (2018). Capital project planning for a circular economy. *Construction Management and Economics, 36*(6), 303–312.

Styles, D., Gálvez-Martos, J. L., Schoenberger, H., & Zeschmar-Lahl, B. (2018). Construction and demolition waste best management practice in Europe. *Resources, Conservation and Recycling, 136*, 166–178.

Thuesen, C., & Hvam, L. (2011). Efficient on-site construction: Learning points from a German platform for housing. *Construction Innovation, 11*(3), 338–355.

UNEP. (2019). *2019 global status report for buildings and construction sector*, 11 December 2019.

United Nations Department of Economic and Social Affairs [UN]. (2019). *World urbanization prospects. The 2018 revision, demographic research*. New York. Available at: https://population.un.org/wup/Publications/Files/WUP2018-Report.pdf

Wörösch, M., Bonev, M., & Mortensen, N. H. (2013). Product platform considerations on a project that develops sustainable low-cost housing for townships. In *CIB world building congress*, Brisbane.

Chapter 8
Blue Ocean Strategy for Business Case of Building Components Designed for Disassembly

Artur Tomczak and Ole Jonny Klakegg

Abstract Sustainability requires us to show careful consideration for nature and future generations. When designing new structures, along with optimisation and cautious material selection, we should also ensure their long-term usefulness. One way to do this is to reuse whole building components, which is known from history to be practical. Life-cycle assessments prove this circular practice to be more environmentally friendly than recycling. Designing adaptable building components for disassembly and reassembly is feasible but not popular. This paper looks at the viability of such a product offering and conceptualises a business case using the Blue Ocean Strategy framework. The analysis is based on data coming from literature, case studies and interviews with practitioners. The business case of adaptable building components not only is built on the premise of subsequent uses of the products but also shows immediate benefits such as a fast assembly process. From a solely economic perspective, such products bring primary value in attracting more clients willing to pay an additional price for more sustainable buildings. Such an offering also helps to form a circular economy market of reusable products, which is desired by European incentives. The results compare and distinguish the circular business case with contemporary alternatives – monolithic and prefabricated structures. The paper provides guidelines for harnessing the value of prefabricated building elements designed with the intention of multiple applications and developing a circular economy business strategy in the built environment. Necessary preconditions, limitations and barriers are also discussed.

Keywords Circular economy · Business case · Blue Ocean Strategy · Adaptable building elements · Design for disassembly (DfD)

A. Tomczak (✉) · O. J. Klakegg
Department of Civil and Environmental Engineering, Norwegian University of Science and Technology (NTNU), Trondheim, Norway
e-mail: artur.b.tomczak@ntnu.no; ole.jonny.klakegg@ntnu.no

G. Lindahl, S. C. Gottlieb (eds.), *SDGs in Construction Economics and Organization*, Springer Proceedings in Business and Economics,
https://doi.org/10.1007/978-3-031-25498-7_8

8.1 Introduction

The Industrial Revolution has brought us significant advancements, allowing us to produce goods in high volume efficiently. High performance and energy-intensive manufacturing processes also have side effects – businesses optimise for short-term profits at the expense of society and the environment. Nowadays, with raising awareness of anthropogenic influence on climate change, sustainability gains attention also in construction. To minimise the greenhouse gas (GHG) emissions, designers righteously try to improve buildings' energy efficiency, but as they progress and as countries shift towards renewable energy sources, the embodied emissions of materials begin to play a very important role (Röck et al., 2020; Wiik et al., 2018).

Fivet and Brütting (2020) name three ways to address the issue of embodied GHG: (1) optimise the design to be material-efficient, (2) employ low-carbon materials, and (3) ensure long-term usefulness of elements. The first two are usually considered by the designers, but the third is overlooked, causing a mismatch between designed and actual lifespans. Speaking of time aspect, Wilkinson et al. (2014) describe three types of building lifespans: (1) the technical, driven by safety, dependent on the physical condition, that can be prolonged by maintenance; (2) functional, which ends when the building limits the use and is no longer fit for requirements and needs of the users; and (3) economic, as long as buildings generate more income than costs. Authors also list other factors that affect building usefulness – social, like fashion or demographic; legal, when a building is not compliant with regulations; and political such as zoning or heritage. When one of the lifespans ends, the building becomes obsolete or even demolished, causing underutilised materials to become waste.

Separation of the technical lifespan from the economic and functional enables the utilisation of materials' full potential. Such a breakup is a key topic of literature on circular practices. European Commission (2015) describes circular economy (CE) as an approach 'where the value of products, materials and resources is maintained in the economy for as long as possible, and the generation of waste minimised'. Bocken et al. (2016) further extend the definition by providing three types of circular economy practices: (1) slowing resource loops by design of long-life goods and product life extension (repair, reuse, remanufacturing), (2) closing resource loops by using resources over and over again, avoiding taking out virgin resources (recycling), and (3) narrowing resource loops, or increasing resource efficiency, by using fewer resources per product.

The revolution in space exploration with reusing rockets, caused what is described as the 'SpaceX Effect' (Reddy, 2018). What used to be perceived as impossible became almost a standard, with international companies practicing the reuse of spaceship parts. Similarly, we have reasons to foresee such turn of events in the construction industry, which is simpler than space travel, but because the economic scale can cause even higher disruption. In fact, reuse of building components is nothing new and was practised for centuries (Fivet & Brütting, 2020). It used to be common for Norway to construct buildings with logs and stones that previously served as parts of other buildings. Prefabrication, modularity, reversible connections, and standardisation of elements are among the factors that enable the reuse of structural components (ISO, 2020) that are often already applied irrespective of future reuse.

Despite vast scientific literature on circular economy, there is little visible conversion to viable businesses, and over 97% of materials consumed in Norway are not cycled back to the economy (de Wit et al., 2020). This conceptual paper aims to reflect on the viability of prefabricated building elements designed for disassembly and reusability. Since the highest embodied carbon is attributed to a building's structure (Wiik et al., 2018), the paper focuses on structural elements. We explore the hypothesis that such components are an unexploited niche with profit potential, which we investigate with help of Blue Ocean Strategy tools from Kim and Mauborgne (2015). The paper intends to contribute with business advice for new or existing enterprises by blending economics and construction engineering domains. This paper is not offering a business model, understood as the blueprint for generating profit off a product or service, strategic choices, or value creation (Jensen, 2013). Instead, it is about the business case to explain why the investment is worth doing.

8.2 Methodology

This is a conceptual paper with theory adaptation (Jaakkola, 2020). Our data is coming from existing literature, documented case studies, national statistics, as well as qualitative interviews with convenient sample of 12 industry representatives. The structure of the paper starts with an explanation of the theoretical framework, followed by the analysis of aspects influencing the business case of component reuse: feasibility, sustainability, and economic consideration. Based on the analysis, we formulate the Blue Ocean Strategy proposal. The complete picture is put together in a discussion and concluding part. Due to the limitations of a conference paper, this study will be derived from existing numerical cost estimates and conceptual logic rather than quantifying each element of the business case.

8.3 Theoretical Framework

8.3.1 *Business Case*

The key function of a business case (BC), according to OGC (2009b), is to justify the effort to be invested in the project. According to Gambles (2009) it is '*a recommendation to decision makers to take a particular course of action for the organisation, supported by an analysis of its benefits, costs and risks compared the realistic alternatives, with an explanation of how it can best be implemented*'. The BC defines why the work needs to be done and provides a crucial baseline for the project (OGC, 2009a). It is typically prepared at an early stage, where information is scarce, of low precision and with high levels of uncertainty.

The process of developing a BC includes understanding the needs and priorities of stakeholders, collecting relevant, correct, and updated data, modelling and processing those data, analysing the contents, quality and consequences of the data, facilitating workshops when needed, project planning, and finally preparing the information for decision-makers. Benefits and other consequences should be expressed as tangible as possible. Respecting that some effects are controversial or not possible to specify and quantify, the BC will be much more useful and likely to lead to the right decision and action if it is explicitly clear about consequences and quantifies effects.

Considering whether to invest or not, the first question is if the benefits outweigh the cost. Only if the benefits are more significant than the cost would a rational actor choose to invest. Investors normally define profit margins for estimates to meet. However, the decision is not as simple as it looks, given that different aspects of benefit weigh differently for different stakeholders. Benefits are open for interpretation by the decision-makers. The next question is if this course of action is better than the realistic alternatives. Investment funds are limited, so any rational actor needs to consider the best investment at hand. The most fundamental test is whether the investment is better than not investing. The reference alternative or 'zero-alternative' needs to be tested, as this decides whether an investment should be made or not. Then there are other investment alternatives. There is no need for a BC if there is really no alternative. Gambles (2009) points out that authors of BC often pretend there are no alternatives, but this usually is not the case. In the Norwegian public sector, the requirement is presenting minimum two real investment alternatives in addition to the reference alternative (Norwegian Ministry of Finance, 2019). This is intended to make the project proponents consider the whole space of opportunities.

The time perspective is another question. Ideally, one would look at all consequences following the investment – for all times. Practically, however, it is usual to define a limited timeframe. The BC needs to consider the short- and long-term perspectives. The limited knowledge of consequences far into the future and the fact that it is hardly possible to connect those future effects to the specific project are two reasons. Other reasons include the technical lifetime of physical assets and the requirement for nondiscrimination of alternatives. Another aspect of a long timeframe is the effect of discounting. Calculations of net present value reduce future effects by interest effect and make the real long-term effects disappear from the basis for decision-making. As demonstrated by the Norwegian Ministry of Finance (2014), one practical way to handle this is to define a standard timeframe to be used in a comparison of alternatives and an interest rate that diminishes with a longer time horizon to reduce the effect of calculating away future consequences. However, these solutions are not ideal, and discussions about how to make sustainable decisions are still ongoing.

8.3.2 *Blue Ocean Strategy*

Since we are conceptualizing an unexploited market opportunity, we apply an analytical framework called Blue Ocean Strategy, developed by Kim and Mauborgne (2015). Blue Ocean Strategy is the desired situation with no direct competition, as opposed to a metaphorical red ocean full of competitive predators. The red ocean is characterised by high saturation of similar profile enterprises, already significantly optimised, and focused on benefit-cost trade-offs. In red oceans, businesses tend to focus on competition rather than a customer. It is usually the known market space, while blue oceans represent the undiscovered market. Blue Ocean Strategy is not focused on outperforming competitors in what they are good at but offering a quantum leap – value that is not addressed at the known market. It is not about achieving a state of monopoly but rather staying ahead of the market.

The key tool we apply from the work of Kim and Mauborgne (2015) is 'the strategy canvas' – a chart with aspects that a product competes in presented on a horizontal axis. The vertical axis represents the offering level that a client receives. The higher the score, the more company offers to clients with their product. The strategy canvas is about showing the relative performance in various fields of competition. To develop the chart, we implement the 'Four Actions Framework', where each action requires answering the related question: (1) eliminate (which of the aspects taken for granted in the industry need to be eliminated?), (2) reduce (which of the aspects should be reduced below the existing industry standard?), (3) create (what are the aspects that need to be created hitherto never offered by the industry?), and (4) raise (which of the aspects should be elevated above the current industry standard?).

8.4 Analysis

8.4.1 *Feasibility of Building Product Reuse*

The first question of feasibility is if the structural building products, such as columns, beams, walls, and floors, allow being taken apart and constructed into a new building. Fivet and Brütting (2020) provide examples of structures successfully built with steel and timber reclaimed components. Two interview respondents mentioned that it is easier to refurbish or reuse buildings made over half a century ago than those more modern ones. 'We all agree that [buildings from before 70s] are easier to adapt than ones from the 80s and 90s; they are over-designed. You can put a building on top, or a roof garden. (...) Even today, we're designing buildings that just meet requirements, because that's more economical.' Another interviewee adds: 'They should not be old enough to tear down, but those are the ones we really want to tear down because they are hard to get quality in refurbishment, but when you have old buildings it is easier actually'. Lindheim (2021) in her study on buildings as 'material banks' concluded that some building owners already design for reusability, especially where there is frequent inventory replacement and need for flexibility so that reuse pays off quickly.

The complexity of such reuse depends on the individual product. Ceramic bricks are quite easily retrievable from walls, but that is not the case for modern porous ceramic blocks, which crush on separation. As we hear from the interviews, timber products – lumber, CLT, glulam – are usually screwed together, but whether they can be unscrewed depends on screw type and length. Both timber and steel products can be cut out and reused as shorter elements. If bolted connections are applied, elements can be reused without processing. Concrete is harder to reuse since most of the applications use monolithic connections. Reinforced concrete can be cut, but not without loss of reinforcement continuity and cover. Rather exceptional is the use of prefabricated concrete with bolted connections, easily accessible but covered with removable lime mortar (Paananen & Suur-Askola, 2018). Nonpermanent connections are more expensive but easier and faster to perform on-site. The case study of 'KA13' project confirms that reuse of building products is feasible and shows the practical side of reclamation (FutureBuilt et al., 2021). The building's foundations, usually made of concrete and bound to the ground, are the hardest to prefabricate and connect reversibly. The solution for reusing foundations could be to make them reusable where they already are in situ. To enable their reuse, designers should provide connection points and specify the allowance for future loads.

ISO 20887 (ISO, 2020) provides guidance on design for disassembly (DfD) and adaptability, showing how to increase reuse suitability. Among recommendations are standardisations of elements, exposing joints for easier access, preparation of a disassembly plan, and digitalisation of all relevant information for future users. Other standards, such as the recent Hollow Core Slabs for Reuse (NS 3682, 2022), go even further and provide detailed procedures for processing, assessment, and documentation of used products.

Another aspect is the lifespan of such products. In Europe, most buildings are designed for 50 years, according to the Eurocodes. However, there is a mismatch between the designed lifespan and the actual. In Denmark, residential lifespan is on average 120 years (Aagaard et al., as cited in Marsh, 2017), while in the United States, it is closer to 60 years (Aktas & Bilec, 2012), and in China and Japan, the average is even less than three decades (Wang et al., 2018; Wuyts et al., 2019). Statistical data for Norway show that buildings, on average, have a lifespan of three to six decades (Todsen, 2014). Wilkinson et al. (2014) describe three types of building lifespans: (1) the technical, driven by safety, dependent on the physical condition, that can be prolonged by maintenance; (2) functional, which ends when the building limits the use and is no longer fit for demand; and (3) economic, as long as buildings generate more income than costs. Rarely structural material degradation is the first reason for a building's unusability. The oldest wooden church in Norway, Urnes stavkirk, which dates back to the twelfth century (UNESCO, n.d.), proves that materials can serve us much longer than we usually designed them for. When an economic or functional lifespan ends while the technical conditions are still good, it causes a mismatch, leading to the waste of valuable materials. As temporary structures prove, with appropriate design, technical lifespan can be separated from the other lifespans, preventing the waste of quality materials.

Although technically possible, there are many barriers to reuse. According to research done by Lindheim (2021), key barriers are (1) compliance with rigid rules and regulations, (2) lack of financial incentives in the market, (3) lack of a system for reuse of building materials and components to make the outcome predictable for everyone involved, and (4) knowledge gaps.

8.4.2 *Environmental Sustainability*

Second is the environmental impact on the natural environment by energy consumption, greenhouse gas emissions, deforestation, pollution, land occupation, etc. Reusable components are not neutral to the environment; they still affect it negatively, but they prevent waste production, and it can be argued that also the production of a new component that would double the environmental impact.

The benefits of reuse should be reflected in the Life-Cycle Assessment (LCA) – the process of measuring a building's environmental performance. How to account for the benefits of reuse in the LCA is not straightforward. In some studies, reuse and recycling are incorporated at phase D of the LCA, called 'benefits and loads beyond the system boundary'. Robert Crawford (2011) writes that the initial use of such elements should not be credited because no guarantee can be made that they will actually be reused. His approach is more realistic than the first one but also means that an element that will be used twice (or more) will have multiple times higher LCA values on first use than on the following ones. On the other hand, people might take advantage of this assumption and sell components that were in use for a short time as low embodied carbon products, arguing that product production stages (A1–A3 of LCA) were already addressed in the first use cycle. The difference is significant since the product stage is responsible for most of the GHG emissions of building materials – for prefabricated steel and concrete even exceeds 95% (Norwegian EPD Foundation, 2021).

The case study project 'KA13' compares GHG emissions of used products with newly produced and achieves 97% reduction for steel, 89% for hollow core slabs, and 92% for windows (FutureBuilt et al., 2021, after Walter & Høydahl, 2020). In research by Brütting et al. (2020), more conservative assumptions are taken, with more energy demanding disassembly and post-processing, and reuse is compared to new steel of much lower embodied carbon. Still, the result shows reduction of over half of GHG emissions when elements are reused.

8.4.3 Economic Consideration

The report by 3XN, supported by the Danish Environmental Agency, shows a case study of a representative office building in Denmark with a new built value of DKK 860 million (Sommer et al., 2016). The study compares traditional demolition with the circular approach. The added costs of investing in reversible elements compliant with CE model are estimated to be only 0.35% of the total value of the building. If such a building needed to be demolished, it would cost additional DKK 16 million (1.9%), but if the disassembly was possible, DKK 35 million (−4.1%) could be claimed from the resale of the superstructure components, 8% on entire buildings, and even 16% in 50 years considering projected material prices.

Previously mentioned 'KA13' study which achieved significant GHG reduction shows that while reuse of some products is cheaper (e.g. windows ~60% cheaper), steel building products are 49% more expensive than buying new material (FutureBuilt et al., 2021, after Walter & Høydahl, 2020). Several interviewees blame this on low price of virgin materials: 'It's much too cheap to tear down the building, and we're not paying enough for landfill' – says one of them – 'I can't understand how virgin steel from China can outcompete steal from local steel providers in Norway, made with renewable energy and fully from recycled steel'.

In a short-time perspective, the proposed products are more expensive than alternatives. However, when the total cost of ownership is considered, adaptable components gain an advantage because apart from the product price, this measure incorporates the product's quality, delivery (including assembly), and maintenance costs. For a higher price, they provide better quality, durability, and reduce uncertainty by allowing for inexpensive building adjustments.

The cost ratio between reuse and new is also foreseen to drop with improved circular processes, especially access to information, establishment of circular market, and experience. The trends on Fig. 8.1 show that in case of Norway, each year more and more demolition waste is being produced. With raising raw material prices due to their scarcity (sand, cement, iron ore, asphalt) and emission trading systems,

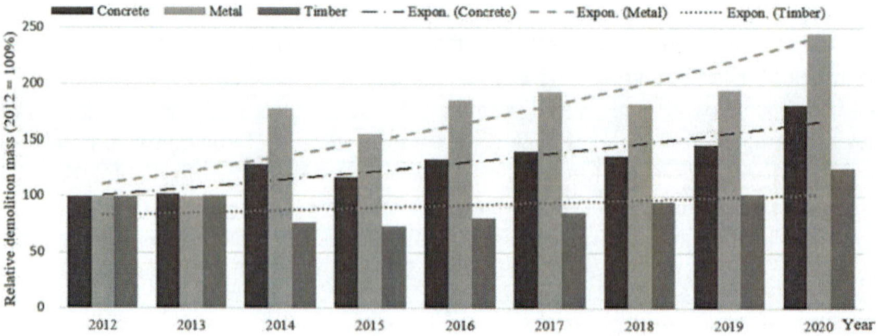

Fig. 8.1 Relative mass demolition for Norway. (Source: SSB, 2021)

as well as raising prices for waste disposal, reuse of building components is expected to become cheaper alternative in near future.

The business case of adaptable building components is built on reduced embodied carbon emissions. Having optimised building stock with extended material flows before they are discarded have a less destructive impact on the environment. From a solely economical perspective, this brings value in attracting more clients who are willing to pay an additional price for more sustainable buildings. Certification systems like BREEAM and DGNB also assign points for reuse practices, which are translated to higher real estate value. Larry Fink, Chief Executive Officer at BlackRock company managing the largest assets globally, writes in his annual letter 'We focus on sustainability not because we're environmentalists, but because we are capitalists and fiduciaries to our clients. (...) Every company and every industry will be transformed by the transition to a net zero world. The question is, will you lead, or will you be led?' (Fink, 2022).

In economics, the term 'Betongold' describes the supposed security of real estate ('concrete') from falling in value, especially in times of crisis. Real estate value is usually linked to the usable floor area and the building's location. While traditional building components are single-use, adaptable components preserve value independently of the building's lifetime and location. For that reason, they can be considered separate financial instruments that gain value with the rise of the circular economy market and provide secure material stock, independent of uncertain raw material prices. Some even conceptualise the idea of 'Building Components as a Service' (Wainwright, 2020), where elements are rented in a take-back system, and the manufacturer or retailer remains the sole owner.

Another economic aspect of adaptable building components is that financial institutions are willing to offer better conditions to sustainable investments, which is already a case for refurbishments loans or mortgages on timber buildings. This use of economic incentives to drive sustainable investments will be even stronger in the future. In June 2020, the EU commission published the EU taxonomy (European Commission, 2019), a classification system for investments. The purpose is to move capital to sustainable investments, which is expected to have considerable effect on the building and construction industry. According to the 'Circular Economy – Principles for Building Design' (EU, 2020), enhancing durability will decrease the financial risk and suggest that finance/insurance companies could specify in requirements standards for due diligence when assessing the circularity of the project. Authors also advice looking at investments from the life-cycle costing (LCC) perspective to grasp the increased revenue streams and capitalise future risks of difficulty to deconstruct buildings and cost of waste management.

The 'triple bottom line' has, for decades, defined sustainability. The first, economic, and, second, environmental, have already been described. The third bottom line is the social impact. Adaptable components address that aspect by opening a new market for disassembly, diagnostics, storage, transportation, and assembly, providing job and development opportunities. Among other social benefits is better adaptation to temporary needs, leading to less obsolete buildings, better land utilisation, and easier recovery after destruction due to natural or anthropocentric causes.

As urbanisation progresses, so does increase the need for change in our physical built environment. Example of shopping malls becoming obsolete after customers switch to online shopping shows how building stock does not reflect changes in people's lifestyles fast enough. Adaptable building components are a means to improve a building's flexibility. Another example is building according to the actual demand, not predictions. Instead of building a school with a constant capacity for decades, the space could be extended or reduced according to the actual demographics. Considering social benefits, second-hand components should be cheaper than brand new, making it more affordable for young and low-earning social group. Ability to modify the building for a reasonable price would allow customers to build houses they can afford without much mortgage and extend it later. It would also work in the other direction; they could reduce the size of property if no longer needed by selling its parts. Altogether it provides better conditions for the future generations.

Sustainability (the triple bottom line) also needs to be supported by a fourth leg – governance – the structures and rules by which society and organisations uphold and develop its existence. Governance includes developing strategies, making decisions, and managing resources on a society or organisational level. The business case is a part of the governance theme and must be informed by all aspects of the triple bottom line. This is the only way strategic decisions on investments can lead to sustainable outcomes in the short, medium, and long range.

8.4.4 Blue Ocean Strategy for Reusable Building Products

The first question from the 'Four Actions Framework' asks what could be eliminated from the existing, competitive market. In the case of adaptive buildings, highly optimised, tailored components limit flexibility, so design optimation is not desirable. Instead, universal design and standardisation are recommended by industry experts and standards (ISO, 2020). Spending less time at this step of design reduces engineering costs. Standardisation also helps to automate the manufacturing and storage of products. Reusability prevents waste generation, so waste-related costs and issues are eliminated. The strategy canvas presenting the described aspects is shown in Fig. 8.2, on the example of three reinforced concrete products: (1) monolithic, (2) prefabricated, and (3) designed for disassembly.

The second action from the framework is to reduce factors predominant in the industry which are not much needed. One of them is the required amount of skilled in situ workforce. In countries such as Norway, labour costs drive the prices of construction. DfD elements with simple, reversible connections can be assembled by a small group of visiting workers and lifting equipment. Tasks requiring specialists, such as fixing reinforcement, formwork, welding, and carpentry, are all done at the prefabrication plant. The product's initial price and carbon footprint might be higher than non-DfD products, thus having a lower offering level. The initial uncompetitiveness in market price comes in exchange for long-term investment in much lower

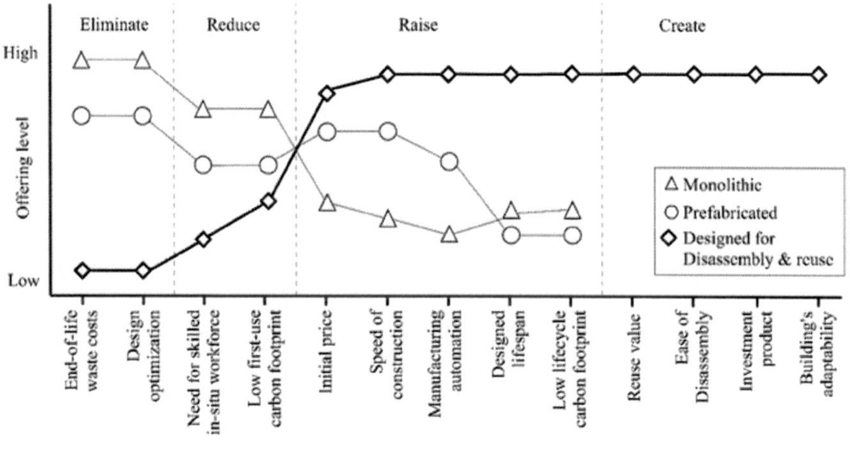

Fig. 8.2 Strategy canvas implementation for structural building elements for reuse – reinforced concrete case. (Source: own graphic)

whole-life carbon footprint products with high reuse value, not susceptible to raw material price and availability.

Another aspect, already mentioned in previous chapters, is the designed lifespan of 50–60 years. This assumption does not match the material's capacity and thus could be favourably restated into shorter predictions of individual use cycles and much longer product lifespan. The already mentioned separation of the technical lifespan of an element from the functional and economic lifespan of a building is necessary to achieve it.

Finally, there are questions that help create entirely new demands and make strategic price shifts: what factors should be elevated or created that are not addressed by existing business models? The adaptable components rely on the quality of reversible connections. Making all designs reversible makes the distinction between temporary and permanent obsolete. Other needs are designing for disassembly, developing proper end-of-life scenario plans, and establishing the reuse market. Remaining factors have already been covered while analysing the sustainability considerations.

8.5 Discussion

The first consideration should always be if the building needs to be deconstructed. As long as there are plenty of underperforming or even obsolete buildings not worthy of refurbishment, the reuse is next best solution. Reuse only extends the lifespan of objects, not neutralising their influence on the environment. At some point, adaptable components will also need to be discarded. That is not the ultimate solution to the environmental threat of construction, but as long as we rely on nonregenerative materials, it is a sustainable step towards it. If possible, the ideally circular, regenerative approach should be pursued (also called 'cradle-to-cradle'), in which non-harmful materials flow in closed loops with minimum loss.

Generalisations of design cause individual elements to be not fully optimised. This stems from the fact that universal elements need higher capacity by applying higher safety and uncertainty factors. Oversized elements have higher volume, thus embodied emissions, which is the opposite of intended results. Similarly to the presented economic aspects, the analysed case requires an upfront investment of resources and greenhouse gas emissions now to be able to save more resources and emissions later.

The business case of adaptable building components is not only built on the premise of subsequent uses of the products but also shows immediate benefits such as a fast assembly process. At the moment, the reuse of structural building components takes place mainly on pilot projects and requires additional investment compared to alternatives. High uncertainty about products history, lack of experience, and established procedures all raise the costs.

The presented business case could be more favourable if encouraged by regulations and subsidies for reusable components (Lindheim, 2021). This is already beginning to take place with regulation of prices of materials causing high emissions, such as cement, for example, in the European Emission Trading System. The EU Taxonomy mentioned above is another example of strong incentives that will force investors in a more sustainable direction. However, it does not necessarily make them go for DfD components. Sustainably produced timber also performs well with regard to GHG emissions.

Key barrier to enabling the reuse workflow is to close the circularity loop, but since the market with used products is not established, the demand for such products is low. Even if the stock of reclaimed products would be available, it requires a functioning information system where designers could search for suitable products.

8.6 Conclusions

The paper presents how reusable building components can be a viable business case. Usually, the value of structural components is wasted when the building ends its life, or only a part of the value is recovered by downcycling. The basis of this business case is to separate the technical lifespan of a component from the functional and economic lifespan of a building, by applying existing industry practices like prefabrication and design for disassembly. This way, one can harness the full value of elements.

The price of adaptable components is higher than of other products on the market, but there are immediate benefits that discount for it. One of them is the speed of construction with bolted connections. Another advantage is the relatively lower values of GHG emissions considering whole life cycle of a component. This translates to more sustainable profile of such products, which can be documented with sustainability certificates, that leads to more sustainable investments.

Full return on investment is realised at moment of building teardown or partial reconstruction: firstly, because the disassembly is cheaper than demolition of monolithic structures, and, secondly, because the salvaged elements are of much higher value if they are ready to be reused without much post-processing. Their value could even exceed that of virgin products, as their environmental analysis no longer includes the production phase.

This paper also displays the indirect benefits of the proposed model, such as increased value of buildings designed for disassembly due to its adaptability. Such structures can be extended, reduced, or modified to fit the given demand, preventing inefficiency. They could also be built temporarily and displaced after in places that would not be profitable otherwise.

The research reported here is only conceptual at this stage. The potential for reusable building components is illustrated, and the necessary conditions to make it possible are identified. In addition, barriers and real-life problems that need to be overcome are described. More thorough analysis is needed to be able to specify a business case that holds as a basis for decisions to choose this path.

The work primarily addresses the 12th Sustainable Development Goal about responsible consumption and production. It shows a path forward for the industry that contributes about 11% of global emissions to the atmosphere just from the use of materials (IEA & UNEP, 2018). This is not the ultimate solution to environmental threat of construction, but as long as humanity relies on virgin resources, it is a sustainable step towards it. Presented approach to building design could be encouraged by regulations and subsidies favouring reusable components. The paper also touches Sustainable Development Goals 8, 9, and 11, by opening a circular economy market with reusable components and makes our built environment more resilient. The scientific contribution of the paper comes from spanning the often-siloed fields of construction and economics, with the help of tools previously not applied in this context.

The blue ocean approach confirms that there is a business case for DfD building components, in particular for those making the first move.

References

Aktas, C. B., & Bilec, M. M. (2012). Impact of lifetime on US residential building LCA results. *International Journal of Life Cycle Assessment, 17*(3), 337–349. https://doi.org/10.1007/S11367-011-0363-X

Bocken, N. M. P., de Pauw, I., Bakker, C., & van der Grinten, B. (2016). Product design and business model strategies for a circular economy. *Journal of Industrial and Production Engineering, 33*(5), 308–320. https://doi.org/10.1080/21681015.2016.1172124

Brütting, J., Senatore, G., Schevenels, M., & Fivet, C. (2020). Optimum design of frame structures from a stock of reclaimed elements. *Frontiers in Built Environment, 6*, 57. https://doi.org/10.3389/FBUIL.2020.00057/FULL

Crawford, R. (2011). *Life cycle assessment in the built environment* (1st ed.). Spon Press.

de Wit, M., Haigh, L., von Daniels, C., Christiansen, A. F., et al. (2020). *Circularity gap report – Closing the circularity gap in Norway*.

EU. (2020). *Circular economy – Principles for building design*.

European Commission. (2015). Closing the loop – An EU action plan for the circular economy. In *Communication from the Commission to the European Parliament, the Council, the Europena Economic and Social Committee and the Committee of the Regions*. European Commision. https://eur-lex.europa.eu/legal-content/EN/TXT/?uri=CELEX:52015DC0614

European Commission. (2019, December 18). *Questions and Answers: political agreement on an EU-wide*. https://ec.europa.eu/commission/presscorner/detail/en/qanda_19_6804

Fink, L. (2022). *The power of capitalism – Larry Fink's annual 2022 letter to CEOs*. BlackRock. https://www.blackrock.com/no/privat/2022-larry-fink-ceo-letter

Fivet, C., & Brütting, J. (2020). Nothing is lost , nothing is created, everything is reused: structural design for a circular economy. *The Structural Engineer, 98*(1), 74–81.

FutureBuilt, Spaces, & Entra. (2021). *Erfaringsrapport ombruk Kristian Augustus gate 13*.

Gambles, I. (2009). *Making the business case. Proposals that succeed for projects that work*. Gower Publishing Ltd.

IEA & UNEP. (2018). *2018 Global status report towards a zero-emission, efficient and resilient buildings and construction sector*. www.iea.org

ISO. (2020). ISO 20887 Sustainability in buildings and civil engineering works — Design for disassembly and adaptability. In *International standard* (Vol. 2020).

Jaakkola, E. (2020). Designing conceptual articles: Four approaches. *AMS Review, 10*(1–2), 18–26. https://doi.org/10.1007/S13162-020-00161-0/TABLES/2

Jensen, A. B. (2013). Do we need one business model definition? *Journal of Business Models, 1*(1), 61–84. https://doi.org/10.5278/ojs.jbm.v1i1.705

Kim, W. C., & Mauborgne, R. (2015). *Blue ocean strategy: How to create uncontested market space and make competition irrelevant*. Harvard Business Review Press (Expanded).

Lindheim, K. (2021). *Lønnsomhet ved bygninger som materialbanker* [Profitability of buildings as material banks] (Masters' thesis). Norwegian University of Science and Technology.

Marsh, R. (2017). Building lifespan: Effect on the environmental impact of building components in a Danish perspective. *Architectural Engineering and Design Management, 13*(2), 80–100. https://doi.org/10.1080/17452007.2016.1205471

Norwegian Ministry of Finance. (2014). *Prinsipper og krav ved utarbeidelse av samfunnsøkonomiske analyser mv* [Principles and requirements for preparing social economic analysis etc].

Norwegian Ministry of Finance. (2019). *Statens prosjektmodell – Krav til utredning, planlegging og kvalitetssikring av store investeringsprosjekter i staten* [The State Project Model – Requirements for elucidation, planning and quality assurance of large investment projects in the state] (p. 7).

NS 3682: Hollow core slabs for reuse. (2022). https://www.standard.no/no/Nettbutikk/produktkatalogen/Produktpresentasjon/?ProductID=1407012

OGC. (2009a). *Directing successful projects with PRINCE2* (1st ed.). The Stationery Office.

OGC. (2009b). *Managing successful projects with PRINCE2* (5th ed.). The Stationery Office.

Paananen, T., & Suur-Askola, P. (2018). *In search of a bright, circular future! Peikko and the circular economy, practical considerations*.

Reddy, V. S. (2018). The SpaceX Effect. *New Space, 6*(2), 125–134. https://doi.org/10.1089/SPACE.2017.0032

Röck, M., Saade, M. R. M., Balouktsi, M., Rasmussen, F. N., Birgisdottir, H., Frischknecht, R., Habert, G., Lützkendorf, T., & Passer, A. (2020). Embodied GHG emissions of buildings – The hidden challenge for effective climate change mitigation. *Applied Energy, 258*(November 2019), 114107. https://doi.org/10.1016/j.apenergy.2019.114107

Sommer, J., Jansen, K. G., 3XN Architects, & GXN Innovation. (2016). *Building a circular future*. Danish Environmental Protection Agency (3rd ed., p. 312). https://issuu.com/3xnarchitects/docs/buildingacircularfuture

SSB. (2021). *09247: Generated waste amounts from construction, rehabilitation and demolition of buildings (tonnes), by material 2004–2020*. https://www.ssb.no/en/statbank/table/09247

The Norwegian EPD Foundation. (2021). *EPD-Norge.no*. https://www.epd-norge.no/

Todsen, S. (2014). *Ny kartlegging av økonomisk levetid og verdifall for realkapital* [New survey of economic life and impairment of real capital]. Statistisk sentralbyrå. https://www.ssb.no/nasjonalregnskap-og-konjunkturer/artikler-og-publikasjoner/ny-kartlegging-av-okonomisk-levetid-og-verdifall-for-realkapital

UNESCO. (n.d.). *Urnes Stave Church*. https://whc.unesco.org/en/list/58

Wainwright, O. (2020). The case for... never demolishing another building. *The Guardian*. https://www.theguardian.com/cities/2020/jan/13/the-case-for-never-demolishing-another-building

Walter, H. K., & Høydahl, V. V. (2020, June). *Ombruk av byggematerialer og – produkter i et bærekraftperspektiv. Vurdering av miljøeffekt og kartlegging av potensialet for en oppskalering av ombruksmarkedet.*

Wang, J., Zhang, Y., & Wang, Y. (2018). Environmental impacts of short building lifespans in China considering time value. *Journal of Cleaner Production, 203*, 696–707. https://doi.org/10.1016/J.JCLEPRO.2018.08.314

Wiik, M. K., Fufa, S. M., Kristjansdottir, T., & Andresen, I. (2018). Lessons learnt from embodied GHG emission calculations in zero emission buildings (ZEBs) from the Norwegian ZEB research centre. *Energy and Buildings, 165*, 25–34. https://doi.org/10.1016/J.ENBUILD.2018.01.025

Wilkinson, S. J., Remøy, H., & Langston, C. (2014). *Sustainable building adaptation: Innovations in decision-making* (p. 296).

Wuyts, W., Miatto, A., Sedlitzky, R., & Tanikawa, H. (2019). Extending or ending the life of residential buildings in Japan: A social circular economy approach to the problem of short-lived constructions. *Journal of Cleaner Production, 231*, 660–670. https://doi.org/10.1016/J.JCLEPRO.2019.05.258

Chapter 9
Ecosystems and Reuse of Building Materials: An Exploratory Study

Francesca Vergani, Rikard Sundling, and Carlos Martinez

Absract The implementation of circular economy strategies, such as the reuse of building materials, represents a valid opportunity for the building and construction sector to diminish the consumption of raw materials and generate revenues and, thus, to contribute to the Sustainable Development Goals 11 and 12. However, the reuse of building materials is still limited. Previous studies suggest that one reason is the lack of communication and collaboration among the actors involved. By using an ecosystem perspective, this study focuses on the characteristics of the ecosystem of actors involved with reuse of building materials. An exploratory research approach was used, which consisted of scientific literature and six semi-structured interviews with actors in the Öresund region. The interviews were conducted to elaborate on the characteristics that best represent this ecosystem. The results suggest that these characteristics are (a) complementariness and collaboration, (b) capability to evolve, (c) willingness to align to circularity, and (d) platformization. In particular, these characteristics highlight the needs of (1) collaboration among the actors already in the early stages of a project, (2) more actors and new roles in the ecosystem, (3) awareness of the objectives and values which guides the ecosystem, and (4) better knowledge and use of the ecosystem platforms.

Keywords Circular economy · Building and construction sector · Reuse of building materials · Ecosystem · Actors

---

F. Vergani (✉) · R. Sundling · C. Martinez
Department of Building and Environmental Technology, Lund University, Lund, Sweden
e-mail: francesca.vergani@construction.lth.se; rikard.sundling@construction.lth.se; carlos.martinez@construction.lth.se

G. Lindahl, S. C. Gottlieb (eds.), *SDGs in Construction Economics and Organization*, Springer Proceedings in Business and Economics,
https://doi.org/10.1007/978-3-031-25498-7_9

9.1 Introduction

The building and construction sector is currently responsible for the consumption of 35% of total energy use and 35–45% of raw materials and generates up to 38% of global carbon dioxide and circa 36–40% of landfill waste (UNEP, 2020). Previous studies show how the circular economy (CE) represents a valid opportunity for both diminishing the use of raw materials and creating economic revenue (Gerhardsson et al., 2020; Munaro et al., 2021; Nußholz et al., 2020; Pomponi & Moncaster, 2017; Sezer & Bosch-Sijtsema, 2020; Whicher et al., 2018). However, the implementation of CE strategies, such as the reuse of building materials, is still limited. The reuse of building materials is a key part of the circular economy in the building and construction sector since reuse is necessary for reducing the use of virgin materials in new developments. The slow progress of the reuse of building materials, as a strategy to achieve circularity in construction, can be attributable to several reasons such as the following:

- The lack of commitment from the authorities that leads to a lack of policies, regulations, and institutional procurement that favor the diffusion of CE (Hart et al., 2019; Knoth et al., 2022)
- The practical barriers inherent in the implementation of circular practices such as the lack of knowledge, technologies, and appropriate business models that would support the transition of organizations (Gerhardsson et al., 2020; Hart et al., 2019; Kanters, 2020; Knoth et al., 2022)
- The inclination of the actors in the building and construction sector to follow the dynamics and processes of the traditional linear method of production (Knoth et al., 2022; Kooter et al., 2021)

The last reason is the focus of this study. The reuse of building materials redefines relationships and establishes new ways of working; thus, it challenges the existing knowledge and roles of the actors. Furthermore, the involvement of new actors is often required to fill the missing knowledge and skills. This is an additional element that challenges the relationships and dynamics established within the traditional linear process. The literature suggests that for the successful implementation of reuse of building materials, a further development of interactions and collaborations is required (Hart et al., 2019; Kooter et al., 2021). With the help of an ecosystem perspective, this study investigates the collaborations and interactions that arise with the reuse of building materials in the building and construction sector.

In the literature, the concept of ecosystem is usually used as a lens to examine the relationship between different, mutually dependent actors who coevolve to respond to a shared purpose to obtain a competitive advantage (Aksenova et al., 2019; Bosch-Sijtsema & Bosch, 2015; Moore, 1993; Vargo & Lusch, 2014). The concept of ecosystem has also been applied to the building and construction sector (Pulkka et al., 2016); however, there are still few studies that connect this concept to circular strategies, and none of them is focusing on the reuse of building materials. Moreover, the characteristics with which ecosystems are usually described do not seem to be sufficient to explain the interactions between the different actors involved with the reuse of building materials.

The aim of this paper is therefore to fill this gap and elaborate on the characteristics of the ecosystem of the reuse of building materials. With the support from both scientific literature and an exploratory interview study, the goal is to argue for the relevance of these characteristics. The interview study consisted of six interviews, with actors in the field of reuse of building materials, all active in the Öresund region (Sweden and Denmark). To promote the transition to the reuse of building materials in the building and construction sector, this study targets, namely, the Sustainable Development Goal (SDG) 11 which seeks to create sustainable cities and communities that are inclusive, safe, and resilient and SDG 12 that promotes sustainable consumption and production.

The rest of this paper is structured as follows. In the next section, the method for the exploratory interviews and the analysis are explained. In the theoretical framework section, the concepts and characteristics of ecosystems and the applicability of ecosystem concepts in construction, as well as literature on circular economy and reuse of building materials, are presented. In the results and discussion section, the data obtained from the interviews are combined with the literature to elaborate on four characteristics of the ecosystem of the reuse of building materials, which were identified. In the conclusion section, the key findings from the study are highlighted, and future research is proposed.

9.2 Method

This study was based on abductive research where the researchers attempted to explore the importance of the ecosystems concept in relation to the reuse of building materials in the building and construction sector. The study relied on scientific literature on the topics of ecosystems and reuse of building materials, as well as data gathering in the form of interviews.

The first step was to search for literature regarding the concepts and characteristics of ecosystems and the applicability of ecosystem concepts in construction, as well as literature on the CE and reuse of building materials. Both the field of reuse of building materials and the literature on the application of ecosystems in the building and construction sector have lately been gaining traction in academia. Nevertheless, the practical application of the reuse in the building and construction sector is still in its infancy and the literature on the use of ecosystem in relation to the implementation of CE strategies is scarce. Therefore, this study examined how the characteristics of ecosystems can be applied to the reuse of building materials.

The second step was to gather data from actors involved with the reuse of building materials. Semi-structured interviews were chosen as the data collecting method since interviews offer a flexible approach where the interviewer gathers experiences and knowledge from the interviewee. As suggested by Alvesson (2011), this was a flexible approach since the interview questions changed and were adapted to the specific interviewee in order to extract more data. On the other hand, interviews were also flexible in the sense that the researchers had to interpret the data from the interviews, meaning that other researchers might draw other conclusions from the same data.

A relevant aspect to define is the ecosystem boundaries since they depend on the goal of the study (Bröchner, 2016). For this study, the circular ecosystem is assumed to have regional boundaries. In particular, we looked at the Öresund region and the actors involved, who were both Danish and Swedish. In this study, six actors were interviewed: two suppliers of reused materials, two developers, and two consultants; see Table 9.1. The suppliers acquire their reused materials mainly from two sources: unused materials from contractors and materials which need upcycling from demolition firms. Both developers identified significant cultural value in the local building materials and wanted to embrace reuse throughout their projects. In both cases, these projects are their first with a focus on reuse. The consultants have been active in the reuse of building materials for several years and thus have significant experience in the field. All the interviewees were reached by an email which provided information about the study. Since the collection of data was carried out during the COVID-19 pandemic, almost all the interviews were conducted online and, with the consensus of the interviewees, were recorded and transcribed. The language used for the interviews was Swedish, apart from the interview with the Danish actor which was held in English. The interviewees were asked questions about previous experiences with the reuse of material; the type, time, and management of the collaborations and relationship with the actors involved in reuse; the essential knowledge and skills needed in the reuse; the challenges the actors had to overcome individually as an organization or as an ecosystem; and the main value of working with the reuse of building materials.

The structure of the analysis was as follows. First, an initial framework of ecosystem characteristics based on the review of the literature was made. The design of the interviews followed this initial framework. The transcribed interviews were then assessed and categorized. Subsequently, the initial framework was then adapted to suit the data from the interviews and literature on the topics of circular economy and reuse of building materials. Finally, the adapted set of characteristics was used as headings in the results and discussion section; under each heading, the categorized data from the interviews are discussed together with relevant literature.

Table 9.1 The interviewees participating in the study

Interviewee (nr)	Type of organization	Role	Nationality
1	Private supplier of reused materials	CEO	Sweden
2	Municipal supplier of reused materials	Foreman	Sweden
3	Private developer	Project manager	Sweden
4	Municipal developer	Project manager	Sweden
5	Sustainability consultant	Consultant	Denmark
6	Reuse consultant	CEO	Sweden

9.3 Theoretical Framework

9.3.1 Circular Economy

Circular economy (CE) is described by Korhonen et al. (2018) as "a sustainable development initiative with the objective of reducing the societal production-consumption system's linear material and energy throughput flows to the linear system. CE promotes high value material cycles alongside more traditional recycling and develops systems approaches to the cooperation of producers, consumers, and other societal actors in sustainable development work." In addition to highlighting the role of the CE for the creation of economic value by exploiting cyclical and regenerative materials flows, this definition also emphasizes that CE should be a process that favors sustainability in all its forms and that requires the contribution of all the actors involved. CE is a broad concept that includes sublevels of definition and strategies such as recycling, reuse, remanufacturing, refurbishment, repair, cascade, and upgrade of products, components, and materials (Mhatre et al., 2021). The building and construction sector contributes with 12% of the global GDP (Crosthwaite, 2000) and plays an important role in the transition from the linear to the circular economy. In order to move toward CE strategies and minimize the production of waste while reducing costs, it is however necessary to reconsider current construction practices, change and improve methods and organizational resources, and allow the reuse of building components and materials (Bertino et al., 2021; Konietzko et al., 2020; Pomponi & Moncaster, 2017). Furthermore, to produce buildings that are circular for their entire life cycle, it is necessary to implement circular strategies and practices in all phases of the building process (Gerhardsson et al., 2020; Mhatre et al., 2021; Pomponi & Moncaster, 2017), thus involving all the relevant actors in this transformation (Konietzko et al., 2020; Pomponi & Moncaster, 2017). The ability to accept and embrace these organizational and role changes is necessary for the transition to circular economy in the building and construction sector.

Currently, the implementation of circular practices in the building and construction sector is still limited due to both a conservative mindset and the intrinsic characteristics of the sector (Munaro et al., 2021; Pomponi & Moncaster, 2017; Hart et al., 2019; Knoth et al., 2022) and to the several practical barriers that are interconnected with circularity (Gerhardsson et al., 2020; Kanters, 2020; Hart et al., 2019; Knoth et al., 2022). Furthermore, the implementation of circular processes in construction involves both radical and incremental innovations throughout the entire life cycle of a building, and for the effective success of this transition, it is essential that these innovations are also supported by further innovations in policies, business models, procurement, and the interrelation of actors (Whicher et al., 2018).

9.3.2 Reuse of Building Materials

The entire life cycle of a building can be summarized in three stages: the pre-use stage (i.e., initiation and preparation, design, and construction), the use and operation of the building, and the stage following the use, usually called end of life (Ghisellini et al., 2018). Mhatre et al. (2021) argue that the circularity of a project depends on the sourcing of building materials. The selective demolition and deconstruction of a building at its end of life contribute to reintroduce building materials in the loop (Ghisellini et al., 2018), and thus, the reuse of building materials can be considered as the key for a successful transition to a more developed CE in the building and construction sector. The European Commission Waste Framework Directive (EU, 2008) presents a five-step hierarchy where reuse is the second-best option in terms of waste prevention. Reuse is typically defined as the use of products in their original form or subject to minimal recovery activities (Ghisellini et al., 2018). Although the reuse of building materials is of critical importance, there are many barriers hindering its implementation on a significant scale. The first barrier is that less than 1% of the current building stock is fully deconstructable (Ghisellini et al., 2018), and thus, there is a need not only to produce new building with reused materials but also to design with the deconstruction in mind.

A recent study conducted in Sweden (Gerhardsson et al., 2020) shows how the reuse of materials in the Swedish construction sector is mainly prevented by the lack of knowledge of all the processes and practices to be implemented and integrated. Starting from the analysis of projects already carried out, Gerhardsson et al. (2020) propose a series of work practices that favor the reuse of building materials: *a materials inventory, targets for reuse, circular building design, planning for new processes resulting from reuse, incentives for reuse in procurement, and long-term documentation strategies enabling future reuse*. Moreover, Knoth et al. (2022) identified several barriers in their interview study and argued for three ways of addressing these barriers, namely, *establishing reuse infrastructure and knowledge base, getting manufacturers onboard, and enabling reuse through regulations and increased reputation*. Both the barriers identified and the practices proposed by Gerhardsson et al. (2020) and Knoth et al. (2022) focus mainly on the pre-use stage of a building revealing that the actors' main efforts should converge during the initiation and preparation, design, production, and construction phases.

It is worth noting that for the implementation of any kind of innovation to create value, all the actors involved must provide a solution to their innovation-related struggles (Adner & Kapoor, 2010), in this case, to the problems related to the reuse of building materials. Working with new practices and following new processes goes, in fact, beyond the capabilities and knowledge of individual organizations and requires effort from all actors involved (Adner & Kapoor, 2010; Vosman et al., 2021). It is therefore essential to use a theoretical framework that helps to understand both the complex interactions that are established in circular projects and the attitudes that can favor the diffusion of the reuse of materials outside the boundaries of a single project or organization.

9.3.3 Ecosystem: Concepts and Characteristics

The concepts of ecosystem and circular economy (CE) were combined before the ecosystem metaphor was introduced into strategic management research to explain the escalation in cooperation and collaboration between interdependent (but also competing) organizations (Aksenova et al., 2019; Bosch-Sijtsema & Bosch, 2015). The interest of researchers in responding adequately to the growing awareness of the environmental, social, and economic consequences that the current economic paradigm was producing dates to the 1970s and 1980s (Pomponi & Moncaster, 2017). As a response to this need, Frosch and Gallopoulos (1989) proposed a new industrial ecosystem that imitated the characteristics and dynamics of biological ecosystems in order to transform the linear model of production and thus limit the consumption of energy and raw materials. The authors hence suggested that the effluents of industrial processes were used as raw materials for other processes, thereby exploiting their intrinsic circularity. It is worth noting that the commitment of both producers and consumers in changing their attitudes and behaviors was highlighted by Frosch and Gallopoulos (1989) as an important aspect to implement a paradigm shift and ensure the success of the new industrial ecosystem.

The involvement of actors, such as producers and consumers or other organizations, is further emphasized in the ecosystem concept developed by Moore in 1993 (Aksenova et al., 2019; Bosch-Sijtsema & Bosch, 2015; Moore, 1993; Pulkka et al., 2016). This ecosystem, however, no longer looks at industrial process as a metaphor for the natural ecosystem, since the focus shifts on to the relationships between the actors involved in the process (Moore, 1993). The characteristics of both ecosystems as presented above envisage (a) a form of symbiosis between the elements of the ecosystem, (b) the coevolution of the actors involved around an innovation, and (c) and the ability to create collective value (Aksenova et al., 2019; Bosch-Sijtsema & Bosch, 2015; Pulkka et al., 2016). Another aspect that is relevant for the success of an ecosystem is the *platform* (Bosch-Sijtsema & Bosch, 2015; Moore, 1993). The latter is considered as the shared tool used to sustain the ecosystem activities and novelties (Bosch-Sijtsema & Bosch, 2015).

9.3.4 Applicability of Ecosystem Concepts in Construction

In the application of the ecosystem concept to the building and construction sector, the studies are divided mainly between those who see similarities between the construction and the service sector and therefore use the lens of the *service ecosystem* (Bröchner, 2016; Sezer & Bosch-Sijtsema, 2020) and those who look at the modalities of innovation in construction and therefore shift the emphasis on to the *innovation ecosystem* (Vosman et al., 2021; Yang et al., 2021; Whicher et al., 2018). Moreover, the concept of *business ecosystem* is also used to understand the need for network-based collaboration and shared logic to approach complex and risky projects (Toppinen et al., 2019).

Sezer and Bosch-Sijtsema (2020) used the concept of service ecosystem to study the barriers and tensions between the actors involved in construction and demolition waste refurbishment projects in Sweden. Vosman et al. (2021) employed the concept of innovation ecosystem as an approach to facilitate long-term collaborations that favor the necessary level of innovation and change to meet the needs and challenges of contemporary society. A couple of studies have focused their analysis on the relation between information and communication technologies (ICT) innovation in construction and the ecosystem concept (Aksenova et al., 2019; Yang et al., 2021). Moreover, Whicher et al. (2018) suggested how ecosystems theory could be used to identify all actors affected by a circular economy action plan, such as the one proposed and implemented for Scotland. The aim of identifying the affected actors is to be able to involve and include all of them in the creation and implementation of circular economy action plans. Finally, Toppinen et al. (2019) scrutinized the fitness of the business ecosystem concept in the context of multistory timber buildings, paying particular attention to the creation of value and benefits for the actors involved in the business. In all these studies, the four aforementioned characteristics return, sometimes taking on slightly different aspects and sometimes complementing themselves with further characteristics.

9.4 Results and Discussion

For this study, literature on the subject of ecosystems and the reuse of building materials has been examined. Definitions of ecosystem concepts and their characteristics were sought and categorized; see Table 9.2. Subsequently, an attempt was made to apply each of the characteristics identified for the reuse of building materials based on both the literature and the six interviews. This attempt showed that in the case of actors involved with reuse of building materials, none of the previous ecosystem definitions in the literature fully explain their interactions and struggles. It was therefore decided to adapt the characteristics presented in previous studies and to use four main characteristics in an attempt to cover all the aspects within an ecosystem of actors working with the reuse of building materials. These four characteristics are (a) complementariness and collaboration, (b) capability to evolve, (c) willingness to align to circularity, and (d) platformization.

9.4.1 Complementariness and Collaboration

As stated by Pulkka et al. (2016), the construction sector is based on a high level of compliance among organizations as each individual actor becomes less and less capable of providing all the technical and process knowledge necessary to develop and complete a given project. The need for collaboration among different actors is a recurring aspect in the literature about the reuse of building materials (Gerhardsson

Table 9.2 Identifying the characteristics of ecosystems

Type of ecosystems	Definition	Characteristics	Literature
Ecosystem (general description applied in different fields)	From an ecosystem perspective, a company can be viewed "not as a member of a single industry but as part of a business ecosystem that crosses a variety of industries and to coevolve capabilities around a new innovation. Companies work cooperatively and competitively to support new products, satisfy customer needs, and eventually incorporate the next round of innovations" (Moore, 1993)	1. Symbiosis 2. Coevolution 3. Platform	Bosch-Sijtsema and Bosch (2015), Frosch and Gallopoulos (1989), and Moore (1993)
Business ecosystem	It is a set of core features which includes nonlinear value creation, interdependency of participants, substantial knowledge exchange, and nonmarket governance mechanisms, and coevolution of capabilities (Pulkka et al., 2016). "The ecosystem is focused on collective value creation as the recognized area of institutional life" (Pulkka et al., 2016)	1. Network of participants 2. Governance system 3. Shared logic	Pulkka et al. (2016) and Toppinen et al. (2019)
Service ecosystem	Service ecosystems are defined as "relatively self-contained self-adjusting systems of resource-integrating actors connected by shared institutional logics and mutual value creation through service exchange" (Vargo & Lusch, 2014)	1. Institutional arrangements and mutual value creation 2. Ability to self-adapt to changes	Bröchner (2016), Sezer and Bosch-Sijtsema (2020), Trischler et al. (2020), and Vargo and Lusch (2014)
Innovation ecosystem	The innovation ecosystem can be understood as "a multi-stakeholder network around certain innovative value propositions" (Vosman et al., 2021)	1. Heterogeneity of actors 2. Strategic alignment of actors 3. Alignment with respect to the value proposition 4. Nonformal governance	Vosman et al. (2021) and Whicher et al. (2018)
Circular innovation ecosystem	"A number of products, business model and ecosystem principles that when combined enable firms to take an ecosystem perspective on the circular economy and work towards higher circularity" (Konietzko et al., 2020)	Collaboration Experimentation Platformization	Konietzko et al. (2020)

et al., 2020; Kanters, 2020; Knoth et al., 2022). In fact, the complementarity between actors is often insufficient to reach the right level of knowledge for CE implementation in building projects; additionally, it is necessary that the actors interact and collaborate even in phases of the project in which they were not traditionally involved (Gerhardsson et al., 2020; Knoth et al., 2022). It was stressed by the private developer and the sustainability consultant that actors such as architects, consultants, demolition and deconstruction firms, recycling companies, and contractors need to collaborate and be more involved throughout the project stages, especially in the earlier stages. Moreover, both the sustainability consultant and the municipal developer emphasized the need to include more actors in the early strategic work "to make them (the other actors) co-owners of the ambition and to understand what we want to do" (Interviewee 5).

All the interviewees mentioned at least once that frustration is one of the most recurrent moods among actors. In collaboration, a winning way to achieve circularity is to manage the relationships through increasing communication, involvement, and motivation of all the actors. This will also prevent frustration (Hart et al., 2019; Knoth et al., 2022; Konietzko et al., 2020). The sustainability consultant explained how collaborating with architects has enabled them to generate design ideas for reusing building materials and thus create new knowledge that can be used in various projects. The municipal developer describes its approach to the engagement and motivation of other actors; it is one of recognizing the importance of stimulating collaboration and making room for other actors' new ideas. The aim is to ensure that the knowledge of the contractors is integrated with that of the architects or with that of the demolition contractor, for example. This type of collaborative attitude is considered as one of the facilitators for the creation of long-term relationships, which in turn represent one of the conditions for the diffusion of innovation (Bosch-Sijtsema & Bosch, 2015). Moreover, as suggested by Sezer and Bosch-Sijtsema (2020), the actions of the actors within a network reflect, directly or indirectly, upon all other actors. For example, according to the sustainability consultant, developing close relationships with clients enables them to influence the sustainability objectives of the project "and push the limits of their boundaries" (Interviewee 5). The same kind of effect was described by the private developer when referring to how the collaboration between the developer and the sustainability strategist helped to achieve a reuse target of 80% for their project.

Moreover, it is worth noting that the interviewees highlighted the need to collaborate not only with complementary actors but also with competing organizations. Collaboration between competing organizations is needed to be able to respond to the market, to grow the market itself, to improve pricing of materials and services, to improve procedures, and to develop certification (Gerhardsson et al., 2020; Knoth et al., 2022). The need to collaborate is in line with previous literature based on competitive engagement approaches (Bosch-Sijtsema & Bosch, 2015).

9.4.2 Capability to Evolve

Previous studies show that one of the major barriers to the implementation of the reuse of building materials in the Swedish building and construction sector is the lack of knowledge and experience in reuse practices (Gerhardsson et al., 2020; Kanters, 2020; Knoth et al., 2022). While Kanters (2020) focuses on the need for architects to acquire both technical skills and greater flexibility, the interviews reveal the necessity of bridging the knowledge gap and encourage the flexibility of all the actors involved in the reuse of building materials. This flexibility is perceived as the ability of the actors to redefine their roles and responsibilities. As explained by some of the interviewees, actors who are new to reuse practices often remain skeptical "just because they are not used to it" (Interviewee 4). The interviewees stressed that, during the design phase, the architect and the client must cooperate with contractors, consultants, demolition contractors, and suppliers for successfully implementing the reuse of building materials. In this process, a new responsibility is added to the role of architects; they are no longer just designers, but they also become the organizers of a logistics process (Kanters, 2020). According to the sustainability consultant, inflexible architects may perceive this new responsibility as a threat not only to their creativity but to their profession.

The transition from a traditional building process to one focusing on the reuse of building materials leads to the creation of new roles and the involvement of actors from other sectors (Gerhardsson et al., 2020; Konietzko et al., 2020; Vosman et al., 2021). The interviewees underline, for example, the lack of actors who can classify and certify the reused materials. Some interviewees believe that public procurement and legislation could push organizations to fill these new roles. Thus, the ability of the actors to respond to the need of the ecosystem by evolving, transforming, and filling new roles is one of the main characteristics of the ecosystem of actors involved with reuse of building materials.

9.4.3 Willingness to Align with Circularity

Whether or not an actor belongs to a certain ecosystem is defined both by the ability of each actor to evolve and coevolve but also by the willingness to align with certain targets (Konietzko et al., 2020; Vosman et al., 2021) and according to certain engagements or interdependencies (Bosch-Sijtsema & Bosch, 2015; Bröchner, 2016; Konietzko et al., 2020; Vargo & Lusch, 2014). As highlighted by Bröchner (2016), ecosystems in construction are often dominated by contractors and by their ability to select procedures that favor the collaboration between all the actors involved in a project. Gerhardsson et al. (2020) add that the clients also play a relevant role for the transition to the reuse of building materials because they can include reuse as a requirement in the procurement process. All interviewees agreed with this last statement; for example, the private developer suggested that the creation of a sustainability action plan can be used to facilitate the selection of the other actors in a project. In their project, those actors who lacked the skills needed or the willingness to strive for certain goals were removed, thereby reducing the number of potential candidates. In this sense, the sustainability action plan can also be considered as an institutional arrangement as described by Bröchner (2016). In line with what stated by Pulkka et al. (2016), all the interviewees emphasized that mutual awareness and alignment with certain values encouraged relationships between the different actors.

Moreover, framing joint strategies, such as the sustainability action plan presented by the private developer, can help the actors to understand that an efficient reuse of building materials can only be reached if all the actors work together. For this purpose, both the private developer and the sustainability consultants declared that defining the objectives and goals of a project should be its very first step. On a similar note, both consultants highlighted the importance of their roles in promoting sustainability since they are often the initiators for the transition to the reuse of building materials. These consultants do not only advise developers, municipalities, and small companies; they also set sustainability strategies for specific projects. A different situation was presented by the municipal developer. In its project, the reuse of materials was first considered after the actor selection process had finished. In this case, the trust between the actors (built on long-term collaboration), the positive attitude to reuse, and the willingness to align with circular practices proved to be the winning factors for the successful implementation of reused building materials.

An important aspect that hinders the willingness to align to circularity is the lack of legislation and regulations related to the reuse of building materials (Gerhardsson et al., 2020; Hart et al., 2019; Knoth et al., 2022). Sometimes, the current regulations constitute an obstacle for the reuse of building materials (Knoth et al., 2022). This was highlighted by almost all the interviewees, especially referring to risk management and to the management of material transportation and ownership from one project to another one.

9.4.4 Platformization

The creation of an online platform that guarantees collaboration, an efficient level of communication between the actors, and the sharing of knowledge and information is a fundamental aspect within the ecosystem (Bosch-Sijtsema & Bosch, 2015; Gerhardsson et al., 2020; Knoth et al., 2022; Konietzko et al., 2020; Toppinen et al., 2019). Both the literature and the interviewees suggest that the presence of a platform is fundamental for the success of a circular project and for implementation of the reuse of building materials. Even so, four out of six interviewees revealed a form of skepticism or lack of knowledge toward the most common platform in Sweden (CCBuild) in favor of platforms within the project or developed by individual actors. In a couple of interviews, the role of a platform was reduced to that of a materials inventory, which is only one aspect a platform can fulfill. The private developer said that their material inventory is an open-access platform which is able to share materials among other projects; however, they later revealed that they want to keep all the materials within the project and will not sell anything. One of the suppliers revealed a lack of a digital system for cataloging materials, stating that all information relating to quantities and costs is better kept in the foreman's head and the heads of a few other employees. In the first of the two examples, it was not made explicit if the material inventory also reports the materials' characteristics, potential for reuse, quality, or environmental potential as suggested by Knoth et al. (2022); in the second example, it was clearly stated that none of these data were collected. This confusion regarding platformization can be a hindrance to further implementation of the reuse of building materials (Knoth et al., 2022). An open-access material inventory is one of the first steps to initiate the reuse of building materials, but it is not sufficient to stimulate and diffuse the knowledge about the building materials alone. The two consultant organizations were well aware of this and, therefore, had developed an information sharing platform to facilitate interaction with other actors and increase the knowledge of the members of the ecosystem.

9.5 Conclusions

If the building and construction sector aims to reach a more sustainable development, the increased reuse of building materials will be necessary. The further implementation of the reused of building materials could contribute to the Sustainable Development Goal 11 (sustainable cities and communities) and Sustainable Development Goal 12 (responsible consumption and production). In this paper, four characteristics of the reuse of building materials ecosystems are presented. These are complementariness and collaboration, capability to evolve, ability to align to circularity, and platformization. Each of these four characteristics is elaborated in the results and discussion and is supported by literature and an exploratory study. Six actors involved with the reuse of building materials were interviewed: all are active in the Öresund region.

There are four overarching results from this study about the successful implementation of the reuse of building materials. First, there is a need for collaboration between multiple actors from an early stage. Since there is a lack of knowledge in the sector, all key actors, in a project, have to contribute with their expertise in order for them to identify collectively how much material reuse is realistic to implement for each project. Secondly, since the field is under development, more actors are needed; in addition, some actors have to take on new roles. Third, there is a need for increased awareness of the project objectives and values. It is essential that actors are willing to align themselves to circularity and that they remain flexible in their roles. Finally, there are a few platforms under development which will accelerate the implementation of the reuse of building materials and increase knowledge sharing; however, better platformization of the knowledge and tools than are currently available is needed.

Suggestions for future research on this topic include expanding the interview study presented here and mapping the actors in the ecosystem. Even though the findings are consistent with the literature, six interviews are not enough to explain the full ecosystem. There are likely, therefore, to be more findings in an extended study. Another suggestion is mapping actors in other regions, where implementation of the reuse of building materials is more advanced. It could be fruitful to compare these different regions by examining their strengths and weaknesses. An in-depth case study is also proposed since thorough data gathering concerning the actors and their roles could give insights into how the reuse of building materials is implemented in practice.

Acknowledgments This research was part of the Reconcile project, funded by the EU Interreg (project grant # NYPS 20204782).

References

Adner, R., & Kapoor, R. (2010). Value creation in innovation ecosystems: How the structure of technological interdependence affects firm performance in new technology generations. *Strategic Management Journal, 31*, 306–333. https://doi.org/10.1002/smj.821

Aksenova, G., Kiviniemi, A., Kocaturk, T., & Lejeune, A. (2019). From Finnish AEC knowledge ecosystem to business ecosystem: Lessons learned from the national deployment of BIM. *Construction Management and Economics, 37*(6), 317–335. https://doi.org/10.1080/01446193.2018.1481985

Alvesson, M. (2011). *Interpreting interviews*. SAGE.

Bertino, G., Kisser, J., Zeilinger, J., Langergraber, G., Fischer, T., & Österreicher, D. (2021). Fundamentals of building deconstruction as a circular economy strategy for the reuse of construction materials. *Applied Sciences (Switzerland), 11*(3), 1–31. https://doi.org/10.3390/app11030939

Bosch-Sijtsema, P. M., & Bosch, J. (2015). Plays nice with others? Multiple ecosystems, various roles and divergent engagement models. *Technology Analysis and Strategic Management, 27*(8), 960–974. https://doi.org/10.1080/09537325.2015.1038231

Bröchner, J. (2016). Coordination in slowly emerging service ecosystems. Construction history. *Revue Europeenne d' Economie et management des services, 2*(2), 17–41). Classiques Garnier. https://doi.org/10.15122/isbn.978-2-406-06930-0.p.0017

Crosthwaite, D. (2000). The global construction market: A cross-sectional analysis. *Construction Management and Economics, 18*, 619. https://doi.org/10.1080/014461900407428

EU. (2008). *Directive 2008/98/EC of the European Parliament and of the Council.* Available at: https://eur-lex.europa.eu/legal-content/EN/TXT/?uri=CELEX%3A02008L0098-20180705. Accessed 08 Feb 2022.

Frosch, R. A., & Gallopoulos, N. E. (1989). *Strategies for Manufacturing, 261*(3), 144–153. https://doi.org/10.2307/24987406

Gerhardsson, H., Lindholm, C. L., Andersson, J., Kronberg, A., Wennesjö, M., & Shadram, F. (2020). Transitioning the Swedish building sector toward reuse and circularity. *IOP Conference Series: Earth and Environmental Science, 588*(4), 042036. https://doi.org/10.1088/1755-1315/588/4/042036

Ghisellini, P., Ripa, M., & Ulgiati, S. (2018). Exploring environmental and economic costs and benefits of a circular economy approach to the construction and demolition sector. A literature review. *Journal of Cleaner Production, 178*, 618–643. https://doi.org/10.1016/J.JCLEPRO.2017.11.207

Hart, J., Adams, K., Giesekam, J., Tingley, D. D., & Pomponi, F. (2019). Barriers and drivers in a circular economy: The case of the built environment. *Procedia CIRP, 80*, 619–624. https://doi.org/10.1016/J.PROCIR.2018.12.015

Kanters, J. (2020). Circular building design: An analysis of barriers and drivers for a circular building sector. *Buildings, 10*. https://doi.org/10.3390/buildings10040077

Knoth, K., Fufa, S. M., & Seilskjær, E. (2022). Barriers, success factors, and perspectives for the reuse of construction products in Norway. *Journal of Cleaner Production, 337*, 130494. https://doi.org/10.1016/J.JCLEPRO.2022.130494

Konietzko, J., Bocken, N., & Hultink, E. J. (2020). Circular ecosystem innovation: An initial set of principles. *Journal of Cleaner Production, 253*, 119942. https://doi.org/10.1016/j.jclepro.2019.119942

Kooter, E., van Uden, M., van Marrewijk, A., Wamelink, H., van Bueren, E., & Heurkens, E. (2021). Sustainability transition through dynamics of circular construction projects. *Sustainability (Switzerland), 13*(21), 1–19. https://doi.org/10.3390/su132112101

Korhonen, J., Nuur, C., Feldmann, A., & Birkie, S. E. (2018). Circular economy as an essentially contested concept. *Journal of Cleaner Production, 175*, 544–552. https://doi.org/10.1016/j.jclepro.2017.12.111

Mhatre, P., Gedam, V., Unnikrishnan, S., & Verma, S. (2021). Circular economy in built environment – Literature review and theory development. *Journal of Building Engineering, 35*. Elsevier Ltd. https://doi.org/10.1016/j.jobe.2020.101995

Moore, J. F. (1993). Predators and prey: A new ecology of competition. *Harvard Business Review, 71*(3), 75–86.

Munaro, M. R., Freitas, M. D. C. D., Tavares, S. F., & Bragança, L. (2021). Circular business models: Current state and framework to achieve sustainable buildings. *Journal of Construction Engineering and Management, 147*(12), 04021164. https://doi.org/10.1061/(asce)co.1943-7862.0002184

Nußholz, J. L. K., Rasmussen, F. N., Whalen, K., & Plepys, A. (2020). Material reuse in buildings: Implications of a circular business model for sustainable value creation. *Journal of Cleaner Production, 245*, 118546. https://doi.org/10.1016/j.jclepro.2019.118546

Pomponi, F., & Moncaster, A. (2017). Circular economy for the built environment: A research framework. *Journal of Cleaner Production, 143*, 710–718. https://doi.org/10.1016/j.jclepro.2016.12.055

Pulkka, L., Ristimäki, M., Rajakallio, K., & Junnila, S. (2016). Applicability and benefits of the ecosystem concept in the construction industry. *Construction Management and Economics, 34*(2), 129–144. https://doi.org/10.1080/01446193.2016.1179773

Sezer, A. A., & Bosch-Sijtsema, P. (2020). Actor-to-actor tensions influencing waste management in building refurbishment projects: A service ecosystem perspective. *International Journal of Construction Management*, 1–10. https://doi.org/10.1080/15623599.2020.1741493

Toppinen, A., Miilumäki, N., Vihemäki, H., Toivonen, R., & Lähtinen, K. (2019). Collaboration and shared logic for creating value-added in three Finnish wooden multi-storey building projects. *Wood Material Science & Engineering*. https://doi.org/10.1080/17480272.2019.1653365

Trischler, J., Johnson, M., & Kristensson, P. (2020). A service ecosystem perspective on the diffusion of sustainability-oriented user innovations. *Journal of Business Research, 116*, 552–560. https://doi.org/10.1016/J.JBUSRES.2020.01.011

UNEP. (2020). Global ABC Roadmap for Buildings and Construction Towards a zero-emission, efficient, and resilient buildings and construction sector. www.iea.org

Vargo, S. L., & Lusch, R. F. (2014). Inversions of service-dominant logic. *Marketing Theory, 14*(3), 239–248. https://doi.org/10.1177/1470593114534339

Vosman, L., Coenen, T. B. J., Volker, L., & Visscher, K. (2021). *Exploring the innovation ecosystem concept for a construction industry in transition.*

Whicher, A., Harris, C., Beverley, K., & Swiatek, P. (2018). Design for circular economy: Developing an action plan for Scotland. *Journal of Cleaner Production, 172*, 3237–3248. https://doi.org/10.1016/J.JCLEPRO.2017.11.009

Yang, Y., Zhang, Y., & Xie, H. (2021). Exploring cultivation path of building information modelling in China: An analysis from the perspective of an innovation ecosystem. *Engineering, Construction and Architectural Management, 28*(3), 662–680. https://doi.org/10.3390/su12176902

Part III
Renovation and Resource Efficiency in the Built Environment

Chapter 10
Replication and Upscaling of Energy-Efficient Solutions in the Building Sector

Anandasivakumar Ekambaram, Daniela Baer, Lillian S. Rokseth, Pål C. B. Vågbø, Håvard Bergsdal, and Nina H. Sandberg

Absract This paper presents relevant theories, approaches and concepts that focus on replication and upscaling of energy-efficient solutions in the building sector. It describes an analytical framework that can be used as an overall guidance/reference for replication and upscaling of solutions from the building level to the district (or neighbourhood) level. The framework consists of multi-level perspective (MLP) and technology innovation systems (TIS). Furthermore, this paper presents two more topics – (1) long-tern dynamic modelling of building stocks and (2) a socio-economic view on the demand side – to provide a better understanding of the dynamics that are involved in the replication and upscaling initiatives. These two topics would provide additional, relevant information and insight that could strengthen the application of the framework mentioned above. Within the framework and the two supporting topics, this paper highlights three aspects that should be specially considered when it comes to replication and upscaling of solutions from the building level to the district (or neighbourhood) level. They are stakeholders, contextual factors and knowledge (training and knowledge development). These three aspects are related to each other. This paper would contribute to provide support/guidance to replicate and upscale energy-efficient solutions in the building sector at the district (or neighbourhood) level. This is a conceptual paper based on narrative literature study and the authors' project experience.

Keywords Renovation · Stakeholder · Contextual factors · Knowledge · Sustainability

A. Ekambaram (✉) · D. Baer · L. S. Rokseth · P. C. B. Vågbø · H. Bergsdal · N. H. Sandberg
SINTEF, Trondheim, Norway
e-mail: siva@sintef.no; Daniela.Baer@sintef.no; lillian.rokseth@sintef.no; pal.vagbo@sintef.no; Havard.Bergsdal@sintef.no; nina.h.sandberg@sintef.no

G. Lindahl, S. C. Gottlieb (eds.), *SDGs in Construction Economics and Organization*, Springer Proceedings in Business and Economics,
https://doi.org/10.1007/978-3-031-25498-7_10

10.1 Introduction

Sustainability and energy efficiency are topics that have gained much attention recently in many sectors, including the building sector. Several authors point out that buildings are responsible for 30–40% of global carbon emissions and that there is hence a significant improvement potential for energy saving in the building sector – e.g. Johansson et al. (2017) and Baumhof et al. (2018). In addition to the existing buildings, there is new construction. New construction efforts are expected to increase the current stock at a rate of 1% per year (Vilches et al., 2017; Xing et al., 2011). In this context, what will be the status of building renovation? A description provided by European Commission (2015, page 11) says: 'The biggest challenge when reducing energy use in buildings is to increase the rate, quality and effectiveness of building renovation (currently only at 1.2%/year). To do this, it is necessary to reduce renovation costs and also to increase the speed at which it can be carried out in order to minimise disturbance for occupiers. To achieve an ambitious increase of the renovation rate (up to 2–3% per year), effective solutions need to be widely demonstrated and replicated.'

The above description not only points out the importance of building renovations in the future of the European building sector but also sets the premise of this paper, that is, replicating and upscaling of effective solutions and contributing to greater benefits for the society.

This paper is connected to an EU project called REZBUILD (https://rezbuildproject.eu/). The main objective of the REZBUILD project is to develop one refurbishment ecosystem based on the integration of cost-effective technologies, business models and life cycle interaction for deep nZEB renovation to diverse residential renovation typologies and interconnecting both building renovation stages and stakeholders. Several solutions were developed in this project such as 3D printing facades, BIPV (building-integrated photovoltaics) integration, super thermal insulation materials, HVAC (heating, ventilation and air conditioning) systems by radiant floor and solar heat pump solution (solar-assisted heat pump, SAHP). A part of the study in the REZBUILD project was to look at the potential for replicating and upscaling the REZBUILD solutions in the building sector at the district level.

This paper aims to find out how to approach such replication and upscaling efforts. Here, we first look at completed or ongoing energy-efficient building (EEB) projects to identify their experiences (lessons learned) on replication and upscaling efforts. This study and the study on sustainability related concepts and approaches provide us with a background knowledge to develop and present an analytical framework that can be used as an overall guidance/reference for replication and upscaling of energy-efficient solutions in the building sector. The contribution of this paper can be considered in connection with Sustainable Development Goals 11 (sustainable cities and communities) and 13 (climate action), at least to a certain extent.

This paper is based on narrative literature review (NLR) and the authors' experience. A brief description on NLR is provided first, followed by the summary of the results from the study on the selected EEB projects, and then concepts and approaches related to sustainability, which are key building blocks of the framework, are described. Finally, concluding remarks wind up the whole discussion.

10.2 Research Method

This paper is based on a narrative literature review (NLR). NLR takes into consideration various studies of a topic and allows the reviewer to gain an understanding of various views associated with the topic and to make a holistic interpretation of the studies by using his/her experience as well as existing theories and models (Campbell Collaboration, 2001; Kirkevold, 1997). Jahan et al. (2016) say that NLR does not necessarily require to report more rigorous aspects that characterise structured literature review – aspects such as research methodology, search term, database that was used and inclusion as well as exclusion criteria.

10.3 Study on EEB Projects

10.3.1 Example Projects

We have chosen a narrative literature review approach including document analysis in this study as it enables for a holistic interpretation of existing EEB projects within H2020 and respond to the complexity of the research topic. We identified 39 completed and ongoing H2020, FB 7[1] and IEE[2] projects and did a literature and document analysis to describe their experiences/reflections (lessons learned) on replication and upscaling efforts. Here are some existing R&D projects addressing the potential for replication and upscaling:

- Pro-GET-OnE (2017–2021): https://www.progetone.eu/
- 4RinEU (2016–2020): http://4RinEU.eu/
- TP2Endure (2016–2020): https://www.p2endure-project.eu/en
- MORE-CONNECT (2014–2018): www.more-connect.eu/
- STUNNING (2016–2019): https://renovation-hub.eu/
- ConZEBs (2017–2019): https://www.conzebs.eu/
- ZEBRA2020 (2014–2016): http://zebra2020.eu/

In our study, we have found that only a few (seven) of the reviewed projects published documents or papers that thematise the potential for replication and

[1] FB7 refers to the EU 7th Framework Program for Research and Technical Development covering the timespan from 2007 to 2013.

[2] IEE refers to projects funded by the Intelligent Energy Europe Program.

upscaling of the tested processes and products. Amongst these are projects dealing with deep renovation processes focussing on consumer- or user-centred approaches (Pro-GET-OnE), robust and reliable technology concepts and business models (4RinEU), prefabricated plug and play (PnP) systems (P2Endure) and prefabricated, multifunctional renovation elements for the entire building envelope (façades and roof) (MORE-CONNECT). Other projects aim at developing a stakeholder community built around a knowledge-sharing platform with information on innovative solutions for building renovation and novel business models (STUNNING), solution sets for cost reduction of new nearly zero-energy buildings[3] (nZEBs) (ConZEBs) or monitoring systems to track the market uptake of nZEBs (ZEBRA2020).

The results in this section are based on documents published as deliverables in the reviewed projects (STUNNING, ConZEBs, ZEBRA2020) and a published paper addressing technical, financial and social barriers and challenges in deep renovation project defined through a joint workshop of four H2020 cluster projects (Pro-GET-OnE, 4RinEU, P2Endure and MORE-CONNECT). We present the results in a chronically order.

10.3.1.1 P2Endure

P2Endure has implemented upscaling as a central part of the project, and this is the focus of one of the work packages: WP 5 Exploitation, standardisation and market upscaling. The work package '[…] will deliver generic exploitation, business and replication plans that are eligible for adjustments and implementations in different market networks' (P2Endure, 2022). In close collaboration between the four H2020 innovation actions P2Endure, 4RinEU, Pro-GET-OnE and MORE-CONNECT, a workshop was organised with the focus on summarizing the experiences from the projects. One of the objectives of this event 'was to support the market upscaling of P2Endure solutions' (ibid.). The main barriers found in deep renovation processes were divided into the following macro-groups: technical, financial and social barriers (D'Oca et al., 2018) (cf. Table 10.1).

10.3.1.2 ZEBRA2020

Based on the experiences in the ZEBRA2020 project, Toleikyte et al. (2016) describe four overarching conditions that should be addressed to ensure effective policy processes that are needed for the market uptake of nZEBs:

- Involvement from a broad set of stakeholders
- Long-term strategies to upgrade the building stock

[3] A nearly zero-energy building is defined as a building that 'has a very high energy performance' where the energy required should be covered to a 'significant extent by energy from renewable sources' (BPIE, 2011, p. 5).

Table 10.1 Identified barriers for upscaling

Barrier type	Description
Technical barriers	A lack of consistent and standardised solutions or integrated solutions to comply with new and different building standards requirements on energy saving Lack of skilled workers to carry out the work Shortcomings in technical solutions and long processes discouraging owners Safety/seismic risk connected with the deep renovation processes (damages can be done to the homes whilst retrofitting or unsure perception of the current safety of the existing buildings) End-users' and owners' lack of technical expertise and trust in effective energy renovation savings
Financial barriers	High up-front costs and owners reluctant to borrow funds for energy renovation purposes Long pay-back times of retrofitting interventions Lack of confidence of the potential investors Insufficient and instable available funding Lack of attractive financing for homeowners with low to medium incomes who are usually not eligible for regular bank loans The fact that existing financial tools are insufficient and unattractive
Social barriers	Decision-making processes that are long and complex, especially in cases of multi-owner houses (condominiums) The lack of consensus, understanding and support from the inhabitants that often hinder the effective approval of the interventions The problem of disturbance during site works and/or relocation (in case owners/users need to leave their homes during the process) Low awareness about energy efficiency and nonenergy benefits of renovation Lack of dialogue between the different stakeholders.

Based on D'Oca et al. (2018)

- Continuous assessment and review including data collection and quality assurance to ensure and monitor progress
- Empowering local level or private initiatives to go beyond the set goals and lead by example to help accelerate the rate and depth of nZEBs (Ibid., p. 28)

The ZEBRA2020 project forwards a total of 35 recommendations for EU Member States categorised within six topics (Toleikyte et al., 2016, p. 29):

- Legislative and regulatory (e.g. regulating minimum standards for building performance through building codes, improvements to the use of energy performance certificates, providing tailored advice to building owners and investors and implementing standardised methodologies for data gathering)
- Economic (e.g. incentivising the market uptake through active price signals, providing financial support for renovation according to long-term benchmarks, clever legislation to mitigate the problem of split incentives)
- Communication (e.g. branding of nZEBs, promotion of demonstration projects that exemplify benefits of high-performance buildings, facilitation of knowledge sharing)
- Quality of action (e.g. development of quality frameworks for nZEB techniques and technologies, training of building professionals, enhancement of the expertise of certifiers)

- New business models (e.g. fostering the uptake of industrialised renovation, encouragement of new business models, enabling the market to support new features of buildings as micro energy-hubs)
- Social measures (e.g. defining energy poverty, specification and increase in support measures for vulnerable target groups, improvement of all social housing to nZEB standards)

The overarching conditions described in ZEBRA2020 are relevant to look at reported experiences from other projects. In the final report documenting the lessons learned from the STUNNING project, the learning outcomes and guidelines for replications are described for each of the five demos included in the project. Amongst the lessons learned concerning technical issues, they find that the innovation potential of the construction methods and processes is case-specific and demands a step-by-step approach, depending on the competence of contractors. The stakeholders involved (building owners, tenants and planning partners) must all be involved in the development of innovative methods. Regarding social issues, the experiences with the demos demonstrate that there is a need for a fundamental change in both awareness and quality of communication and the need for increased involvement amongst inhabitants. Another learning outcome is that the reduced construction time caused by a high degree of prefabrication had a positive effect on all participants, including the inhabitants who experienced less disturbance (Laffont-Eloire et al., 2019). The positive effect of time reductions in the construction period is also amongst the key added social values from the innovation actions P2Endure, 4RinEU, Pro-GET-OnE and MORE-CONNECT (D'Oca et al., 2018).

10.3.2 Discussion

The study on the 39 R&D projects in the field of EEB shows that only a limited number of projects have published and shared their experiences of upscaling. The seven projects that present their lessons learned approach different levels of transition. Whilst the P2Endure and the STUNNING project focus on the market uptake of specific technologies as prefabricated building elements, the ZEBRA project approaches the nZEB concept and its market penetration as a whole with a focus on policymaking on the landscape level.

P2Enduro and STUNNING follow a bottom-up approach by investigating the drivers and barriers of upscaling in the contextual setting of the demo sites and identify the three thematic areas of technical, finance and social aspects. Their contribution is thereby to deliver findings on practical implementation of the technologies. In all of the three dimensions identified, the end-user plays a prominent role.

Findings from the ZEBRA project addressing the upscaling of the nZEB concept are more general and on the policy level as they are not addressing a specific technology but an interplay between different technologies that in combination enable the realisation of environmental ambitious goals as the nZEB concept. The measures identified in ZEBRA on the policy level match with barriers identified in the STUNNING project – e.g. the providing of financial support for renovation fits with the identified lack of insufficient and instable available funding. Additionally, we see that the integration of contextual factors of the specific demo sites contributes to identify specific end-user needs – e.g. the homeowners with low to medium incomes who are usually not eligible for regular bank loans.

Besides the necessity of identification of contextual factors and their incorporation in respective policy measures, our results point the attention to stakeholders and their specific roles in the upscaling process.

The results show that stakeholders play an important role as contributors to open innovation processes and that therefore a broad involvement of all stakeholders should be realised. This is in line with academic literature and the concept of open innovation processes (Chesborough, 2003). The results also stress the importance of knowledge of the diverse stakeholders and the diffusion of knowledge to diverse stakeholder groups.

10.4 Developing a Framework

10.4.1 Key Parts of the Framework

Academia has produced several theories and approaches that explain the process of upscaling and that point to different factors that influence the process. We have chosen the multi-level perspective (MLP) in Transition Theory (Geels, 2002) to elaborate on the process from the primary invention (niche) of a new energy efficiency building technology to market application. MLP enables a deeper understanding of the coevolution of technologies and the incorporation of different stakeholders' role and their interplay as a system-wide interaction. The theoretical approach of Technology Innovation Systems (TIS) (Hekkert et al., 2007) identifies seven functions of these systems and its inherent influence factors for their performance. Thereby, the concept of TIS helps us to identify influencing factors of upscaling for energy-efficient building technologies within the broader framework of sustainable transition of the built environment as elaborated in Transition Theory. We present Transition Theory first, followed by TIS, and then we combine these two approaches in an analytical framework.

10.4.1.1 Socio-technical Transition and the Multi-level Perspective

The socio-technical transition approach is an umbrella term that includes the multi-level perspective (MLP), transition management (TM) and strategic niche management (SNM) and is mainly explored by Dutch researchers. A socio-technical system comprises of three interrelated elements: a network of actors and social groups;

formal, cognitive and normative rules that guide their activities; and material and technical elements as artefacts and infrastructures (Geels & Schot, 2007). MLP is a prominent transition framework to analyse and indicate appropriate transition pathways. The MLP posits that transitions will be enabled through interaction processes within and amongst three analytical levels: niches (micro-level), socio-technical regimes (meso-level) and a socio-technical landscape (macro-level).

The niche level (micro) is the level in which space is created for experimentation and technology development as done in the REZBUILD project. On this level, the strategic niche management (SNM) highlights the importance of protected 'incubator' spaces, user involvement as well as learning to develop new technologies that can either be an incremental or radical innovation. The regime level (meso) presents the current structures and practises in the market, characterised by established institutions, technologies and rules and that provide stability and reinforcement to the current socio-technical systems. The landscape level (macro) is the overall socio-technical setting that encompasses the dynamics of deep cultural patterns, macro-economics and macro-political developments that make up the environment or context of socio-technical transition (Twomey & Gaziulusoy, 2014).

In transition studies, three success factors are generally recognised when it comes to market uptake of technological inventions from the niche to the regime level (van den Heiligenberg et al., 2017). Successful niches have the following:

- Visions which are shared by many actors, specific and of high-quality building of social networks,
- Broad and deep networks
- Learning processes at various dimensions, i.e. first-order learning (maintenance learning) and second-order learning (reframing, or reordering of assumptions; Sterling, 2007). Raven (2005) emphasises that learning processes that take place align the technical features of the niche experiment with its social dimensions (e.g. regulation, user preferences) and that induce the actors to reflect about their underlying norms and values about the niche experiment

10.4.1.2 Technology Innovation Systems (TIS)

Many authors see technological change as one of the major determinants of economic growth, and the concept of sustainable technology development contributes to understand technological development within a bigger system approach. The system and its inherent components must change to enable for sustainable development (Weaver et al., 2000). Technological development and change are analysed in a heuristic approach within the field of innovation systems. The concept of technology innovation systems (TIS) is developed to analyse all societal subsystems, actors and institutions contributing in one way or the other, directly or indirectly, intentionally or not, to the emergence or production of new technologies and the upscaling of them (Hekkert et al., 2007). As technology development is not an autonomous and self-going process, management of the process is necessary. The specific emphasis of TIS lies in the identification of activities within the system – Hekkert et al. (2007) call them functions of the system. An analysis of the TIS and its functions will deepen the understanding of what kind of activities that fosters or hampers innovation.

We present the seven functions of TIS developed by Hekkert et al. (2007) in a chronological order:

1. Knowledge development: creation of new knowledge, which include different types as, e.g. science-based research activities as well as experience-based knowledge
2. Knowledge diffusion: diffusion of knowledge within the innovation system and abroad
3. Guidance of the search: selection of a development pathway of the technological development as management asset
4. Entrepreneurial experimentation: exploration and exploitation of business models and opportunities on the basis of new technologies and applications
5. Market formation: interplay between the type of innovation within the spectrum from incremental to radical and the formation of the market
6. Resource mobilisation: availability and usability of different resources (e.g. financial and human capital) to enable innovation system activities
7. Creation of legitimacy: creation of legitimacy through diverse activities (e.g. lobbying, marketing, capacity building) to counteract to resistance to change

10.4.1.3 The Combination of MLP and TIS

Our analytical framework is based on the MLP approach within transition studies and the concept of TIS and its functions. The TIS approach focuses on the technology itself and how an upscaling process can lift the technology from the niche level to market uptake on the regime level. The multi-level perspective (MLP) is another heuristic framework, which takes a broader approach than TIS theory by looking at transformative societal processes (Twomey & Gaziulusoy, 2014). The MLP frames a broad understanding of sustainable transition by technological innovations within the socio-technical system and points the attention to the different levels and its inherent stakeholders.

We therefore combine both approaches to develop an analytical framework. The MLP contribute with the multi-level perspective and its specific stakeholders as well as the focus on policy measures, whilst TIS gives us a deeper understanding of the necessary functions/activities to consider to upscale a technology from the niche level to the market performance.

10.4.2 Two Supporting Topics

The two topics that we will present below (long-term dynamic modelling of building blocks and a socio-economic view on the demand side) provide additional, relevant information and insight that could strengthen the application of the key parts of the framework mentioned above.

10.4.2.1 Long-Term Dynamic Modelling of Building Stocks

The REZBUILD project has developed several innovative technologies for enabling energy-efficient refurbishment solutions. To understand the possible upscaling and wider implementation of new solutions, it is important to have some understanding of the dynamics of building stocks, their composition and their likely development.

A dynamic building stock model developed primarily for Norway, but applied also to a selection of other European countries, is called RE-BUILDS 2.0 model (Sandberg et al., 2021). RE-BUILDS 2.0 uses material flow analysis (MFA) methodology to predict future developments in stock size and composition and associated flows of demolition and construction. In addition, different renovation scenarios have been implemented in the dynamic modelling to understand the impacts of various renovation and upgrading measures. In the dynamic building stock model for residential housing, the key driver for determining the size of the building stock is the population's need to reside and associated lifestyle parameters. This includes population size and demand for number and floor area of dwellings.

10.4.2.2 A Socio-economic View on the Demand Side

Several of the EU projects that we have looked at focused mainly on the supply side of the renovation market whilst making assumptions about the demand side. Similarly, research focussing on the barriers for sustainable retrofitting highlights that the problems are informational and economic and not necessarily technical (Ástmarsson et al., 2013). In short, without demand there will be no market for the developed technical solutions. Thus, in order to understand how to facilitate a large-scale system-wide deep renovation adoption, the underlying mechanics of demand for such products should be understood.

There has been some research into the demand side of deep renovation. For example, Ástmarsson et al. (2013) point out some demand-side barriers: political consciousness, lack of information and understanding amongst main stakeholders regarding the technical solutions available in the market, the property itself and which solutions should be prioritised. Moreover, in the EU projects 4RinEU, ProGETonE and P2Endure, the residents themselves were considered a barrier to upscaling.

There can be many possible variables that can affect the deep renovation market, including the demand side of it, in the future. It is thus important to find out what factors that have been historically important and would mostly likely be central for explaining future variation within the respective market. This can be achieved by utilizing machine learning (ML) methods to analyse time series data that are relevant to the green transition for the property market.

10.4.3 Highlighting Three Aspects/Areas

Within the framework and the two supporting topics, we highlight three aspects/areas that should be specially considered when it comes to replication and upscaling of energy-efficient solutions:

1. Stakeholders: replication and upscaling of the solutions at the district level involve many actors with varying degree of knowledge, interest and influencing power. And, the number, nature dynamics of stakeholder tend to change over time. Hence, it is important to map, involve, engage and manage stakeholders in project that focus on replication and upscaling of solutions. The importance of good stakeholder management and project success is widely mentioned in project management literature – for example, please refer Joslin and Müller (2016).
2. Contextual factors: focus on contextual elements can be seen in two dimensions:
    - Micro-level: here, the focus is on the context of the renovation project. In this regard, technical, organisational and people issues (knowledge transfer between individuals) are specially taken into consideration.
    - Macro-level: it deals with national and international standards, as well as regulations and long-term strategies that various countries have when it comes to building renovation.
3. Knowledge (training and knowledge development): Learning, training and knowledge sharing play a prominent role to make sure that the requirements, functioning and effects of the solutions are correctly understood and that the installation is properly done with a systemic understanding of the dynamics of the district (or neighbourhood) where the solutions are replicated and upscaled. In addition, there is a need to ensure that the users have adequate knowledge and information to use the solutions in an appropriate way so that the intended positive effects can be accomplished effectively.

   These three aspects are related to each other.

10.5 Concluding Remarks

As a summary of the discussion so far, we present the following model (cf. Fig. 10.1). The framework/model would contribute to provide support and guidance to replicate and upscale energy-efficient solutions in the building sector at the district level (or neighbourhood level).

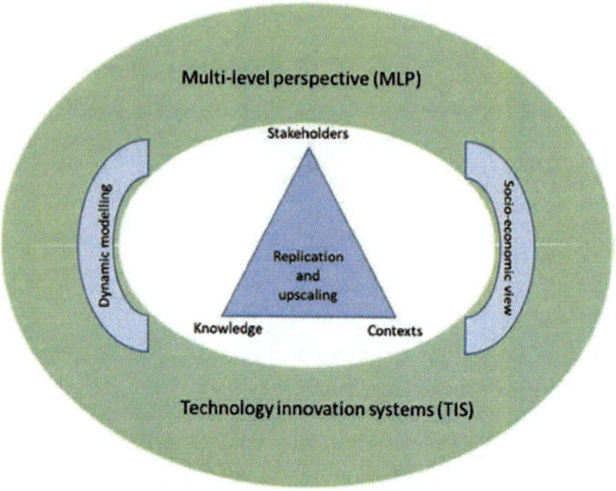

Fig. 10.1 A model for replication and upscaling

Our study of lessons learned of upscaling processes within R&D projects and the incorporation of learnings from the REZBUILD project has highlighted the three areas that play a specific role in the upscaling process: stakeholders, contextual factors and knowledge. These findings are in general in line with existing literature studies but also stress the interplay between these three areas. Especially findings in the REZBUILD project have shown that different types of knowledge are located at different stakeholder groups and need specific measures to develop within the specific contextual setting of a geographical area. This poses the question to the management of learning activities and stakeholder cooperation in the uptaking of EEB technologies from the niche level to the market. Little is said about the governance structures behind upscaling activities of EEB technologies. Whilst specific stakeholders' roles and respective activities are identified at different phases of the innovation process including upscaling, little could be found on the leadership of the process as a whole and the specific management approach. We therefore pose the questions of responsibility and leadership of upscaling of EEB technologies from the early invention to market penetration. The management of upscaling needs to be addressed in all phases of the innovation process and needs to relate to all presented functions of a TIS, the incorporation of policies on the landscape level and with a specific eye on the three elements of stakeholder and knowledge management as well as contextual factors since no 'one solution fits all'.

Acknowledgements This paper is connected to a research project called REZBUILD. This project has received funding from the European Union's Horizon 2020 research and innovation programme under grant agreement No 768623. Any dissemination of results must indicate that it reflects only the author's view and that the Commission is not responsible for any use that may be made of the information it contains.

References

Ástmarsson, B., Jensen, P. A., & Maslesa, E. (2013). Sustainable renovation of residential buildings and the landlord/tenant dilemma. *Energy Policy, 63*, 355–362.

Baumhof, R., Decker, T., Röder, H., & Menrad, K. (2018). Which factors determine the extent of house owners' energy-related refurbishment projects? A Motivation-Opportunity-Ability Approach. *Sustainable Cities and Society, 36*, 33–41.

BPIE – Buildings Performance Institute Europe. (2011). *Europe's buildings under the microscope. A country-by-country review of the energy performance of buildings*. BPIE.

Campbell Collaboration. (2001). *An author's guide to writing articles and reviews for Educational Research review [pdf]*. https://www.huidziekten.nl/diversen/opleiding/CATDatabase/guidetowritingreviews.pdf. Accessed: 20.01.2022.

Chesborough, H. W. (2003). *Open innovation – The new imperative for creating and profiting from technology*. Harvard Business School Publishing Corporation.

D'Oca, S., Ferrante, A., Ferrer, C., Pernetti, R., Gralka, A., Sebastian, R., & op 't Veld, P. (2018). Technical, financial, and social barriers and challenges in deep building renovation: Integration of lessons learned from the H2020 cluster projects. *Buildings, 8*, 12.

European Commission. (2015). *Draft Horizon 2020 Work Programme 2014–2015 in the area of Secure, Clean and Efficient Energy*. http://ec.europa.eu/research/horizon2020/pdf/work-programmes/secure_clean_and_efficient_energy_draft_work_programme.pdf. Referred 18th April 2018.

Geels, F. W. (2002). Technological transitions as evolutionary reconfiguration processes: A multi-level perspective and a case-study. *Research Policy, 31*, 1257.

Geels, F. W., & Schot, J. (2007). Typology of sociotechnical transition pathways. *Research Policy, 36*(3), 399–417.

Hekkert, M. P., Suurs, R. A. A., Negro, S. O., Kuhlmann, S., & Smits, R. E. H. M. (2007). Functions of innovation systems: A new approach for analysing technological change. *Technological Forecasting and Social Change, 74*, 413–432.

Jahan, N., Naveed, S., & Zeshan, M. (2016). How to conduct a systematic review: A narrative literature review. *Cureus, 8*(11).

Johansson, T., Olofsson, T., & Mangold, M. (2017). Development of an energy atlas for renovation of the multifamily building stock in Sweden. *Applied Energy, 203*, 723–736.

Joslin, R., & Müller, R. (2016). The relationship between project governance and project success. *International Journal of Project Management, 34*, 613–626.

Kirkevold, M. (1997). Integrative nursing research – An important strategy to further the development of nursing science and practice. *Journal of Advanced Nursing, 25*, 977–984.

Laffont-Eloire, K., Peraudeu, N., Petit, S., Bordeau, M., Joumni, H., Belaid, F., … La, X. W. (2019). *Sustainable business models for the deep renovation of buildings*. EU Horizon 2020 project STUNNING (GA No. 768287). Available at: https://renovation-hub.eu/downloads/. Accessed: 24.01.2022.

P2Endure. (2022). *P2Endure – Work Programme*. Available at: https://www.p2endure-project.eu/en/project/work-programme. Accessed: 24.01.2022.

Raven, R. P. J. M. (2005). *Strategic niche management for biomass: A comparative study on the experimental introduction of bioenergy technologies in The Netherlands and Denmark*. Technische Universiteit Eindhoven. https://doi.org/10.6100/IR590593

Sandberg, N. H., Næss, J. S., Brattebø, H., Andresen, I, & Gustavsen, A. (2021). Large potentials for energy saving and greenhouse gas emission reductions from large-scale deployment of zero emission building technologies in a national building stock. *Energy Policy, 152*.

Sterling, S. (2007). Riding the storm: Towards a connective cultural consciousness. In A. E. J. Wals (Ed.), *Riding the storm: Towards a connective cultural consciousness, social learning towards a sustainable world* (pp. 63–82). Wageningen Academic Publishers.

Toleikyte, A., Kranzl, L., Bointner, R., Bean, F., Cipriano, J., De Groote, M., & Volt, J. (2016). *ZEBRA 2020 – Nearly zero-energy building strategy 2020. Strategies for a nearly Zero-Energy Building market transition in the European Union.* Available at: https://zebra2020.eu/publications/an-overview-of-zebra2020-publications-and-tools/. Accessed: 25.01.2022.

Twomey, P., & Gaziulusoy, A. I. (2014). *Review of system innovation and transitions theories concepts and frameworks for understanding and enabling transitions to a low carbon built environment.* Working paper for the Visions & Pathways project [pdf]. Available at: https://www.researchgate.net/publication/306119135_Review_of_System_Innovation_and_Transitions_Theories_Concepts_and_frameworks_for_understanding_and_enabling_transitions_to_a_low_carbon_built_environment. Accessed: 25.01.2022.

van den Heiligenberg, H. A., Heimeriks, G. J., Hekkert, M. P., & van Oort, F. G. (2017). Habitat for sustainability experiments: Success factors for innovations in their local and regional contexts. *Journal of Cleaner Production*, 169 [pdf]. Available at: https://www.sciencedirect.com/science/article/pii/S0959652617313471?via%3Dihub. Accessed: 23.01.2022.

Vilches, A., Garcia-Martinez, A., & Sanchez-Montañes, B. (2017). Life cycle assessment (LCA) of building refurbishment: A literature review. *Energy and Buildings, 135*, 286–301.

Weaver, P., Jansen, L., Grootveld, G., & van Spiegel, P. (2000). *Sustainable technology development*. Greenleaf Publishing.

Xing, Y., Hewitt, N., & Griffiths, P. (2011). Zero carbon buildings refurbishment – A hierarchical pathway. *Renewable and Sustainable Reviews, 15*, 3229–3236.

Chapter 11
A Delphi Pilot Study to Assess the Impact of Location Factors for Hyperscale Data Centres

David King, Nadeeshani Wanigarathna, Keith Jones, and Joseph Ofori-Kuragu

Abstract The rapid expansion of digital technology used by individuals and organisations has resulted in the need for many data centres on a rapid scale. Data centres are high energy-consuming buildings during their operation. Therefore, data centres are increasingly being built in regions with good access to energy infrastructure and cold climates to reduce the cooling energy demands of the facilities. This paper presents location factors that must be considered when estimating and modelling the capital cost of data centres built outside the UK. The research presented here is based on a questionnaire survey conducted as a pilot study of an impending Delphi study. The participant's responses were obtained via an open-ended questionnaire with a cross-sectional timeframe. A thematic analysis of the responses revealed six overarching themes: ease of doing business, design, customer pricing, land, power and fibre, and supply chain. The three most dominant are land, power and fibre and supply chain. This pilot study confirmed the knowledge gap and supported the need that further investigatory work is required. A complete Delphi study will develop the themes identified in this pilot study to achieve consensus on their significance and support an assessment of the impact of location on the modelling and forecasting of capital expenditure for hyperscale data centres.

Keywords Data centres · Location · CAPEX · Delphi

D. King (✉) · N. Wanigarathna · K. Jones · J. Ofori-Kuragu
Anglia Ruskin University, Chelmsford, UK
e-mail: dk555@pgr.aru.ac.uk; nadeeshani.wanigarathna@aru.ac.uk; keith.jones@aru.ac.uk; joseph.ofori@aru.ac.uk

G. Lindahl, S. C. Gottlieb (eds.), *SDGs in Construction Economics and Organization*, Springer Proceedings in Business and Economics,
https://doi.org/10.1007/978-3-031-25498-7_11

11.1 Introduction

Data centres are buildings that house information technology (IT) equipment to store and process data and services such as the internet (Whitehead et al., 2015). The rapid expansion of digital technologies resulted in building many data centres on a rapid scale (Parker, 2019). Hyperscale data centres are purposefully built large-scale data centre facilities above 20 MW, owned, and typically operated by the organisation they support, such as Apple or Facebook (Christensen et al., 2018). They are usually service platforms for social media, search engines, communication, entertainment, virtual reality, artificial intelligence, machine learning and e-commerce (ibid).

Data centres are high energy-consuming buildings during their operation. Therefore, hyperscale data centre site selection is often predicated on access to cheap power and utilities and in locations with substantial energy infrastructures already in place (Hogan, 2015: 4; Rosenwald, 2011; cited in Parker, 2019). Due to IT demands and cooling equipment primarily and lighting, power distribution and other requirements, they are very energy-intensive during their operation (Avgerinou et al., 2017). Therefore, energy efficiency became a primary concern of data centre development (Wahlroos et al., 2018). The Nordic region has attracted significant investment in new data centres with massive investments from the cloud and hyperscale investors like Facebook, Google, AWS and Apple (Christensen et al., 2018). These regions are preferred due to advanced technological progress in the sector and the cold climate conditions that reduce the cooling energy demands of the facilities (Avgerinou et al., 2017). This trend in locating data centres outside of the UK presents a significant challenge to cost consultants within the UK during the estimating and modelling capital cost of these buildings.

Cost planning (or cost modelling) at the feasibility stage is defined by the Royal Institution of Chartered Surveyors (RICS) as 'the determination of possible cost of a building(s) early in the design stage in relation to the employer's fundamental requirements. This takes place before the preparation of a full set of working drawings or bills of quantities and forms the initial build-up to the cost planning process' (RICS, 2012).

During the cost planning in early development phases, cost consulting professionals often use historical cost data as base cases and adjust their costs to suit the circumstances of new projects. Whilst the impact of specific characteristics such as shape, inflation and specifications is easy to adjust based on case-based reasoning, the impact of location is a difficult factor to predict for construction professionals. Therefore, they rely on location cost indices for this purpose. For example, several location cost indices such as Spon's Architects' and Builders' Price Book 2022 (Aecom, 2021) and the Building Cost Information Service (Royal Institution of Chartered Surveyors, 2021) are available for cost consultants to use. Similar indices are available in other European countries, yet they are less relevant for data centres. Likewise, there are often no precedents set to use as a baseline for cost comparisons. An example would be that specific standards and regulations for noise attenuation for hyper-size generators for data centres did not exist in Sweden and had to be modelled on regulations from other countries (Vonderau, 2017). Furthermore, compounding the lack of available cost data is the secrecy surrounding data centres, their operations, cost and locations (Holt & Vonderau, 2015).

Whilst international location cost indices, such as those provided by Eurostat (European Union, 2021), World Bank (World Bank, 2021) and the OECD (Organisation for Economic Co-operation and Development, 2021) otherwise known as purchasing power parities, are broad and primarily model variations at country level, they are less effective during the cost planning for individual projects specific to a particular region as there are many variables ranging from macroeconomic, construction methodology and geographical and geological categories. However, there is a gap in knowledge about what variables and types are relevant to the modelling and forecasting of hyperscale data centres. Through the exploration and literature review, it has been established that a wide range of variables impact construction projects' cost and cost modelling. However, there is no evidence identifying if and how these variables would impact the site location on the cost planning for the capital expenditure of hyperscale data centres. This review identifies that further research is required to establish and define specific variables relevant to hyperscale data centres.

It is recognised above that whilst there is published data on traditional construction costs in the UK along with published location indices. Neither of these provides sufficient information on establishing construction costs overseas. This is further compounded due to the specific design requirements of data centres and identifies a knowledge gap in the existing body of research.

This paper aims to identify consensus and an assessment for probability to identify themes that will inform the first round of a questionnaire of a complete Delphi study to identify the impact of location on the modelling and forecasting of capital expenditure for hyperscale data centres. The research question is to establish the impact of site location on the capital expenditure of hyperscale data centres and the variables that impact the modelling and forecasting as identified in the research gap above. This will assist in selecting the correct location to make informed decisions and reduce the financial risk to capital expenditure. The indicative research question raised is 'the impact of location on the modelling and forecasting capital expenditure for hyperscale data centres'.

To support the research topic, four subtopics have been identified:

- Does the location of a data centre change the risk of overspend? If so, how?
- What variables of capital expenditure are directly affected by location?
- What elements of capital expenditure cost are the most likely to overspend?
- What are the essential items to consider when choosing the right location for a hyperscale data centre?

The research will improve knowledge, linking innovation and infrastructure and the development of resilient infrastructure, which are all items targeted by the United Nations Sustainable Development Goals.

11.2 Methodology

11.2.1 Research Instrument

A modified Delphi technique was used as the research instrument. Delphi is a methodological technique where consensus can be formed from responses arrived at from a panel of experts based on questions that are uncertain (Pill, 1971). This is expanded on further in the seminal book by Linstone and Turoff (1975), where they state 'Delphi may be characterised as a method for structuring a group communication process so that the process is effective in allowing a group of individuals, as a whole, to deal with a complex problem'.

Therefore, the research uses a Delphi technique as the most appropriate method as Keeney et al. (2006) stated a Delphi approach 'is an important method for achieving consensus on issues where none previously existed'. In addition, Hasson et al. (2000) note that 'insufficient information has led to an increased use of consensus method' such as Delphi.

Whilst there are several modified uses of Delphi (Thompson, 2009), this pilot study has used a qualitative approach as it is suggested as the preferred method. Jairath and Weinstein (1994) note that pilot studies should be carried out for all Delphi studies in advance of the leading research questionnaire being developed. As a result, this pilot study creates a robust framework for developing the first-round questionnaire in a complete Delphi study.

11.2.2 Data Collection

To inform this pilot study, a sample (n = 5) of data centre industry experts was selected to identify the topics to inform the complete Delphi study. According to Connelly (2008), experts suggest that a pilot study sample should be 10% of the total Delphi study sample size. It is anticipated that due to the corporate sensitivity of obtaining the data, the number of participants for the full Delphi will be less than 50. Therefore, the number of participants selected for this pilot study is within range. It is known that a fundamental component of Delphi research is the identification of a 'panel of experts' (Baker et al., 2006) as they form an established method for determining consensus (Beech, 2001). Using a Delphi method with industry experts makes it an ideal solution where such information is lacking (Graham et al., 2003). The identification and selection of an expert is determined as a person regarded or consulted as an authority on account of special skill, training or knowledge, a specialist (Stevenson, 2010).

The selection of participants for this pilot study is also experts as identified by their senior position in an organisation (Mead & Mosely, 2001), and as such, the participants selected are all at senior levels within an organisation. The experts were purposefully selected using a multistep iterative approach. Participants were selected with design, development, engineering and commercial management expertise. The experts had similar expertise, experience and senior roles within the data centre industry.

11.2.3 Developing the Questionnaire

The survey was cross-sectional in its timeline, and the questions were open-ended with unlimited free text. The questions contained within the questionnaire are those identified as the research subtopics. One additional question was included within the questionnaire. This question was included to avoid any restrictions that the participants may experience whilst completing the four predetermined questions as above. In addition, it gave them the freedom to identify any other items they considered would impact capital expenditure outside the predetermined questions. This additional question was asked to the participants: what other variables may impact capital expenditure?

11.2.4 Details of the Data Collection

Participants were identified through known contacts based on their experience in cost consultation for data centre construction projects. The other criteria considered during the participant selection were their expertise and experience in hyperscale data centres both inside the UK and globally. The participants were not restricted to where and when the questionnaire was completed. The questionnaire was distributed to the participants utilising the online surveys distribution method via password-protected invites. The survey was open for 14 days to enable the participants' sufficient time to respond. All responses were received within this period.

11.2.5 Data Analysis

According to Kuhn (2021), having a paradigm as guidance for research is key to any research project as it is the 'set of beliefs and agreements shared between scientists about how problems should be understood and addressed'. Therefore, it is critical to establish ontological and epistemological approaches when forming the research methodology.

Williams (2016) states that ontology is 'the branch of philosophy concerned with the nature of things that exist'. The ontological approach taken within this pilot study and the subsequent interpretation and analysis of the participants observations identifies that the research has taken a realistic approach in that 'realism denies that there is any objective knowledge of the world' (Maxwell, 2012).

Epistemology is 'a way of understanding and explaining how we know what we know' (Crotty, 2020). It is also 'concerned with providing a philosophical grounding for deciding what kinds of knowledge are possible and how we can ensure that they are both adequate and legitimate' (Maynard, 1994). This pilot study has adopted a constructivist approach as the thematic analysis of the qualitative data is 'not discovered but constructed' (Crotty, 2020). This constructivist approach to the pilot study has also been supported by Gray (2021), who states that 'truth and meaning do not exist in some external world but are created by the subject's interactions with the world'.

Inductive thematic analysis was used to determine the themes and subsequent categories of the responses arising from the questionnaire as recognised by King (2004) that it assists in producing a concise analysis.

The thematic analysis was produced in NVivo software; the responses were coded to establish themes and analysed for themes and then grouped into categories. The coding and themes reflect the interpretation of the excerpts with the research question in mind.

The item frequency was calculated based on the aggregated items' responses. The two most dominant categories were explored in each question as most responses had a significant break between frequency counts.

11.2.6 Ethical Considerations

It is acknowledged that ethical considerations are fundamental to research for moral and institutional reasons (Farrimond, 2012). Therefore, before the questionnaire issue, written informed consent was obtained in advance from all participants and gatekeepers. Ethical approval for this study was received from Anglia Ruskin University Research School Research Ethics Panel (SREP) under the terms of Anglia Ruskin University's Research Ethics Policy.

11.3 Results and Discussion

11.3.1 Does the Location of a Data Centre Change the Risk of Overspend? If So, How?

The thematic analysis identified both land and the supply chain as the dominant categories in the participant's response when asked how the location of a data centre changes the risk of overspend. The analysis identified that land had a significant impact with overspending due to location choice. Results of the data analysis indicated that poor ground conditions, competing land use pressures and flood plain having significant impact.

In the thematic analysis, land equated to 50% of the themes. However, the perception of this impact varies, and whilst land appears to be the dominant category, there are many facets within this category. Following the data analysis, whilst geotechnical and environmental, land cost and infrastructure are identified, there is no consensus on impact ranking. The supply chain was also recognised as a dominant category in particular a lack of suitable labour, the cost of labour and materials being imported along with climatic conditions (hot/cold) impacting material selection.

These perceptions arise from the supply chain theme whilst showing the effect does not identify the cause. Therefore, there is a consensus that the supply chain does have an effect; there is no consensus identifying the cause, particularly in identifying the elements that could impact the schedule.

11.3.2 What Variables of Capital Expenditure Are Directly Affected by Location?

The thematic analysis identified both land and power and fibre as the two dominant categories in the participant's response when asked to identify what they considered the most significant cost impact to capital expenditure through location choice.

All of the participants of this study held a view that land had a significant impact, with examples from the questionnaire to support the theme categories being land parcel size, neighbouring high-risk activities (above/below ground) and identification of contamination risk profile.

In analysing these references from the questionnaires, land was mentioned 18 times. However, the perception of impact varies significantly from purchase price to contamination. This identifies that land appears to be the dominant category. However, identifying what constitutes a theme within the land category is unclear, and whilst geotechnical, climate and flooding are identified, there is no consensus on the dominant theme. All participants cited power and fibre as a significant theme, referenced 14 times. Results of the data identified specifically that the availability of power, fibre, existing network backbone structure and 'dark' fibre infrastructure have significant impacts. Typical statements being 'how much power is available' along with 'where does the fibre come in and how is it connected to the regional network'.

In analysing this data, most respondents agree that the availability of power and fibre has the most significant impact on this category. However, there is no consensus on whether power or fibre is the dominant theme.

These results largely agree with the existing literature and confirm that infrastructure requirements specific to data centres are a primary determinant of capital cost. Accurately estimating the impact of these factors on capital cost could improve the accuracy of development appraisal exercise and early decisions on location selection.

11.3.3 What Elements of Capital Expenditure Are Most Likely to Overspend?

The thematic analysis identified both design and the supply chain as the dominant categories in the participant's responses when asked to identify what elements of capital expenditure are most likely to overspend.

The design of the data centre was the dominant category at 35%, with response examples of design-related themes being permitting and planning, the method of colling the building and physical security of the site. Elements such as seismic design we also identified with comments such as 'a 20 MW two-story design for a Data Centre in the UK cannot be transposed to a site in Gebze region of Istanbul as no seismic cost/design have been accounted for within the substructure or superstructure'.

Whist this question asks what elements are most likely to overspend, the themes do not form a consensus within the category.

The supply chain was also recognised as a dominant category particularly labour and material costs. Other items resulting from the supply chain themes identified construction delays along with associated preliminary costs as having a significant impact.

Therefore, there is a consensus that the supply chain does have an effect; there is no consensus identifying the cause, particularly in identifying the elements that could impact the schedule.

11.3.4 What Are the Essential Items to Consider When Choosing the Right Location?

The thematic analysis identified land, power and fibre as the dominant categories in the participant's response when asked what essential items to consider when choosing the right location for a hyperscale data centre. The data identified that land had a significant impact on overspending due to location choice, property ownership laws, natural hazard risks, the previous use of the land, proximity to tenant/customer requirements and the accessibility of the sites.

In the thematic analysis, land equated to 43% of the categories. However, the perception of this impact varies, and whilst land appears to be the dominant category, there are many facets to the potential consensus of a category. This has identified that whilst geotechnical and environmental are dominant. Land cost and infrastructure are equally referenced, and subsequently, a consensus has not been formed. Likewise, all participants also cited power and fibre as a significant theme, referenced 11 times. Particular themes identified proximity to the data's point of use (latency), the availability of large utility power supply and redundancy power sources such as hydro, gas, oil, diesel and solar.

The category analysis identifies that most respondents agree that power and fibre have a significant impact. However, there was no consensus on whether power or fibre was the dominant theme when the participants were asked what the essential items are to consider when choosing the right location for a hyperscale data centre.

11.3.5 What Other Variables May Have an Impact on Capital Expenditure?

When asked to identify what variables of capital expenditure are directly affected by location, the most prominent category in the participant's response is design, ease of doing business and power and fibre. The design was the dominant category at 27%, with the key themes being the power and cooling technologies used, the substructure and superstructure design and noise attenuation. Whist this question asks what variables of capital expenditure are directly affected by location, there is no consensus within the category.

Power and fibre were identified as a significant theme, forming 20% of the categories, specifically distances from utilities (substation, fibre, water). Costs associated with incoming services and the need for an onsite substation were also frequent themes. Whilst the analysis identifies that power and fibre has a significant impact, there is not a consensus on whether it is either power or fibre that is the dominant theme.

11.3.6 The Relationship to the United Nations Sustainable Development Goals

By understanding the impact of power, fibre, land and design implications on the development of data centres provides knowledge and a greater understanding on the impact and importance of resilient infrastructure.

11.4 Conclusion

This pilot study was undertaken to seek expert opinion on the key themes impacting location on the modelling and forecasting of capital expenditure for hyperscale data centres, and the results of the data analysis are intended to provide rigour and to inform the questionnaire for the main Delphi study and increase the validity of the proposed questions.

This paper provides some indications of the current understanding of the variables that impact the modelling and forecasting for planning the capital expenditure of hyperscale data centres.

This review identified the gap in knowledge and the need for research in this area to move beyond prediction and forecasting of construction projects to that specifically for hyperscale data centres. Clibbens et al. (2012) state that pilot Delphi studies are rarely reported in the academic literature, making it difficult to establish best practice in this area. For this pilot study, industry expert knowledge was obtained through several expert participants ($n = 5$). The response rate was 100%.

Through having an open-ended questionnaire, experts were able to respond freely and without restriction.

Having completed the thematic analysis of the data arising out of the questionnaire, this pilot study has identified categories and themes that will address and inform the first round of a questionnaire of a complete Delphi study. Six overarching categories have been identified:

- Customer pricing
- Design
- Ease of doing business
- Land
- Power and fibre
- Supply chain

Additionally, the thematic analysis associated with each particular question has identified the themes within the categories as being:

- Access and infrastructure
- Geotechnical and environmental
- Land cost and availability
- Fibre
- Power
- MEPH (mechanical, electrical and public health)
- Substructure
- Superstructure
- General contractors

Therefore, from this pilot study, these categories and themes will inform round one of the main Delphi study. This pilot study has also confirmed the knowledge gap and supported the need that further investigatory work is required. Likewise, the results of this pilot study and the methodological stance taken have demonstrated that a Delphi approach would enable consensus in further research where none previously exist.

The results of this paper have identified that there are variables that impact the modelling and forecasting for planning the capital expenditure of hyperscale data centres that could not be identified through either existing published cost data or location variables thus connecting the results to the existing body of knowledge.

In addition, it is recommended that a further literature review is carried out to include the themes that have been identified through this pilot study, as this will subsequently inform and direct further research.

The results of this paper and the impending future research will provide knowledge to assist in achieving goal Nr 9 of the United Nations Sustainable Development Goals where it is identified that a key component of achievement should be that of a resilient infrastructure, particularly where there is a growing need for data led services.

11.4.1 Limitations

The respondent's location may have been biased as although the participants have experience in other geographic regions, they are all currently based in the UK. This may have led to biases in the answers, as some elements may differ at a country level. Such biases will be acknowledged as part of the main research study. However, it was expected that due to the extensive experience, the participants have the necessary expertise to respond to the questions in this pilot study, excluding their own biases.

Additionally, it is acknowledged that in undertaking this pilot study, the author has leveraged professional relationships, and this may have resulted in participant bias (Moore et al., 2010).

Acknowledgements The author would like to thank the R & D team at Yondr Group for their support in this research along with all of the expert participants for their response to the questionnaire, without whom this research would not be possible.

References

Aecom. (2021). *Spon's architects' and builders' price book 2022*. Taylor & Francis Group.

Avgerinou, M., Bertoldi, P., & Castellazzi, L. (2017). Trends in data centre energy consumption under the European code of conduct for data centre energy efficiency. *Energies, 10*(10), 1470.

Baker, J., Lovell, K., & Harris, N. (2006). How expert are the experts? An exploration of the concept of 'expert within Delphi panel techniques. *Nurse Researcher, 14*(1), 59.

Beech, B. (2001). The Delphi approach: Recent applications in health care. *Nurse Researcher (through 2013), 8*(4), 38.

Christensen, J. D., Therkelsen, J., Georgiev, I., & Sand, H. (2018). *Data centre opportunities in the Nordics: An analysis of the competitive advantages*. Nordic Council of Ministers.

Clibbens, N., Walters, S., & Baird, W. (2012). Delphi research: Issues raised by a pilot study. *Nurse Researcher, 19*(2), 37.

Connelly, L. M. (2008). Pilot studies. *Medsurg Nursing, 17*(6), 411.

Crotty, M. (2020). *The foundations of social research: Meaning and perspective in the research process*. Routledge.

European Union. (2021). Statistics | Eurostat. [Online]. Available at https://ec.europa.eu/eurostat/databrowser/view/prc_ppp_ind/default/table?lang=en. Accessed 19 Dec 2021.

Farrimond, H. (2012). *Doing ethical research*. Macmillan International Higher Education.
Graham, B., Regehr, G., & Wright, J. G. (2003). Delphi as a method to establish consensus for diagnostic criteria. *Journal of Clinical Epidemiology, 56*(12), 1150–1156.
Gray, D. E. (2021). *Doing research in the real world*. Sage.
Hasson, F., Keeney, S., & McKenna, H. (2000). Research guidelines for the Delphi survey technique. *Journal of Advanced Nursing, 32*(4), 1008–1015.
Holt, J., & Vonderau, P. (2015). Where the internet lives. *Signal Traffic*, 71–93.
Jairath, N., & Weinstein, J. (1994). The Delphi methodology (Part one): A useful administrative approach. *Canadian Journal of Nursing Administration, 7*(3), 29–42.
Keeney, S., Hasson, F., & McKenna, H. (2006). Consulting the oracle: Ten lessons from using the Delphi technique in nursing research. *Journal of Advanced Nursing, 53*(2), 205–212.
King, N. (2004). 21: Using templates in the thematic analysis of text. In *Essential guide to qualitative methods in organisational research* (pp. 256). Sage.
Kuhn, T. (2021). *The structure of scientific revolutions*. Princeton University Press.
Linstone, H. A., & Turoff, M. (1975). *The Delphi method*. Addison-Wesley Reading.
Maxwell, J. A. (2012). *A realist approach for qualitative research*. Sage.
Maynard, M., 1994. Methods, practice and epistemology: The debate about feminism and research. Researching women's lives from a feminist perspective Routledge, 10(26), pp.10–26.
Mead, D., & Mosely, L. (2001). The use of the Delphi as a research approach. *Nurse Researcher, 8*(4), 4–23.
Moore, D. A., Tanlu, L., & Bazerman, M. H. (2010). Conflict of interest and the intrusion of bias. *Judgment and Decision Making, 5*(1), 37.
Organisation for Economic Co-operation and Development. (2021). *Prices and purchasing power parities (PPP)*. OECD. [Online]. Available at: https://www.oecd.org/sdd/prices-ppp/. Accessed 19 Dec 2021.
Parker, M. (2019). *An apple a day: Listening to data centre site selection through a ionospheric investigation*.
Pill, J. (1971). The Delphi method: Substance, context, a critique and an annotated bibliography. *Socio-Economic Planning Sciences, 5*(1), 57–71.
RICS. (2012). *New rules of measurement-NRM 1: Order of cost estimating and cost planning for capital building works*. Royal Institution of Chartered Surveyors.
Royal Institution of Chartered Surveyors. (2021). *BCIS Indices – basic parameters*. [Online]. Available at: https://service-bcis-co-uk.proxy-lib.anglia.ac.uk/BCISOnline/Indices. Accessed 19 Dec 2021.
Stevenson, A. (2010). *Oxford dictionary of English*. Oxford University Press.
Thompson, M. (2009). Considering the implication of variations within Delphi research. *Family Practice, 26*(5), 420–424.
Vonderau, A. (2017). Technologies of imagination: Locating the cloud in Sweden's North. *Imaginations: Journal of Cross-Cultural Image Studies, 8*(2), 8–21.
Wahlroos, M., Pärssinen, M., Rinne, S., Syri, S., & Manner, J. (2018). Future views on waste heat utilisation–Case of data centers in Northern Europe. *Renewable and Sustainable Energy Reviews, 82*, 1749–1764.
Whitehead, B., Andrews, D., & Shah, A. (2015). The life cycle assessment of a UK data centre. *The International Journal of Life Cycle Assessment, 20*(3), 332–349.
Williams, M. (2016). *Key concepts in the philosophy of social research*. Sage.
World Bank. (2021). *PPP conversion factor, GDP* (LCU per international $) | Data. [Online]. Available at: https://data.worldbank.org/indicator/PA.NUS.PPP?view=map. Accessed 19 Dec 2021.

Chapter 12
Motivation for Doing a Sustainable Building Refurbishment from a Norwegian Building Owner's Perspective

Marit Støre-Valen, Alenka Temeljotov-Salaj, and Bintang Noor Prabowo

Abstract Norway aims to be a part of the European Green Deal where refurbishing and renovating buildings is an important action towards sustainable development to reach national and global SDG. This paper aims to shed light on what motivates the building owner to do sustainable building refurbishment (SBR) and discuss strategies that promote further SBR. This is examined through a scoping literature review and in-depth interviews of public and private property owners in the southeast of Norway. This study confirms that the first and foremost motivations of doing a SBR are cost-driven, technically, and regulatory-driven. Secondly, environmental aspects have the potential to be a motivator due to future changes in terms of demands of doing climate gas calculations and the implementation of the EU's taxonomy. Social aspects such as user demands and user involvement are discussed but not found as a motivation in itself. However, for historical buildings, the willingness to invest and find sustainable solutions is more likely to be true, as historic buildings are important for the identity and attractiveness in their neighborhood. The findings suggested that stricter regulations and higher demolition fees, climate gas calculations and life cycle cost demands, and EU's taxonomy are likely to further promote SBR.

Keywords Sustainable building renovation · Sustainable refurbishment · Adaptive reuse · Adaptability · Historic buildings

M. Støre-Valen (✉) · A. Temeljotov-Salaj · B. N. Prabowo
Department of Civil and Environmental Engineering, Norwegian University of Science and Technology (NTNU), Trondheim, Norway
e-mail: marit.valen@ntnu.no; alenka.temeljotov-salaj@ntnu.no; bintang.n.prabowo@ntnu.no

G. Lindahl, S. C. Gottlieb (eds.), *SDGs in Construction Economics and Organization*, Springer Proceedings in Business and Economics,
https://doi.org/10.1007/978-3-031-25498-7_12

12.1 Introduction

In the last decades, Norwegian cities put a lot of effort into achieving emissions reductions by setting visions, policies, strategies, and actions. According to the UNFCCC Paris Agreement at COP 21 in 2015, targeting the limitation of the global temperature increase by 2050, a fast transition to renewable and fossil-free energy is required and reducing the energy demand. As the building sector accounts for approximately 28% of total energy-related CO_2 emissions, there is a vast potential for sustainable renovation/refurbishment of existing buildings (Temeljotov & Lindkvist, 2021). The Buildings Performance Institute Europe emphasizes that 97.2% of the building stock is currently not in the energy performance A class and should be upgraded. Even though some barriers and drivers for sustainable renovation/refurbishment (SBR) have been studied broadly (Jensen et al., 2022), there is still a need to develop new instruments to increase the volume of SBR and methods to evaluate such instruments, especially from the perspectives of building users. This social viewpoint is critical as there is a need for more research that can support a pull from the demand side, including building owners, facilities managers, and end-users, to disclose and drive unfulfilled needs and new opportunities (Temeljotov & Lindkvist, 2021).

SBR is relevant to meet the UN Sustainable Development Goals (SDGs) (Jensen & Maslesa, 2015; Jowkar et al., 2022). SBR is an action that supports SDG 11, protection and securing cultural heritage and negative impact on the environment, and SDG 12 by reducing the primary energy consumption and affecting the circular economy.

Although accelerated investment in energy retrofitting of the existing buildings is supported (EPBD 844, 2018), the challenge of climate resilience at the building level is still high (Kristl et al., 2020). Therefore attention should be focused on creating adequate guidelines for SBR of the existing stock adapted to future climatic conditions and user (owner and end-user) inclusion. The EU emphasizes the involvement of citizens in the generation of renewable energy, which can increase social acceptance and thus enable the low carbon energy transition (Hauge et al., 2019). In Norway, the White Paper 13 (2010–2011)-Active Ownership, mentioned that "Utilization, renovation, and refurbishment of existing buildings rather than the construction of new buildings have a most significant effect on the environmental carbon footprint."

A dilemma to renovate or demolish was discussed in many articles. The conclusion is similar that three primary decision-making criteria still exist: cost-driven, technical building conditions, and regulatory aspects (Alba-Rodríguez et al., 2017; Bullen & Love, 2011; Shah, 2012). Hagen and Sørstrøm (2021) consider the higher investment risks of many refurbishment projects due to unknown factors revealed. While the economic aspects are still a significant driver (high risk and high economic profit), the ecological and societal impact assessments are becoming more critical in reducing the carbon footprint and developing facilities that add value to the owner and the end-user's well-being. SBR is a necessary action to lower the building's carbon footprint. According to Hopkinson et al. (2020), building products associated with environmental benefits have a massive potential to reuse and redesign existing building products.

The European Green Deal focuses on transforming into a sustainable and circular economy by reducing greenhouse gas emissions and pollution. Furthermore, health, life quality, and the generation of new workplaces will also be focus areas (The European Green Deal, 2019). The taxonomy is one of the cornerstones of the EU Action Plan for Sustainable Financing is the introduction of EU taxonomy, a classification system with criteria that define whether an action or a project investment is considered green or sustainable, thereby reducing the risk of greenwashing and steering private capital in a green direction. The taxonomy is also introduced to Norway by January 2022 in large enterprises. This year will be profound to implement knowledge, a standard syllabus, and adjusted terminology relevant to Norwegian property owners, investors, and developers.

Although the taxonomy is initially aimed at the financial sector, it will impact large parts of the market, including buildings and construction (Finansdepartementet, 2021). In terms of SBR and energy efficiency, it is mentioned in the report *that* taxonomy should focus more on the "production phase" and the "recycle phase" of the life cycle to improve carbon print and achieve better total energy efficiency through the entire life cycle. To overcome the cost-driven approach of developing a building to adaptive reuse or a SBR, the taxonomy is seen as a solution to prioritize the upgrading and modernization of existing buildings.

One of the measures discussed in The European Green Deal is to initiate a wave of renovation in the construction sector, referred to as a "renovation wave" (The European Green Deal, 2019). This is due to the great need to take measures on existing and inefficient buildings. Today, the energy-efficient renovation rate in Europe is 1%. In addition, only 0.2% of European building stock undergoes deep renovations, where energy consumption is reduced by at least 60%. The goal is to double annual energy-efficient renovations in Europe by 2030 (European Commission, A Renovation Wave for Europe - greening our buildings, creating jobs, improving lives, 2020).

This paper looks at Norwegian property owners' motivation to do a sustainable building refurbishment (SBR) or adaptive reuse supporting sustainable development. We focus on the owner's decision base before doing a SBR and discuss the need for new strategies to be taken in the future in terms of achieving SDGs 11 and 12 at a national level.

12.2 Research Methodology

The aim of this paper is to understand the motivation and decision base of Norwegian building owners' motivation of doing a SBR. This was done in two steps. First, a scoping literature review was conducted looking at the drivers of doing an adaptive reuse of buildings in terms of sustainability. Secondly, the list was tested on a

Table 12.1 Characteristics of informants interviewed

Role	Informants	$N = 18$
Private property owners	Norwegian private property owners that develop properties for rent in Southeast Norway	14
Public property owners	Oslo Municipality, Oslo University and Statsbygg	5

selection of property owners, the Norwegian Green Building Council members. As members, they are first and foremost interested in the Green Shift and contribute toward sustainable development in society. The information among the property owners was collected by semi-structured in-depth interviews. The informants were asked questions about what motivates them to do SBR and confronted with the list found from the literature. Their attitude toward doing SBR, as well as the challenges that they experienced, was explored. In this paper, we focus on the motivation of doing SBR and discuss what is suggested to be important strategies to promote SBR in the future. The findings from the interviews were analyzed and discussed compared to the findings in the literature. The interviews were conducted over a 3-month period in Spring 2021. The following characteristics of the informants are presented in Table 12.1.

12.3 Literature Review

12.3.1 Sustainable Building Renovation and Refurbishment (SBR)

ISO's definition of a sustainable building is "a building that creates the required performance and functionality with minimum environmental impact and at the same time encouraging improvements in economic and social as well as cultural aspects at local, regional and global levels" (ISO 15392:2008, *Sustainability in Building Construction — General Principles*, 2008).

So, when sustainable building renovation or refurbishment (SBR) is discussed, it addresses the renovation or refurbishment actions that will improve the building performance to satisfy the users' needs both socially, functionally, and environmentally. The actions taken should lower the carbon footprint and optimize the building performance in terms of usability, functionality, and energy efficiency.

Graabak et al. (2014) highlighted the importance of increasing the focus on the principle of sustainable development of today's buildings as the building technical requirements for the buildings continue to increase. The main purpose of the principle of sustainable development is to look at the building's opportunities to meet the current and future needs for the construction's purpose. At the same time, they are having positive impacts on the building's stakeholders, such as building owners and users, by looking at the conditions of the environment, the economy, and the social developments in context. By putting these three areas in context, one can ensure that important aspects are considered when designing and building and when the building is used, which will prevent a unilateral focus on, for example, conservation in the rehabilitation of protected buildings (Graabak et al., 2014).

According to the report prepared by Selvig (2011), listed inhabitants were having difficulties rehabilitating their assets to the current energy standard and achieving as good energy efficiency as new buildings. The purpose of the report is to show the impracticality to require equal energy requirements for new and older houses to achieve reduced energy demand from the building stock. This is illustrated by comparing greenhouse gas emissions from an older protected building with high energy consumption in the operational phase with a new low-energy building where greenhouse gas emissions from the production of the building are included. The report showed which of the buildings contributes to the highest greenhouse gas emissions. Selvig (2011) mentions that experience in this area indicated that energy use in the operational phase of a protected rehabilitated building is higher than that of a low-energy house. However, existing buildings are already built, and emissions per year from building components are thereby low, seen over the life of the building. In addition, rehabilitation of an existing building will result in lower energy use than when new buildings are built, as emissions from the production of the new materials are already included and negligible emissions from material transport. One way to reduce greenhouse gas emissions in the protected building can be by changing the energy used in the operational phase or reducing or changing the energy supply. The report also sheds light on the level of energy efficiency that the building must have for greenhouse gas emissions for both types of buildings to be as equal as possible.

The study showed that greenhouse gas emissions due to the operation of the building could be compensated by the low emissions from material use when renovating the listed building. If a protected building is required to achieve the equivalent amount of greenhouse gas emissions as for a low-energy house, or lower, energy sources that affect greenhouse gas emissions must be used to a lesser extent. This can be done by replacing energy sources that use oil, gas, and electricity with renewable energy sources that use bioenergy, solar heat, and heat pumps (Flyen et al., 2019).

The criterion for constructing new buildings depends on the national definition of nZEB. The threshold values for nZEB vary within the EU by between 20 and 100 kWh/m2 per year. This can cause significant differences in how many buildings can fulfill the activity from one country to another. An unfortunate effect of this may be that investors will add investments to countries where the requirements are less stringent since it will provide "greener" portfolios for their own business (Schütze & Stede, 2020). There is no national definition of nZEB for all EU member states (Raux & Fischer, 2021), which is a major challenge. Soares et al. (2017) argue that the biggest challenge associated with developing the nZEB definition lies in the balance between reducing energy consumption and efficient energy systems and suitability for renewable energy production.

12.3.2 The Circular Economy

The scoping literature review searched for what drives building owners to SBR. The literature says it is necessary to transition to a circular economy to achieve the UN Sustainable Development Goals (SDG). The circular economy is about the more prudent and efficient use of resources (Kvale & Norang, 2021). On 11 March 2020, the EU's new circular economy action plan was introduced and became an integral part of the European Green Deal to ensure a cleaner and more competitive Europe. The construction industry is highlighted as one of the most critical sectors for achieving these goals, and a separate strategy for the sector is addressed. More efficient use of materials, longer life spans, more recycling, and better waste sorting were decided to be implemented (The European Green Deal, 2019).

Sandberg and Kvellheim (2021) reviewed several research articles and documents to understand the circular economy In Norway. They concluded that new buildings must be designed with a high ability to handle changes using materials and building components that easily can be reused and demounted. One of the challenges is that the contractors do not consider that the component should be dismantled and reused one day. However, one practice that has changed in this concern is that it is common to screw the beams together rather than nail them (Sandberg & Kvellheim, 2021). As long as components and building parts are not designed for reuse, there will be a cost related to demounting the parts of the building. Another challenge is the logistics and storage for those components, which require new knowledge and competencies in the whole value chain. Sandberg and Kvellheim (2021) believe there is a future market for environmental consultants and dismantling assessments and demolition contractors.

12.3.3 Adaptive Reuse

Refurbishment is about putting older buildings, building parts, technical facilities, and objects in usable condition, adapted to current regulatory and user requirements, but without changing functionality, including repair, restoration, upgrade, and floor plan construction. Adaptability is the characteristics of a building to meet the requirements for functionality and the building's authenticity flexibility (Finansdepartementet, 2008). Adaptability and usability are important elements in sustainable upgrading and rebuilding. Buildings with poor functional suitability combined with poor adaptability constitute a significant challenge. Many such buildings may need to be demolished. A small, adaptable building, which is suitable today, will develop into unsuitable and can hardly be justified to upgrade from a sustainability perspective. According to Finansdepartementet (2008), one can make a technical upgrade in a short-term outlook for such buildings. However, an incompatible building with high adaptability can be refurbished sustainably.

The reuse of building parts and materials can be cost-saving. There may be a profit potential in reuse, but at the same time, it is pointed out that anything that delays the construction process has little value. The price and availability of matching building materials/parts have been identified by Bullen (2007) as a barrier in the conversion project. This is also in line with the findings of Sandberg & Kvellheim (2021), as the prices of new building materials have stayed relatively low. This view is also supported by the fact that several problematic technical regulations stand in the way of reuse and that strict technical requirements are often disproportionately resourced intensive. The decision-making authorities' priority between safety considerations and technical requirements appears particularly unclear in buildings with protective status.

Before deciding on redevelopment, builders are advised to strategically assess whether the building should be refurbished in the first place. The principles of sustainability should form the basis for this. High rebuilding costs may indicate that the building is not adaptable and therefore not suitable for sustainable redevelopment. Possible ways to regulate this can be by increasing demolition fees, requiring a demolition plan to document the reused materials and components, stricter regulations for sorting waste, and economic incitements for reusing materials in other projects (Klungerbo & Sørland, 2021; Prabowo et al., 2021). According to Ali et al. (2018), redevelopments' economic, social, and environmental benefits have made it an increasingly popular alternative to demolition and new construction. However, rehabilitation work is often risky and uncertain, and the result is generally less planned and more challenging to control than is the case for new construction.

Consequently, more coordination and different planning and control methods, tools, and techniques must be established (Jensen et al., 2022). Ali et al. (2009) note that rehabilitation work is often completed at high cost and time variations. One of the main reasons for this is the late discovery of design information. The building owner should strengthen the information base before construction starts (Ali et al., 2009). One familiar reuse project in Norway, called KA13, is seen as a role model project with great success. They identified that the major challenge is time and resource-demanding to collect information about the products' properties, qualities, and hazardous environment content. They also experienced that the components and building parts are not designed for dismantling and that it is too demanding to plan and demount components and products for reuse. They experienced that the industry is not mature enough to handle circularity reuse processes (Hagen & Sørstrøm, 2021; Klungerbo & Sørland, 2021; Kvale & Norang, 2021).

Bullen (2007) has identified barriers to adaptive reuse and concludes that the barriers always revolve around costs since (1) conversion is only considered sustainable when costs and benefits are included over the lifetime (life cycle perspective), (2) the cost of redevelopment may be high and the construction work, (3) building owners do not consider redevelopment to be economically beneficial, (4) performance of older buildings and ability to meet current building requirements are uncertain, (5) maintenance costs may be higher than for new buildings, and (6) price and availability of matching building materials/parts are uncertain.

It is believed that the safety authorities should be more flexible in measures on protected buildings. It is also mentioned that protected buildings often have good architectural and material qualities and that it is thus worth taking care of these. “The protection in Norway is perceived as conservative, limiting how much can be built on or on listed buildings. It can be reversed, but there is little willingness for the buildings to be transformed from a future perspective. This can prevent increased utilization, active first floors, etc.” (Klungerbo & Sørland, 2021; Kvale & Norang, 2021).

12.3.4 The Motivation of Doing Adaptive Reuse of an Existing Building

As stated among the researchers, the motivation and drivers of doing adaptive reuse of an existing building are presented in Table 12.2, based on the findings from the master thesis of Klungerbo & Sørland (2021) and further developed adapted by the authors.

12.4 Results and Discussions

This chapter presents the findings from the interviews of the property owners. The discussion follows, comparing the results from practice with the findings from the literature.

12.4.1 Motivation for Doing SBR

The master thesis from NTNU examined the Norwegian property owner’s motivation for sustainable refurbishment and adaptive reuse projects (Klungerbo & Sørland, 2021). They found that property owners will always be cost-driven, so the main focus is to search for other parameters and motivators to make a project more sustainable without increasing the cost. The findings are based on interviews among 18 property owners and are presented in Table 12.3.

12.4.1.1 Environmental Impact

Baker et al. (2017) and Fufa et al. (2020) stated that environmental assessments are important information in decision-making. It is evidenced that property managers are motivated by how adaptive reuse projects reduce climate gas emissions.

Table 12.2 Motivation and drivers of doing a SBR, according to the literature

Sustainability issues	Drivers	Sources
Environmental	Environmentally friendly and reduced carbon footprint	Ball (2002), Bullen (2007), Wilkinson (2011), Foster (2020), Fufa et al. (2020)
	Less resource consumed	Douglas (2006), Highfield & Gorse (2006), Bullen (2007), Power (2008), Shah (2012), Baker, et al. (2017), Alba-Rodríguez et al. (2017)
	Branding/symbolism	Ball (2002), Shipley et al. (2006)
Social	Conserve the social value and cultural value of the building	Douglas (2006), Bullen (2007)
	Positive urban development and individual development	Ball (2002), Highfield & Gorse (2006), Bullen (2007), Bullen & Love (2011), Flyen, et al. (2019)
	Protected buildings	Highfield & Gorse (2006)
Technical	Adaptability is high	Shah (2012)
	Good technical conditions	Alba-Rodríguez et al. (2017), Bullen & Love (2011), Shah (2012)
Economic	Low investment costs	Ball (2002), Douglas (2006), Shipley et al. (2006), Bullen & Love (2010)
	Lower material costs	Highfield & Gorse (2006)
	Shortened construction time	Highfield & Gorse (2006), Douglas (2006), Power (2008), Wilkinson et al. (2009), Shah (2012), Baker (2020)
	The building can be in use under construction.	Shipley et al. (2006), Power (2008), Shah (2012)
	The possibility of adding more floors	Highfield & Gorse (2006), Douglas (2006), Wilkinson (2011)

Adapted from Klungerbo & Sørland (2021)

Table 12.3 The motivation for doing a sustainable building renovation among 15 private and three public property owners (Based on the interviews conducted by Klungerbo and Sørland, 2021)

	Motivation	Responses
Environmental	Reduced climate gas emission Less resource-demanding	14 of 18 5 of 18
Social	Increased attractiveness and identity of a community High user demands	13 of 18 6 of 18
Technical	A high degree of adaptability	6 of 18
Economic	Low cost Short construction time Low cost due to possibility to keep parts of existing buildings	3 of 18 3 of 18 3 of 18

However, few informants believed that environmental aspects are the only motivator for practicing SBR, as the cost is always a more robust driver. Besides, many of them commented that there is an increased awareness of the reduced carbon footprint of reusing structural elements and foundations. Several informants stated that they make greenhouse gas calculations to show the positive or negative impact of different alternatives, but not as part of the decision-making for SBR as such calculations are not mandatory today. However, most of them agreed that this is expected to change in the future due to the implementation of the EU's taxonomy, new regulations, and standards.

While the literature argues that SBR is less resource-demanding (Alba-Rodríguez et al., 2017; Baker et al., 2017; Bullen, 2007; Douglas, 2006; Highfield & Gorse, 2009; Power, 2008; Shah, 2012), only 5 of 18 interviewees agreed with that in practice. Their opinion is that mostly this depends on the nature of the project. Often, an SBR project combines reusing some part of a building and demolishing and extending other parts. Some property owners view a SBR project as complex that requires knowledge of utilizing existing structures and components and good competencies to find reasonable solutions. The informants do not necessarily consider a sustainability strategy in all projects, especially in the case of terrible technical conditions.

Similar to the literature by Eray et al. (2019), the property managers mentioned refurbishment of historical buildings regarding the requirement to protect heritage components. However, in this case, the involvements of experts become natural and are doable. Many argue for the complexity in the need to involve both expertise and stakeholders, which is time-consuming in conceptualizing the project. Eray et al. (2019) discussed this, pinpointing the scares research of framework and interface management models of adaptive reuse projects. Their framework focused on circular economy was tested on adaptive reuse projects, focusing on the dialogue and exchange of information between the conceptualization, planning, and execution phases of the projects.

Overall, both the literature and the informants agreed that climate gas calculation in a particular project could clarify the effect of refurbishment versus demolishing and add valuable information that can promote sustainable refurbishment.

12.4.1.2 Social Impact

Douglas (2006) and Bullen (2007) argued that transforming an existing building would add social value and be attractive in their neighborhood. In terms of the social impact, more than 13 of 18 agreed that SBR leads to the increased attractiveness and identity of the community. The informants stated that especially in the case of the projects that include historical buildings, it could be linked with the identity value for individuals in the area, community, and other users, similarly mentioned in Kristl et al. (2019) that heritage buildings are vital in terms of transferring cultural and historical memory. This creates extra motivation for building owners to invest in the SBR process. In addition, if the site allows an extension, it could provide a sustainable solution that uptakes the users' historical values and needs. This is in line with the findings of Highfield and Gorse (2009).

SBR leads to a positive individual and urban development (Ball, 2002; Bullen, 2007; Bullen & Love, 2011; Highfield & Gorse, 2009). Considering sociocultural sustainability, Murphy (2012) looked at the participation, equity, and awareness as part of the conservation of sociocultural patterns. Even though the goals of conservation, restoration, and renovation look more for the preservation of authenticity to ensure historical, cultural, and social values (Li et al., 2022), SBR or adaptive reuse of buildings in an area with poor performance will increase in value and attractivity in terms of willingness to invest and establish new businesses. This is relevant for residential spaces, as renovation helps renew poor neighborhoods (Power, 2008).

The respondents agreed that the user demand is important, but only 6 of 18 respondents said that the user demands are a drive for doing SBR. They emphasize that the users/renter's attitude is that they require new and fresh workplace facilities and must be convinced that an SBR project can have the same qualities as a new building. However, some tenants are more aware of the sustainability aspects as working in a building that has been through a sustainable refurbishment/development will align with the company's sustainability profile.

This is not found highlighted in the literature as an important driver. However, Bullen and Love (2011) found that while the building owners emphasize commercial performance, the users focus on the usability of the building (well-being and productivity). Here, we see the benefit of including the users in finding reasonable profitable and functional solutions. The informants do not mention this in this research. However, the property owners are aware of this, as one of them said: "It's the tenants that govern the market. If we offer a product that the tenants do not ask for, we have missed an important point."

12.4.1.3 Technical Aspects

The technical aspects are always relevant before doing an SBR. However, only 6 of 18 say that the high adaptability is a motivation of itself. In many ways, this reflects the fact that often the low adaptability, like, low ceiling height, is a limitation that can lead to demolition instead of SBR. Both Shah (2012) and findings from the study confirm this. Adaptability is becoming more relevant, but the need for adaptability depends on the building category. In office buildings that are internally renovated every 10 years, when the tenant contract is renewed, there is important to think about flexible and detachable interior walls and technical solutions that are not integrated with the walls.

12.4.1.4 Economic Impact

Only 3 of 18 said that an SBR project would reduce cost or shorten construction time. They argue that in practice, the design phase in an SBR project will take more time than a new building project as there are many modifications during the process that takes time to find good solutions. Also, their experience is that the cost typically increases in an SBR because of many unforeseen actions revealed during the process. Another argument is that building materials are reasonably low cost while labor payments are costly.

Bullen and Love (2011) argued that it will almost always be a more viable option to use an adaptive reuse strategy from a sustainability perspective. Highfield and Gorse (2009) also argued that SBR projects give lower costs when considering whole life cycle costing, including the sustainable outcome in terms of increased value, energy efficiency, social attractiveness, and well-being. However, Bullen and Love (2011) also point out the lack of accurate sustainability assessment measures that hinder making an SBR. Douglas (2006) and Power (2008) argued that demolition is costly and that SBR will benefit. However, they looked at the main renewal of residential housing.

Hagen and Sørstrøm (2021) investigated why refurbishment projects often have a cost overrun. In their study, they discovered that the building owner lacked information about the technical condition, lack of drawings, and information about the adaptability of the building and lacking strategies for sustainability development. They recommended early involvement of the contractor and expertise at the strategic phase to improve the decision base before deciding on SBR.

12.5 Conclusions

Motivation for doing an SBR varies in terms of the benefits and added value of the project. Low competencies of sustainability assessments and cost overrun risk are the main barriers to doing so. However, the informants confirm earlier findings that the environmental aspects are important in terms of reuse materials and climate gas emission. However, there is still a need to develop further incentives and strategies to increase the profitability and sustainability of adaptive reuse projects. Based on the Norwegian property owners' views and what is emphasized in the literature, it is evident that the EU's taxonomy, future regulatory requirements, and economic incentives will be relevant to stimulating and promoting SBR.

The future focus on the circular economy will also promote further SBR projects. To sum up, the findings that motivate the building owner to do an SBR are (1) user involvement to increase well-being, (2) technicalities and adaptability, (3) regulatory requirements, and (4) climate gas calculations.

References

Alba-Rodríguez, M. D., Martínez-Rocamora, A., González-Vallejo, P., Ferreira-Sánchez, A., & Marrero, M. (2017). Building rehabilitation versus demolition and new construction: Economic and environmental assessment. *Environmental Impact Assessment Review, 66*, 115.

Ali, A. S., Kamaruzzaman, S. N., & Salleh, H. (2009). The characteristics of refurbishment projects in Malaysia. *Facilities, 27*(1/2), 56–65.

Ali, A. S., Azmi, N. F., & Baaki, T. K. (2018). Cost performance of building refurbishment works: The case of Malaysia. *International Journal of Building Pathology and Adaptation, 36*(1), 41–62.
Baker, H. (2020). *The adaptation and demolition of existing buildings on masterplan sites.* University of Cambridge.
Baker, H., Moncaster, A., & Al-Tabbaa, A. (2017). Decision-making for the demolition or adaptation of buildings. *Proceedings of the Institution of Civil Engineers-Forensic Engineering, 170*(3), 144–156.
Ball, J. (2002). Can ISO 14000 and eco-labelling turn the construction industry green? *Building and Environment, 37*(4), 421–428.
Bullen, P. A. (2007). Adaptive reuse and sustainability of commercial buildings. *Facilities, 25*, 20.
Bullen, P. A., & Love, P. E. (2010). The rhetoric of adaptive reuse or reality of demolition: Views from the field. *Cities, 27*(4), 215–224.
Bullen, P., & Love, P. (2011). A new future for the past: A model for adaptive reuse decision-making. *Built Environment Project and Asset Management, 1*, 32.
Douglas, J. (2006). *Building adaptation.* Routledge.
EC, The European Green Deal, 640 European Commission (2019). https://doi.org/10.1017/CBO9781107415324.004.
EC, A Renovation Wave for Europe - greening our buildings, creating jobs, improving lives, 53 European Commission 1689 (2020).
Eray, E., Sanchez, B., & Haas, C. (2019). Usage of interface management system in adaptive reuse of buildings. *Buildings, 9*(5), 105.
Finansdepartementet. (2008). *Veileder nr. 2 Felles begrepsapparat KS 2.* 1–6.
Finansdepartementet. (2021). *Finansmarkedsmeldingen 2021 - Meld. St. 31 (2020–2021).*
Flyen, C., Flyen, A. C., & Fufa, S. M. (2019). *Miljøvurdering ved oppgradering av verneverdig bebyggelse.* SINTEF akademiske forlag.
Foster, G. (2020). Circular economy strategies for adaptive reuse of cultural heritage buildings to reduce environmental impacts. *Resources, Conservation and Recycling, 152*, 104507.
Fufa, S. M., Flyen, C., & Venås, C. (2020). *Grønt er ikke bare en farge: Bærekraftige bygninger eksisterer allerede.* SINTEF akademiske forlag.
Graabak, I., Bakken, B. H., & Feilberg, N. (2014). Zero emission building and conversion factors between electricity consumption and emissions of greenhouse gases in a long term perspective. *Environmental and Climate Technologies, 13*(1), 12–19.
Hagen, H. F., & Sørstrøm, K. (2021). *Ombyggingsprosjekter - En studie av hva byggherrer bør prioritere i tidligfase for å unngå kostnadssprekk.* Master thesis, NTNU.
Hauge, Å. L., Hanssen, G. S., & Flyen, C. (2019). Multilevel networks for climate change adaptation – What works? *International Journal of Climate Change Strategies and Management, 11*, 215–234.
Highfield, D., & Gorse, C. (2006). *Refurbishment and upgrading of buildings.* Routledge.
Highfield, D., & Gorse, C. (2009). *Refurbishment and upgrading of buildings.* Spon Press.
Hopkinson, P., De Angelis, R., & Zils, M. (2020). Systemic building blocks for creating and capturing value from circular economy. *Resources, Conservation and Recycling, 155*, 104672.
Jensen, P. A., & Maslesa, E. (2015). Value based building renovation–a tool for decision-making and evaluation. *Building and Environment, 92*, 1–9.
Jensen, P. A., Thuvander, L., Femenias, P., & Visscher, H. (2022). Sustainable building renovation-strategies and processes. *Construction Management and Economics, 40*, 157–162.
Jowkar, M., Temeljotov-Salaj, A., Lindkvist, C. M., & Støre-Valen, M. (2022). Sustainable building renovation in residential buildings: Barriers and potential motivations in Norwegian culture. *Construction Management and Economics, 40*, 161–172.
Klungerbo, M. H., & Sørland, L. M. (2021). *Rive eller bygge om? En studie av hva som kan fremme ombygging i byggenæringen.* Master thesis, NTNU.

Kristl, Ž., Temeljotov Salaj, A., & Roumboutsos, A. (2019). Sustainability and universal design aspects in heritage building refurbishment. *Facilities, 38*(9/10), 599–623.

Kristl, Ž., Senior, C., & Temeljotov Salaj, A. (2020). Key challenges of climate change adaptation in the building sector. *Urbani Izziv, 31*(1), 101–111.

Kvale, N., & Norang, H. (2021). *Grønt er skjønt? En studie av hvordan EUs taksonomi og Level (s) kan konkretisere bærekraft innen bygg og eiendom. Master thesis,* NTNU.

Li, Z., Diao, J., Lu, S., Tao, C., & Krauth, J. (2022). Exploring a sustainable approach to Vernacular Dwelling spaces with a multiple evidence base method: A case study of the Bai People's courtyard houses in China. *Sustainability, 14*, 3856.

Murphy, K. (2012). The social pillar of sustainable development: A literature review and framework for policy analysis. *Sustainability: Science, Practice and Policy, 8*, 15–29.

Power, A. (2008). Does demolition or refurbishment of old and inefficient homes help to increase our environmental, social and economic viability? *Energy Policy, 36*(12), 4487–4501.

Prabowo, B. N., Salaj, A. T., & Lohne, J. (2021). Urban heritage facility management: A scoping review. *Applied Sciences, 11*(20), 9443.

Raux, C., & Fischer, S. (2021). *Testing the application of the EU Taxonomy to core banking products: High level recommendations*. available at https://www.ebf.eu/wp-content/uploads/2021/01/Testing-the-application-of-the-EU-Taxonomy-to-core-banking-products-EBF-UNEPFI-report-January-2021.pdf

Sandberg, E., & Kvellheim, A. K. (2021). *Ombruk av byggematerialer–marked, drivere og barrierer*. SINTEF report., 36.

Schütze, F., & Stede, J. (2020). EU sustainable finance taxonomy–what is its role on the road towards climate neutrality? *DIW Berlin Discussion Paper,* no. 1923, available at SSRN: https://ssrn.com/abstract=3749900

Selvig, E. (2011). *Klimagassberegninger for vernede boligbygg vs. nye lavenergiboliger.* Civitas report., Riksantikvaren, 20.

Shah, S. (2012). *Sustainable refurbishment*. Wiley.

Shipley, R. U. S., & Parsons, M. (2006). Does adaptive reuse pay? A study of the business of building renovation in Ontario, Canada. *International Journal of Heritage Studies, 12*(6), 505–520.

Soares, N., Bastos, J., Pereira, L. D., Soares, A., Amaral, A. R., Asadi, E., Rodrigues, E., Lamas, F. B., Monteiro, H., & Lopes, M. A. R. (2017). A review on current advances in the energy and environmental performance of buildings towards a more sustainable built environment. *Renewable and Sustainable Energy Reviews, 77*, 845–860.

Sustainability in building construction — General principles. (2008). (testimony of ISO 15392:2008).

Temeljotov Salaj, A., & Lindkvist, C. (2021). Urban facility management. *Facilities, 39*(7/8), 525–537. https://doi.org/10.1108/F-06-2020-0078

White Paper 13. (2010–2011). Active Ownership, 43 Norwegian Ministry of Trade and Industry (2011).

Wilkinson, S. J. (2011). *The relationship between building adaptation and property attributes*. Dr.art.thesis. Deakin University.

Wilkinson, S., James, K., & Reed, R. (2009). Delivering sustainability through the adaptive reuse of commercial buildings: the Melbourne CBD challenge. In *Proceedings of the Pacific Rim Real Estate Society 15th Annual Conference* (pp. 1–19).

Chapter 13
The Best Solution for Renovation in Terms of Climate and Economy

Terttu Vainio and Eero Nippala

Abstract EU aims to reach carbon neutrality by 2050. Besides energy consumption reduction, also greenhouse gas emissions have to be cut starting from the production of materials and construction work through the use phase to the end of the use of the building. Existing buildings are estimated to provide a high potential for reducing global warming. This paper focuses on research question, how reasonable are energy efficiency improvements of existing buildings, as the materials used in the process produce CO_2 emissions and increase costs compared with conventional maintenance. This issue is a part of the Sustainable Development Goal 13 Climate Action, which integrates climate change measures into national policies, strategies, and planning and a part of Goal 11 Sustainable cities and communities, which tries to increase the number of cities and human settlements adopting and implementing integrated policies and plans towards inclusion resource efficiency mitigation and adaption to climate change. The carbon footprint of an existing renovated building constitutes mainly from energy consumption emissions. In life cycle costs, the deciding factor is investment. If the building was heated by zero-emission ground source heat, structural renovations would not be worth doing. On the other hand, structural improvement of energy efficiency is recommendable if a building is connected to district heating (DH). Strong reasons, either endogenous or exogenous, must exist for replacing an existing building with a new one. They cannot be justified with the carbon footprint or life cycle costs. These results apply to countries, where the energy efficiency of existing buildings is reasonably good.

Keywords Deep renovation · Rebuilding · Carbon footprint · Life cycle cost · Energy efficiency

T. Vainio (✉)
VTT Technical Research Centre of Finland, Smart Energy and Built Environment, Espoo, Finland
e-mail: terttu.vainio@vtt.fi

E. Nippala
Tampere University of Applied Sciences, Built Environment and Bioeconomy, Tampere, Finland
e-mail: eero.nippala@vtt.fi

G. Lindahl, S. C. Gottlieb (eds.), *SDGs in Construction Economics and Organization*, Springer Proceedings in Business and Economics,
https://doi.org/10.1007/978-3-031-25498-7_13

13.1 Introduction

EU aims to reach carbon neutrality by 2050 (EU, 2019). In practice, this means that the greenhouse gas emissions of a Member State may not exceed the carbon sinks of the country. The EU has made a determined effort to reduce energy consumption in buildings. In the future, low emissions or even the carbon neutrality requirement will be extended to cover the entire life cycle of a building, starting from the manufacture of materials and construction work through the use phase to the demolition of the building. Carbon dioxide emissions from the life cycle phases are referred to as the carbon footprint of a building. The carbon footprint is therefore a narrower concept than life the cycle assessment (LCA) or environmental product declaration (EPD), which also takes into account other greenhouse gases and their impact on the environment.

Economic decision-making has traditionally been strongly based on the comparison of alternative investment costs or the cost-benefit analysis, which takes into account not only the costs but also the achievable benefits. In order to promote energy efficiency, developers have been recommended to use life cycle costs (LCC) as a decision-making tool (EU, 2012). If energy prices increase significantly, it will be worth investing in energy efficiency. Therefore, in addition to investment costs, developers have been encouraged to assess the operating costs of various alternatives. The outcome of the Swedish study came also to this conclusion in economic comparison between a zero alternative (no energy efficiency improvement), renovation to a passive house today, renovation to a passive house in ten years, and rebuilding. The most economically viable alternatives were the ones that contain renovation. Although these alternatives lead to higher investment costs compared with the zero alternatives, relatively soon they were covered by earnings (Dahlöf & Malmros, 2011). Almost a decade later, the Swedish study had the same conclusion. Renovation is more affordable in terms of life cycle costs. The moderate level, as a rule, is more cost-optimal than an ambitious repair. The LCC of the energy renovation is highly dependent on the building type and thermal performance prior to energy renovation (La Fleur et al., 2019). However, renovation is not an option in all cases. The building may simply be of such poor quality that it is not worth making an expensive energy-related investment (Verbeeck & Cornelis, 2011).

The existing building stock is seen as a highly potential target for improving energy efficiency and reducing climate emissions (Bpie, 2011). Besides positive planetary impacts, a Swedish study found the positive social impacts of renovation in terms of the social cost of carbon in the study where renovation and new construction are compared with keeping buildings in their original state (Nydahl et al., 2022). However, developers have a half-hearted approach to energy efficiency on account of the fact that, with the current energy prices, repayment times for measures may amount to dozens of years. For this reason, energy and renovation grants and benefits linked to energy efficiency agreements have been introduced in order to promote improvements in energy efficiency (Bertoldi et al., 2020). The question has also been raised whether low-profit projects should be supported (Dubois & Alleker, 2015).

The assessment of the carbon footprint of the life cycle has thus brought new perspectives to public debate. Can energy efficiency improvements in old buildings be seen as reasonable as the materials used in the process produce CO_2 emissions? How long will it take to compensate for such extra emissions with a more energy-efficient building or low-emission heating? Is it even worth renovating buildings when the energy industry is in the process of giving up fossil fuels in energy production? The energy produced in a purely centralized manner reduces emissions from buildings with poor energy efficiency.

A systematic review of case studies that compared the life cycle carbon footprint of refurbished and new buildings showed that most refurbishments had a lower footprint and some new buildings performed better than refurbished ones. This study also shows that on the basis of current evidence, it is still not possible to conclusively determine which of the alternatives is preferred (Schwartz et al., 2018). Another similar literature study focused on buildings located in northern latitudes and found 15 comparisons between renovation and rebuilding (Huuhka et al., 2021). A summary of the comparisons was that renovation is less carbon over a short time span, but often also over a longer time span. Rebuilding may become more low-emission in the long term if a highly energy-efficient new building and a weak existing building have been paired. Long carbon investment compensation time is problematic from the perspective of combating climate change and mitigating its effects. This is the case also in the Swedish million program era buildings. The study of them came to the conclusion that the renovation of the current building clearly had the least climate impact compared to demolishing and building a new building, assuming 50 years life span (Eskilsson, 2015).

The question is now topical for apartment buildings built in the 1960s and 1970s from precast concrete units. With the society urbanizing at the time, a great number of such buildings were built. Construction techniques that made use of industrial precast concrete units were developed to satisfy high demand, but their quality and, in particular, architecture have proved inferior to those of buildings dating back to the previous periods. Many buildings absolutely need the renewal of their facade and potable water pipelines (Nippala & Vainio, 2017). At the same time, due to energy and climate targets, the energy efficiency of buildings should also be improved. Additional measures will increase repair costs to the extent that demolition and rebuilding will become viable alternatives to renovation.

Upgrading the existing stock to reduce CO_2 emissions cheaply, quickly, and easily would be invaluable in shaping future housing policy. All referred researchers reminded that further analysis is needed for example carbon footprint and building embodied energy content. Instead, centralized energy production, decentralized energy supply, and micro-combined heat and power production (CHP) can change the outcome of energy analyses on what is the best option in terms of costs and climate impacts (Power, 2008).

13.2 Research Question and Method

What is the best development scenario for the economy and the environment regarding apartment buildings dating back to the 1970s: to carry out only the necessary technical repairs or a large-scale renovation which, in addition to technical repairs, would improve energy efficiency, or the demolition and replacement by a new building?

The method of research is a case study. The case is a typical 1970s concrete panel building. Four development options are being tested for it. For alternatives, carbon footprint and life cycle costs are calculated by standardized methods.

Option A: Maintenance. The exterior walls of the building will be subjected to surface finishing, with the water and sewer systems being renewed. Old windows are replaced by new windows. With regard to the heating system, measures are taken to maintain its operation, which include the renewal of the heat exchanger and automation system, as well as the balancing of heat distribution. After these measures are taken, the energy class of the building will be C.

Option B: Deep Renovation. Building exterior walls will be further insulated, with the windows and water and drainage systems being renewed. The energy efficiency of the heating system will be improved with a new heat exchanger and heat recovery for the ventilation system. The building will remain connected to the district heating system because it is the predominant heating method for apartment buildings in urban areas. Renovation meets the energy efficiency requirements and the share of renewable energy. After renovation, the energy class of the building will be B.

Option C: Concrete Rebuilding. The old building will be demolished and a new building of the same size made of precast concrete units will be built. As the new building meets the current energy efficiency requirements set for new A-class construction, it will also be equipped with a heat recovery system and solar panels. The new building will also be connected to the district heating system. Its energy class will be A.

Option D: Wooden Rebuilding. The old building will be demolished, with a new wooden block of flats being built in its place. In terms of energy efficiency, it will be equal to option C, that is, its energy class will be A.

For all four options, the carbon footprint is calculated in accordance with the calculation guidelines issued by the Finnish Ministry of the Environment (Kuittinen, 2017) by using national emission coefficients (CO_2-data, 2021). As emissions are calculated for 30 years, there is no need for the replacement of building parts. The sensitivity of energy production's global warming potential has been tested in four scenarios. Scenario 1 emission coefficients are the same as set in the Finnish long-term Renovation Strategy for 2020 (Ministry of Environment, 2020; Energy year, 2021) and will not change in 30 years. Scenario 2 takes into account in emission coefficients the cleaning of energy production when fossil fuels are replaced by renewals in order to make Finland carbon neutral by 2035 a reality (Koljonen et al., 2019). Scenario 3 is an accelerated scenario 2 to allow district heating production to get out of fossil fuels imported from Russia. In scenario 4, district heating (DH) is replaced by ground source heat (GSHP) in all options. The default coefficient of GSHP performance is 2.9.

Life cycle costs are calculated for 30 years by using EU rules (EU, 2012). Renovation costs and new building costs (investment cost) are typical for this building type (The Housing Finance and Development Centre of Finland (ARA), 2022). The residual cost has been calculated based on the technical life of the buildings (Rakennustieto, 2008). The level of maintenance costs of the property is based on statistics (Finance of housing companies, 2021). The sensitivity of costs is examined using two different energy prices which indicates the price level 2021–2022 (Nordpool, 2022), and three different assumptions of energy price development and alternative discount rates as well. Interest rate options have been set at a level suitable for public investment and private low-risk investments (Streicher et al., 2020). The 30 years chosen for the review period is an EU recommendation for cost-optimality reviews in residential buildings (EU, 2012). The baseline situation of the case under consideration and the significant variables of alternative development strategies are summarized in Table 13.1. The calculation assumptions and variables are summarized in Table 13.2.

This analysis uses constant euro. Prices do not contain the effect of inflation but represent a standard price level, that is, the purchasing power of the selected base year. To discount them, a real interest rate will be used.

Table 13.1 Description of scenarios

	Unit	A. Maintenance	B. Deep renovation	C. Concrete rebuilding	D. Wooden rebuilding
Total floor area	m^2	3000	3000	3000	3000
U-value of windows	W/ m^2K	2.5	1.0	1.0	1.0
U-value of walls	W/ m^2K	0.4	0.29	0.17	0.17
U-value of roof	W/ m^2K	0.4	0.4	0.09	0.09
U-value of basement	W/ m^2K	0.4	0.4	0.16	0.16
Heat recovery	%	–	65	65	65
Heating	–	DH/GSHP	DH/GSHP	DH/GSHP	DH/GSHP
Heat consumption	kWh/ m^2	200/70	70/25	55/20	55/20
Auxiliary electricity	kWh/ m^2	5	5	5	5
Other electricity (exl. from LCC)	kWh/ m^2	30	30	30	30
Energy class DH	–	C	B	A	A
Investment cost	€/m^2	600/700	1000/1100	3000/3100	3300/3400
Residual value	€/m^2	275	425	2350	2350

Table 13.2 Calculation assumptions

Variable	Unit	Value
Lifespan of building/calculation period		
Scenario A, B	a/a	50/30
Scenario C, D	a/a	100/30
Interest rates		
Basic	%/a	3
Sensitivity	%/a	0 and 6
Energy price incl. taxes		
Electricity	cent/kWh	10/20
District heating	cent/kWh	10/20
Energy price development		
Basic	%/a	2
Sensitivity	%/a	0 and 4
CO_2 emissions year 2020/year 2050		
Electricity	g/kWh	65/12
District heating	g/kWh	160/45
Maintenance cost	€/m^2, a	6
Maintenance cost development	%/a	2

Swiss research (Streicher et al., 2020) uses 3% as a discount rate. This rate was considered too low for companies awaiting returns. On the other hand, the study showed that a discount rate of more than 8% caused energy investments to be no longer profitable. The study used 30 years as a time span, justifying it to be typical. Similarly, the same 30 years is the recommendation of the European Commission for the period considered (EU, 2012).

The price of electricity in Finland has been at a level of 30–40 €/MWh for a long time. Electricity transmission and taxes raise the unit price by 10 snt/kWh (Energy, 2022). At the beginning of the year 2022, price level has been stable at 10 €/MWh. In the year 2021, NordPool's price has ranged from zero to 60 €/MWh. At the beginning of the year 2022, the Russian invasion and economic sanctions have not pushed prices as high as did dry summer in the Nordic countries. The price of district heating has changed since 1996 to about 30€/MWh for 2022 approx. 85€/MWh (Energy, 2022). Used district heating price 10 snt/kWh is at the same level as 2022–2023 becoming heating prices. Giving up fossil fuels pressures to raise prices.

13.3 Results

13.3.1 Carbon Footprint

In options A and B, emissions from materials and construction work are low (Table 13.3). Typically, renovation measures can be done without heavy, high-emission materials. Compared with options A and B, in options C and D emissions

Table 13.3 Carbon footprint of options A, B, C, and D over 30 years by energy scenarios 1, 2, 3, and 4

	Unit	Option A	Option B	Option C	Option D
Building materials and construction	$kgCO_2/m^2$	25	40	340	240
Energy scenario 1	$kgCO_2/m^2$	1000	355	285	285
Energy scenario 2	$kgCO_2/m^2$	510	180	145	145
Energy scenario 3	$kgCO_2/m^2$	435	155	125	125
Energy scenario 4	$kgCO_2/m^2$	70	30	25	25
Carbon footprint 1, total	$kgCO_2/m^2$	1025	**395**	625	525
Carbon footprint 2, total	$kgCO_2/m^2$	535	**220**	485	385
Carbon footprint 3, total	$kgCO_2/m^2$	460	**195**	465	365
Carbon footprint 4, total	$kgCO_2/m^2$	100	**75**	370	370

are caused by the demolition of the old building and the construction of new buildings. With the emission coefficients of the current energy generation, described by energy scenario 1, the emissions of materials would be almost irrelevant in renovation options (A, B), and about half the emissions of rebuilding options (C, D). As energy production abandons fossil fuels, either in a planned (energy scenario 2), or accelerated (energy scenario 3) time frame, the materials' proportion of emissions will increase.

Apartment buildings are typically located in cities, and connected to the district heating network (DH). District heating could be replaced by a property-specific ground source heat system (GSHP, energy scenario 4). Heat harvested from the ground is emission-free, and the operation of heat pumps uses electricity. The maximum cumulative emissions have Option A Maintenance, which heating energy consumption is nearly twice as much as other options. In terms of emissions, the order of alternatives from the lowest carbon footprint to the largest would change to B, A, C, and D. In DH energy scenarios 1–3, option A owned the largest global warming potential in a 30-year period.

In DH scenarios 1–3 rebuilding options C and D need 10 years or more to compensate for emissions of demolition and materials of new buildings (Fig. 13.1). GSHP scenario 4 emission savings of A-energy class new buildings are not sufficiently compensated material emissions in a selected 30-year period.

13.3.2 *Life Cycle Costs*

The case building is either connected to district heating (Table 13.4) or heated by ground source heat (Table 13.5).

If there is no return expectation for investment, zero may be used as a discount rate for life cycle calculation. The zero interest rate, but a brisk rise in energy prices makes option A: maintenance the most affordable. Of course, low energy prices or even free energy (GSHP) support option A.

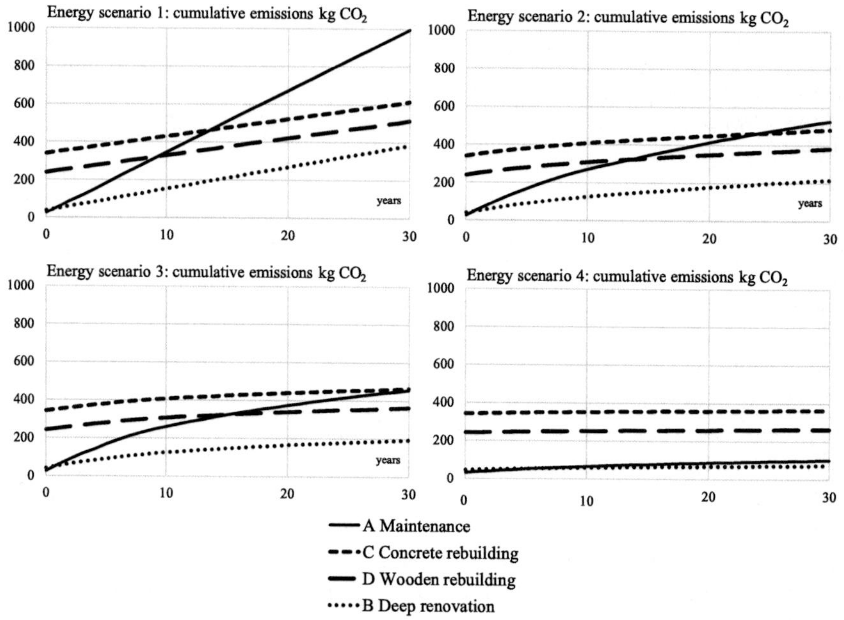

Fig. 13.1 Cumulative emissions of options A, B, C, and D by energy scenarios 1, 2, 3, and 4

Table 13.4 Net present value (€) of options over 30 years if building is connected to DH network

		Energy price 10 snt/kWh			Energy price 20 snt/kWh		
Option	Interest rate	+0%	+3%	+6%	+0%	+3%	+6%
A	0%	1215	**1070**	**966**	**1850**	1493	1268
B		**1062**	1141	1147	1294	**1295**	**1258**
C		1090	2317	2790	1276	2440	2879
D		1390	2617	3090	1576	2740	3179
A	2%	1448	1199	**1041**	2317	1750	1420
B		**1147**	**1188**	1175	1465	1389	1314
C		1159	2354	2812	**1413**	2516	2923
D		1459	2654	3112	1713	2816	3223
A	4%	1796	1385	**1147**	3012	2123	1632
B		1274	**1256**	1214	1719	**1526**	**1391**
C		**1260**	2409	2843	**1616**	2625	2985
D		1560	2709	3143	1916	2925	3285

Higher discount rates are used if there is a return expectation for the investment or a loan has been used for financing, the interest of which should be paid. If the interest rate is higher than zero and the energy price is either 10 snt/kWh 20 snt/kWh or free (GSHP), option B: deep renovation improves its competitiveness. The advantage of renovation options A and B arises from investment costs. Although

Table 13.5 Net present (€) value of options over 30 years if the building is heated by GSHP

		Energy price 10 snt/kWh			Energy price 20 snt/kWh		
Option	Interest rate	+0%	+3%	+6%	+0%	+3%	+6%
A	0%	**912**	**902**	**874**	1144	**1057**	**984**
B		1022	1148	1181	**1115**	1210	1225
C		1082	2345	2839	1159	2396	2875
D		1382	2645	3139	1459	2696	3175
A	2%	**997**	**950**	**901**	1315	**1151**	**1040**
B		1056	1167	1192	**1184**	1247	1247
C		1110	2360	2848	1216	2428	2894
D		1410	2660	3148	1516	2728	3194
A	4%	1124	**1018**	**940**	1569	**1288**	**1117**
B		**1107**	1194	1208	**1285**	1302	1279
C		1153	2383	2861	1301	2473	2920
D		1453	2683	3161	1601	2773	3220

new buildings consume less energy, it is not enough to repay higher investment costs over a period of 30 years. It is the higher investment costs that are the reason why option D: wooden rebuilding remains in last place in comparison to life cycle costs.

13.4 Discussion

For the first time, the Energy Performance of Buildings Directive adopted a position on improving the energy efficiency of old buildings in 2010. Due to this and climate concerns, the environmental perspective has also been strongly involved in the official steering of construction and property management. In the first phase, the focus was on energy consumption and greenhouse gas emissions created during use. For these, the steering emphasized technical feasibility, economy, and functionality. Another guiding principle was combining improvements in energy efficiency improvements with other repairs that would otherwise be made. For example, an intact facade need not to be touched, but if it needs repairs due to damage, at least improving heat insulation should be considered.

The most recent perspective included in the steering is the carbon footprint of the building's life cycle. Based on this study, the carbon footprint and life cycle costs can, but not always, coexist in harmony. Timber construction provides an exception; it has a small carbon footprint but is somewhat more expensive than concrete construction. The cost-effectiveness of concrete in apartment building construction stems from determined development work which began in the 1960s. Regarding the construction of wooden apartment buildings, technological development is lagging behind in industrial-scale construction.

This study has sought to provide an answer to the question of what should be done with apartment buildings dating back to the 1970s, which need renovation. The options examined were A: maintenance, B: deep renovation, C: concrete rebuilding, and D: wooden rebuilding. For these scenarios, carbon footprints and life cycle costs were calculated. Compared with the current new buildings, the energy consumption of a building of this age is three times higher.

Based on the carbon footprint and life cycle costs, option A: maintenance is rather competitive. This option is often excluded in studies if the focus is on the reduction of energy consumption and emissions. In Finland, however, the alternative is relevant because energy efficiency has been taken seriously since the 1970s oil crisis. If the expected life of the building is limited, it is the recommended choice in all respects. However, in terms of the carbon footprint, district heating should be replaced with a ground-source heat pump. Heat pumps are also recommended by other studies (Niemelä et al., 2017; Hirvonen et al., 2021). If the building has an expected long life span, the recommended scenario is B (deep renovation). This option is particularly favored from an environmental point of view, as has been found in other studies (Huuhka et al., 2021; Hasika et al., 2019).

Scenario C (Concrete Rebuilding) has a large carbon footprint. Scenario D (Wooden Rebuilding) has high life cycle costs. The adoption of these solutions requires strong arguments. The change from embodied and operational energy consumption to global warming potential has not changed the outcome of comparisons in 10 years (Nippala & Heljo, 2010; Dahlöf & Malmros, 2011). These naturally include serious damage or interior health hazards. The decision of demolishing a building and build a new, larger one may also be based on the fact that the location has untapped potential. For example, a new building multiplies the amount of floor area in an attractive location. In addition, when taking into account the difference in sales prices or rent levels between old and new apartments, replacing the old building with a new one may be a clearly profitable choice.

Even, if buildings may look similar to each other, they are also individuals. There is a great number of endogenous and exogenous factors. Their combination determines the choice between conservation, renovation, or rebuilding (Verbeeck & Cornelis, 2011; Schwartz et al., 2018; Vainio & Nippala, 2019).

In the spring of 2022, one major exogenous factor is the EU's goal of getting rid of fossil fuels imported from Russia. Fossil fuels are used in combined heat and power production (CHP). District heat is an excellent form of heating but only if the energy source is renewable. The foundation of the Kyoto pyramid is structural improvements before changing the heat source. These good principles may be worth bargaining and betting on the electrification of heating, since it is easier to find zero-emission propulsion for heat pumps than enough renewable energy sources for large-scale district heating. This potential development has been seen many years ago (Power, 2008).

13.5 Conclusion

The main conclusion from the comparisons of carbon footprints and life cycle costs is that the lower energy consumption will not be able to pay back their higher carbon or financial investment compared with the renovation of existing buildings. European Energy Crisis 2022 may not change the mutual competitive situation of alternatives. The increase in energy prices will also increase construction costs as well as the economic and social risks that hold back investment in new construction.

Most of the carbon footprint in the options comes from the emissions of operational energy. The crucial factor in life cycle costs is renovation or construction costs. From both perspectives, option B: deep renovation proved the most advantageous option. Environmentally, it loses its affordability for option A: maintenance if the selected heating means is GSHP. As a result of option B: deep renovation, the building will remain a building from the 1970s. With regard to the housing market, this means that housing sale prices and rents will remain at a significantly lower level compared with renovated or completely new buildings. Strong reasons, either endogenous or exogenous, must exist for demolishing an old building, that is, implementing option C: concrete rebuilding or option D: wooden rebuilding. They cannot be justified with the carbon footprint or life cycle costs.

The study has been carried out in the climate conditions of northern Europe, where energy is consumed for heating buildings. However, this has been taken into account in building technology, because, as a rule, old buildings are heat-insulated and equipped with three glass windows, for example. Such endogenous properties of buildings and, exogenous properties accountable to location, limit the general applicability of the results. As they are, they are best applied to the Baltic Sea countries. One finding that is more broadly applicable to Europe indicates that, in the terms of carbon footprint, renovation is the most profitable option. Strong arguments must exist for demolishing a building and replacing it with a new one.

References

Bertoldi, P., Economidou, M., Palermo, V., Boza-Kiss, B., & Todeschi, V. (2020). How to finance energy renovation of residential buildings: Review of current and emerging financing instruments in the EU. *Energy and Environment, 10*(1), 384.

Bpie. (2011). *Europe's buildings under microscope*. Buildings Performance Institute Europe.

CO2-database. (2021). *Emissions database for construction*. Finnish Environment Institute.

Dahlöf, L., & Malmros, A. (2011). *The million programme - demolish or renovate [Miljonprogrammet - Riva eller renovera]*. Chalmers University of Technology.

Dubois, M., & Allaker, K. (2015). Energy savings from housing: Ineffective renovation subsidies vs efficient demolition and reconstruction incentives. *Energy Policy, 86*, 697–704.

Energy. (2022). *Energy prices*. Statistics Finland.

Energy year 2020. (2021). *Energy and emissions*. Statistics Finland.

Eskilsson, E. (2015). *Renovate or rebuild?* Göteborgs universitet.

EU. (2012). *A comparative methodology framework for calculating cost-optimal levels of minimum energy performance requirements for buildings and building elements*. EUR-lex 32012R0244.
EU. (2019). *The European Green Deal*. EUR-lex 52019DC0640.
Finance of housing companies. (2021). *Income statement statistics of housing companies*. Statistics Finland.
Hasika, V., Escott, E., Bates, R., Carlisle, S., Faircloth, B., & Bilec, M. (2019). Comparative whole-building life cycle assessment of renovation and new construction. *Building and Environment, 161*, 106218.
Hirvonen, J., Heljo, J., Jokisalo, J., Kurvinen, A., Saari, A., Niemelä, T., Sankelo, P., & Kosonen, R. (2021). Emissions and power demand in optimal energy retrofit scenarios of the Finnish building stock by 2050. *Sustainable Cities and Society, 70*, 102896.
Housing Finance and Development Centre of Finland (ARA). (2022). *Prices for new residential buildings and renovations and energy grants*. Statistics and Surveys.
Huuhka, S., Vainio, T., Moisio, M., Lampinen, E., Knuutinen, M., Bashmakov, S., Köliö, A., Lahdensivu, J., Ala-Kotila, P., & Lahdenperä, P. (2021). *To demolish or to repair? Carbon footprint impacts, life cycle costs and steering instruments*. Ministry of the Environment. 2021:9.
Koljonen, T., Soimakallio, S., Lehtilä, A., Similä, L., Honkatukia, J., Hildén, M., Rehunen, A., Saikku, L., Salo, M., Savolahti, M., Tuominen, P., & Vainio, T. (2019). *Long-term development of total emissions*. Prime Minister´s Office, 24/2019.
Kuittinen, M. (2017). Method for the whole life carbon assessment of buildings. Helsinki: Ministry of the. *Environment, 2019*, 22.
La Fleur, L., Rohdin, P., & Mosfegh, B. (2019). Energy renovation versus demolition and construction of a new building - a comparative analysis of a Swedish multi-family building. *Energies, 12*(11), 2218.
Ministry of Environment. (2020). *Long-term renovation strategy 2020–2050 Finland*. EU: Energy Efficient Buildings.
Niemelä, T., Kosonen, R., & Jokisalo, J. (2017). Energy performance and environmental impact analysis of cost-optimal renovation solutions of large panel apartment buildings in Finland. *Sustainable Cities and Society, 32*, 9–30.
Nippala, E., & Heljo, J. (2010). *Comparing life cycle energy, cost and profitability in building retrofit and replacement*. Proceedings of 22nd international housing research conference ENHR 2010 4-7 July 2010.
Nippala, E., & Vainio, T. (2017). *The need for renovation in residential buildings in 2006-2035 [Asuinrakennusten korjaustarve 2006-2035]* (p. 274). VTT Technology.
Nordpool (2022). Market data.
Nydahl, H., Andersson, S., Åstrand, A., & Olofsson, T. (2022). Extended building life cycle cost assessment with the inclusion of monetary evaluation of climate risk and opportunities. *Sustainable Cities and Society, 76*, 103451.
Power, A. (2008). Does demolition or refurbishment of old and inefficient homes help to increase our environmental, social and economic viability? *Energy Policy, 36*, 4487–4501.
Rakennustieto. (2008). *Technical life and maintenance periods of the property [Kiinteistön tekniset käyttöiät ja kunnossapitojaksot]* (pp. 18–10922). RT Building Information File.
Schwartz, Y., Raslan, R., & Mumovic, D. (2018). The life cycle carbon footprint of refurbished and new buildings - a systematic review of case studies. *Renewable and Sustainable Energy Reviews, 81*, 231–241.
Streicher, K. N., Mennel, S., Chambers, J., Parra, D., & Patel, M. (2020). Cost-effectiveness of large-scale deep energy retrofit packages for residential buildings under different economic assessment approaches. *Energy and Buildings, 215*, 109870.
Verbeeck, G., & Cornelius, A. (2011). *Renovation versus demolition of old dwellings Comparative analysis of costs, energy consumption and environmental impact*. Proceedings of PLEA 2011 Architecture and Sustainable Development, 13-15 July 2011.
Vainio, T., & Nippala, E. (2019). *Deep renovations in different business environments* (pp. 88–97). Global Construction Data.

Part IV
Innovations and Digitalization for Sustainable Development

Chapter 14
Multi-objective Genetic Algorithm for the Time, Cost, and Quality Trade-Off Analysis in Construction Projects

Marco Alvise Bragadin, Luca Pozzi, and Kalle Kähkönen

Abstract Project management methods and practice address project success and the well-known "iron triangle" targeting time, cost, and quality trade-off have great importance in this process. Quality optimization, including safety and sustainability, plays a key role in construction project management choices. Since the relationship between quality, time and cost can be different from case to case, an application of artificial intelligence (AI) has been proposed for this purpose. The objective of the research work in this chapter is to demonstrate that AI applications can help project managers the trade-off between time, cost, and quality objectives. A comprehensive approach concerning three estimates of time, cost, and quality of project activities is proposed to optimize project performance in construction. The proposed approach implements a genetic algorithm (GA) to optimize project performances, with the aim of creating a decision support system for construction project managers. Genetic algorithm is an AI application that creates a learning optimization process that discards worse solutions and re-introduces better solutions to search for an optimal or sub-optimal solution. Therefore, time, cost, and quality trade-off can be performed by a multi-objective genetic algorithm that evaluates the effectiveness of various combinations, selecting better solutions with an iterative process. Therefore, the most suitable balance between the three project targets can be achieved. A simple case study of a deep renovation project of two residential is presented to evaluate the proposed approach with a sample application. This study contributes to the understanding of AI applications for construction management.

Keywords Construction · Project Management · Genetic Algorithm · Project planning · time-cost-quality trade-off

---

M. A. Bragadin (✉) · L. Pozzi
Department of Architecture, University of Bologna, Bologna, Italy
e-mail: marcoalvise.bragadin@unibo.it; luca.pozzi2@studio.unibo.it

K. Kähkönen
Faculty of the Built Environment, Tampere University, Tampere, Finland
e-mail: kalle.kahkonen@tuni.fi

G. Lindahl, S. C. Gottlieb (eds.), *SDGs in Construction Economics and Organization*, Springer Proceedings in Business and Economics,
https://doi.org/10.1007/978-3-031-25498-7_14

14.1 Introduction

Construction quality is the essence of a construction contract. Time and cost are other main objectives of project control activities, as clients and contractors in the construction industry are primarily concerned with quality, time, and cost of projects. Nevertheless, quality identifies the overall level of performance of the desired building facility or civil infrastructure, therefore quality can include sustainability items. This is a key component in addressing the SDG – Sustainable Development Goal of the United Nations, no. 9 – Building resilient infrastructure, to promote inclusive and sustainable industrialization and foster innovation (United Nations, 2015). Construction project completion must achieve targets concerning the required level of quality, timely completion, and budget limits. Nevertheless, traditional approaches to construction procurements are based upon cost parameters (i.e., lump-sum contracts) with time and quality constraints, or cost and quality parameters with time constraints, and traditional project control is mainly concerned with time and cost control as addressed by the earned value (EV) methodology (ISO 18045:1, 2010). Competitive bidding for construction works for the Design – Build approaches often ask to minimize construction costs and project durations, and more recently also to maximize construction quality (Anderson & Russel, 2001; Kandil & El-Rayes, 2006). These are the traditional Project Manager's objectives, often described as the "Iron Triangle" (Atkinson, 1999). In terms of construction project success, most of the research works focus on few Key Performance Indicators (KPI), for instance, timely completion, under-budget completion, meeting quality criteria, safely completed work, and client satisfaction. Nevertheless, still most important KPIs are ones concerned with time, cost, and quality (Demirkesen & Ozorhon, 2017). Therefore, construction project managers are used to select a combination of construction technologies and resource usage that minimizes cost and time while maximizing quality. This is termed the time-cost-quality trade-off problem (TCQT) (Zhang & Xing, 2010). Since the relationship between quality, cost, and time is uncertain, and a model of the behavior of these factors can be different from case to case (if exists!), an artificial intelligence (AI) application has been proposed with the purpose of demonstrating that AI applications can help project managers in performing project management processes concerning trade-off between time, cost and quality objectives. As a secondary objective, the positive results of the application confirm that the proposed genetic algorithm-based AI procedure addresses such issues successfully. Artificial intelligence is playing a core role in the Fourth Industrial Revolution, providing significant productivity improvements via analyzing large datasets quickly and accurately, and the optimization of construction management problems via genetic algorithms has been largely addressed by pertinent literature. Genetic algorithms indeed create a learning-based optimization process because better solutions are re-introduced in the iterative optimization process while worse solutions are discarded. Therefore, an optimized solution can be found in a reasonable amount of time, that is, the algorithm converges to better solutions, even if sub-optimal (Darko et al., 2020). This chapter contributes to the understanding of AI applications for construction management. A proposal concerning the application of genetic algorithm-based optimization for time, cost, and quality trade-off has (TCQT) been presented. This chapter is organized as follows. The literature review and the theoretical basis are described in the following sections, addressing genetic algorithms methodology and the time, cost, and quality trade-off (TCQT) problem of construction management. Then the proposed procedure is described within the Pilot study application section. Gained results are discussed in the Sect. 14.5.

14.2 Genetic Algorithms for Time, Cost, and Quality Trade-Off in Construction: A Literature Review

The use of genetic algorithms (GAs) was introduced by J. H. Holland (1975) as a research method based on the mechanics of natural selection and natural genetic of Darwin's Evolutionary Theory. Later, Goldberg (1989) developed further the GAs approach in the field of automation engineering. GAs have been implemented in many engineering, operations research, and optimization problems, for instance, the Travelling Salesman Problem (Mitchell, 1996; Razali & Geraghty, 2011). A genetic algorithm is a global and stochastic research method termed "genetic" because of the mutual terminology from genetics, a branch of biology. It is a probabilistic search procedure designed to work on large spaces involving states that can be represented by mathematical strings (Goldberg & Holland, 1988). A genetic algorithm can be used with the aim of planning and controlling the activities of a project as they are search and optimization tools that assist decision-makers in identifying optimal or near-optimal solutions for problems with large search space. Genetic algorithm is an artificial intelligence application proposed in the literature to optimize construction management problems that creates a learning optimization process that discards worse solutions and reintroduce better solutions to search for an optimal or sub-optimal solution (Mitchell, 1996). Genetic algorithm methods have been used by many researchers in literature as an optimization technology to address Architecture, Engineering, and Construction (AEC) optimization goals as for instance construction scheduling, and cost optimization. Most of the AI techniques used in the AEC sector are GAs (Darko et al., 2020). Construction Engineering and Management benefit from GAs because of Intelligent Optimization, meaning the task of searching for the optimal solution to minimize or maximize an objective function subject to a set of constraints. This problem can be divided into two versions. The simple version is the single objective optimization to identify a single optimal alternative, while the complex one is the multi-objective optimization to simultaneously optimize more than one objective function with a set of feasible solution (Pan & Zhang, 2021). Few researchers focused on the problem of the evaluation of the global quality of a project or a system by means of a quality indicator, and the development of a time-cost-quality trade-off procedure is seldom the objective of research papers. Babu and Suresh (1996) suggested that project quality can be affected by project crashing for minimal cost search. Time–cost trade-off can have effects on quality and therefore a time-cost-quality trade-off is needed. Time is considered the independent variable and quality can be computed with cost constraints. Khang and Myint (1999) tested the Babu and Suresh approach with a case study of the construction of a cement factory in Thailand, highlighting key problems and difficulties faced. A major limitation of the method is that only a very small portion of the overall quality of a work package has a direct relationship with time and cost performances. Quality can be measured through a global quality KPI, termed Quality Index, and based upon a Quality Breakdown Structure (QBS) of the project. The QBS-developed approach builds on the "Quality-Based Performance Rating System" of the American National Cooperative Highway Research Program (NCHRP) (Anderson & Russel, 2001) for contractors' qualification. QBS aims at evaluating the final quality of the products of the construction process, with a performance-based approach. Therefore, a set of quality indicators are detected to evaluate the final product quality. With the aim of improving construction quality, El-Rayes and Kandil (2005) presented a multi-objective optimization model that supports decision-makers in creating an optimal resource optimization plan that minimizes construction cost and time while maximizing its quality. An automated optimization system for construction resources termed MACROS, was implemented (Kandil & El-Rayes, 2006), and the time, cost, and quality trade-off algorithms were developed by genetic algorithms. Following this research line, El Razek et al. (2010) addressed the TCQT problem by the implementation of a java programming code, AMTCROS, based upon a genetic algorithm. San Cristóbal (2009), instead, proposed an Integer Programming model which enables meeting quality output standards and time and cost objectives respectively. Addressing artificial intelligence applications, Zhang and Xing (2010) presented a fuzzy-multi-objective particle swarm optimization to solve the time–cost–quality trade-off (TCQT) problem. Dong et al. (2012) presented a new GA-based method that automates construction schedule generation, with the objective of minimizing time or cost, taking into account engineering and space constraints. Later, Faghihi et al. (2014) developed a computer application that can automatically derive a statically stable construction schedule by data extraction from a BIM model using the concept of genetic algorithms. In the field of Information Technologies, Mishra and Mahanty (2016) indicated that the optimization of project cost, schedule, and quality for a software development project in an outsourcing environment, can be studied with a system dynamics simulation approach. Kyriklidis and Dounias (2014) addressed the resource leveling optimization problem with an evolutionary algorithm (GA) in the project management field, while in the specific construction sector, Monghasemi et al. (2015) proposed a Multi-criterion decision-making approach that identifies all global Pareto optimal solutions by a multi-objective genetic algorithm. Sorrentino (2013) applied GAs to a time, cost, and quality optimization problem for project scheduling of road construction, while Tiene et al. (2018) investigated a similar application for the selection of design alternatives for a building envelope. Liu et al. (2019) presented a GA-based optimization for the Resource-Constrained Project

Scheduling problem that enhances the evolution strategy by proposing modified operators for selection, crossover, and mutation. Hiun et al. (2021) developed a multi-objective optimization tool for modular unit production lines based on genetic algorithms that assumes that the duration of activities on a production line in modular construction depends on the number of workers and reducing construction duration and labor cost will be the optimization objectives. Soman and Molina-Solana (2022) presented a novel Look-Ahead Schedule generation method that uses reinforcement learning algorithms and linked data-based constraint checking to help construction planners as decision support systems. The output schedule is compared with the one generated manually, with the critical path method, and with the modified GA by Liu et al. (2019). Therefore, time, cost, and quality trade-off can be performed by a multi-objective genetic algorithm that evaluates the effectiveness of various combinations, computing better solutions with an iterative process. At the end of the process, the most suitable balance between the three project targets can be selected between the outputs by project managers.

14.3 Proposed Procedure

Genetic algorithms usually start by generating an initial population of possible solutions, called individuals. This generation of individuals is based upon a random approach. Every individual of the population is coded in the form of a string, called chromosome. Then, each chromosome is assessed by calculating its fitness value by the objective function, and chromosomes are sorted depending on their fitness values. The best individuals are selected as parents, creating a sequence of new populations, termed generations (Holland, 1975; Goldberg, 1989; Mitchell, 1996). Therefore, the basic structure of a genetic algorithm involves cyclic operations that simulate the evolutionary process of a population. Each loop represents one generation and each new population generated is formed by better and better individuals. Five phases are considered in a genetic algorithm: initial population; fitness function; selection; mating (crossover/mutation); termination (Mitchell, 1996) (Fig. 14.1). The GA proposed implementation is based upon a table that reports needed data. In the pilot study, the database is a set of construction activities. Each activity can be defined by a set of three parameters, namely time, cost, and quality. The performance of an activity depends on the quality of input products, the fitness of the built component and the building systems, the number and quality of workmanship, and working procedure. The direct cost of the activity and its duration also depend on these variables. Therefore, it is possible to individuate different performing alternatives for each activity, depending on the chosen set of time, cost, and quality indicators. A table summarizes for each project activity the alternatives related to activity duration, activity cost, and quality (Figs. 14.2 and 14.3). Therefore, each project activity includes three possible options for its development that creates a search space of thousands of possible solutions. The definition of different performing alternatives depends on the following set of data:

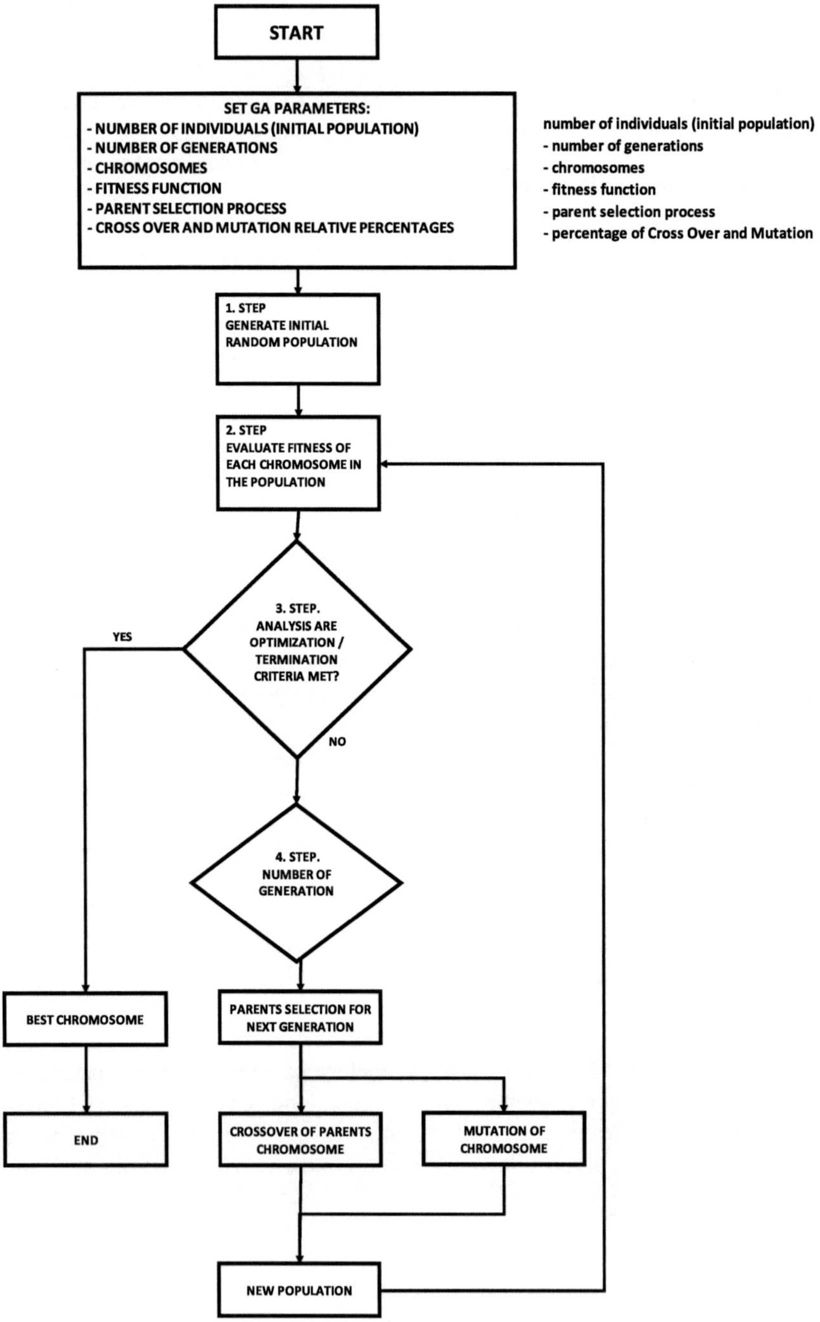

Fig. 14.1 Stages of a genetic algorithm approach

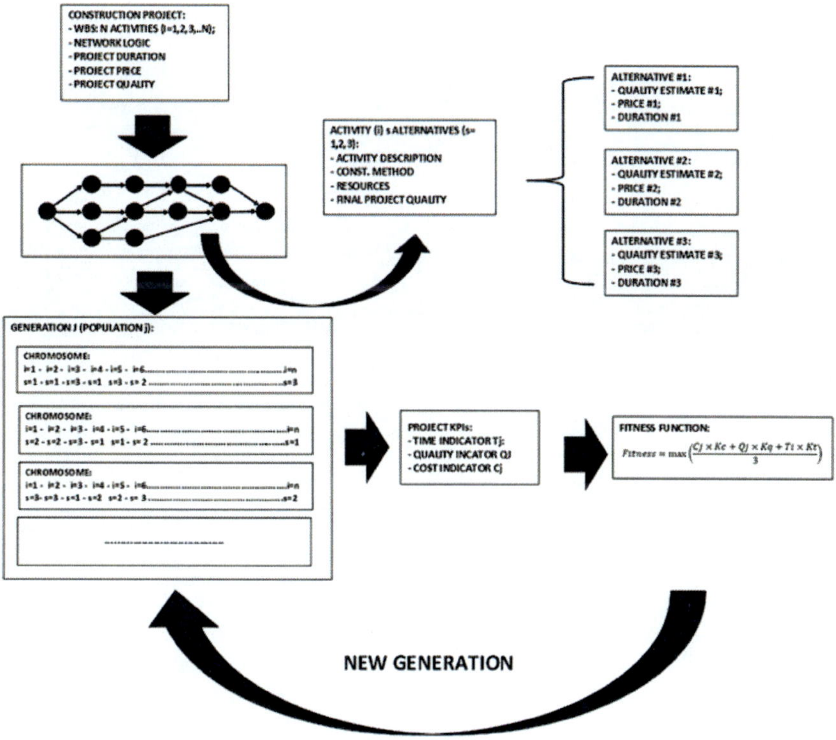

Fig. 14.2 Proposed computation procedure

- Product (input): product quality, product cost, time of installation or transformation;
- Labor: number and qualification of laborers, labor cost, effectiveness of the workmanship, working time;
- Equipment: type and number of required equipment, equipment cost, working capacity.

The following three KPIs have been defined for each activity: quality indicator, time indicator, and cost indicator.

14.3.1 Quality, Time, and Cost Indicators

The Quality Breakdown Structure (QBS) of a construction project enables the objective evaluation of the overall construction quality through quality items measurement (El-Rayes & Kandil, 2005; Kandil and El-Rayes, 2006). The development of a QBS for a construction project provides the capability of estimating the overall construction quality performance at both the activity and the project levels being in

WORK AND QUALITY BREAKDOWN STRUCTURE (1)

A No. / WP alternative	C WBS	D Work Package description	Quality estimate Qei (%)	E Cost Ci (€)	quality weight Qwi	Quality Index Qi	O Duration (h) Di
1	A.01	demolition and removal			0,223		
A		demolition and removal works A	90%	€ 15.174,00		0,201	177
B		demolition and removal works B	100%	€ 15.345,85		0,223	177
C		demolition and removal works C	110%	€ 15.409,35		0,246	179
	A.02	Brickwork					
2	A.02.01	saw-unslich brickwork			0,266		
A		solid bricks	100%	€ 18.226,22		0,268	248
B		old-style solid bricks semi-crafted	110%	€ 30.846,14		0,292	307
C		old-style hand-crafted solid bricks	90%	€ 35.972,58		0,239	306
3	A.02.02	partition walls perforated bricks cm8 thickness			0,008		
A		six hole perforated bricks 8 x 14 x 28	110%	€ 592,08		0,009	8
B		ten hole hollow blocks 8 x 25 x 25	100%	€ 572,64		0,008	7
C		gypsum panels partition wall thickness 8 cm	90%	€ 738,24		0,008	8
4	A.02.03	partition walls perforated bricks cm 10 thickness			0,020		
A		ten hole perforated blocks 8 x 25 x 25	100%	€ 1.393,42		0,020	18
B		gypsum panels partition wall thickness 10 cm	90%	€ 1.839,02		0,018	20
C		gypsum-clay panels partition wall thickness 10 cm	110%	€ 2.304,48		0,022	21
	A.03	Concrete screed					
5	A.03.01	lightened insulating screed			0,075		
A		lightened insulating screed with expanded clay quick	110%	€ 6.664,40		0,083	122
B		lightened insulating screed with natural cork	90%	€ 4.876,40		0,068	154
C		lightened insulating screed with expanded vermiculite	100%	€ 5.198,30		0,075	153
	A.04	Plaster finish					
6	A.04.01	Premixed plaster for interior wall coatings			0,073		
A		Interior plaster with lime - cement mortar	100%	€ 4.994,08		0,073	83
B		Interior plaster with lime mortar	110%	€ 4.877,60		0,080	82
C		Interior plaster with cement mortar	90%	€ 4.994,08		0,065	82
	A.05	Floorings and sheathing/walls					
7	A.05.01	ceramic floor blasted tiles			0,041		
A		ceramic floor blasted tiles 40 x 40	100%	€ 2.837,97		0,041	18
B		ceramic floor blasted tiles 60 x 60	110%	€ 4.874,85		0,045	14
C		ceramic floor blasted tiles 20 x 20	90%	€ 2.497,11		0,037	18
8	A.05.02	Ceramic wall cladding			0,053		
A		ceramic wall tiles 20x20 mono-coloured	90%	€ 6.261,71		0,048	49
B		ceramic wall tiles 20x20 marble-effect	100%	€ 3.692,72		0,053	49
C		ceramic wall tiles 10x10 stone-effect	110%	€ 7.433,79		0,058	60
9	A.05.03	Skirting board			0,028		
A		Stoneware skirting board 10x20	110%	€ 2.504,80		0,027	21
B		Clinker skirting board 8x24 glazed	100%	€ 1.700,00		0,028	17
C		Wooden skirting board - cherry 75x10mm	80%	€ 1.201,60		0,028	9

WORK AND QUALITY BREAKDOWN STRUCTURE (2)

A No. / WP alternative	C WBS	D Work Package description	Quality estimate Qei (%)	E Cost Ci (€)	quality weight Qwi	Quality Index Qi	O Duration (h) Di
	A.06	Windows and doors					
10	A.06.01	counter frame for sliding doors			0,007		
A		metal sub-frame for sliding doors	100%	€ 511,28		0,007	3
11	A.06.02	counter frame for hinged doors max width 11 cm			0,001		
A		wooden counter frame fir depth 2,5 cm width 11 cm	100%	€ 93,66		0,001	1
12	A.06.03	wooden hinged solid door			0,047		
A		Hinged solid interior door - tanganika walnut wood	90%	€ 2.412,84		0,042	8
B		Hinged solid interior door - walnut wood	110%	€ 3.855,00		0,051	7
C		Hinged solid interior door - oak wood	100%	€ 3.209,88		0,047	8
	A.07	Electric system					
13	A.07.01	Electric system for one apartment			0,032		
A		Electric system for one apartment	100%	€ 2.290,00		0,032	21
	A.08	Plumbing and sanitary system					
14	A.08.01	Supply and installation of vitreous china toilet bowl			0,007		
A		Vitreous china toilet bowl	100%	€ 473,50		0,007	5
B		Vitreous china wall-hung toilet bowl	90%	€ 639,10		0,006	5
C		Vitreous china wall-hung chrome finish toilet bowl	110%	€ 1.034,08		0,008	6
15	A.08.02	Supply and installation of vitreous china bidet			0,007		
A		Vitreous china bidet	100%	€ 469,28		0,007	4
B		Vitreous china wall-hung bidet	90%	€ 635,10		0,006	5
C		Vitreous china wall-hung bidet with chrome finish	110%	€ 1.030,08		0,008	6
16	A.08.03	Supply and installation of vitreous china wash basin			0,010		
A		vitreous china wash basin 70x55	100%	€ 687,34		0,010	5
B		vitreous china wash basin 65x50	90%	€ 643,20		0,009	5
C		vitreous china washbasin 70x55 pedestal basin	110%	€ 842,06		0,011	6
17	A.08.04	Sanitary waste water system			0,009		
A		sanitary waste water system for one bathroom PVC	100%	€ 629,12		0,009	8
18	A.08.05	domestic hot/cold water system			0,014		
A		domestic hot/cold water system polybutylene	100%	€ 987,00		0,014	9
B		domestic hot/cold water system cross-linked polyethylene	110%	€ 1.156,64		0,016	18
C		domestic hot/cold water system galvanized steel	90%	€ 1.360,44		0,013	17
	A.09	Building assistance					
19	A.09.01	wall assistance plumbing	100%	€ 640,00	0,009	0,009	11
20	A.09.02	wall assistance electrical system	100%	€ 930,00	0,013	0,013	15
	A.10	Painting					
21	A.10.01	Indoor water-based paint			0,059		
A		Indoor water-based breathable paint	90%	€ 3.648,38		0,053	56
B		Indoor water-based breathable / washable paint	100%	€ 4.077,48		0,059	60
C		Indoor palette knife effect resin-based wall coating	110%	€ 20.839,40		0,065	309

Fig. 14.3 Pilot study quality breakdown structure and data sets

relation to the Work Breakdown Structure WBS (Fig. 14.3). In the pilot study, the quality breakdown structure has been integrated with the Work Breakdown Structure, meaning that for every work package of the WBS, a Quality estimate has been evaluated. Then the quality estimate of the activity has been weighted to find the overall project quality assessment. The quality weighting of the activities of the WBS was developed taking into account the cost proportion of each activity to the overall project cost (El-Rayes & Kandil, 2005). The Quality KPI, representing the overall quality of the Construction Project, can be estimated as the sum of the quality estimate of each activity of the project, weighted to represent the importance and contribution of the quality of every activity to the overall quality of the project (El-Rayes & Kandil, 2005; Kandil and El-Rayes, 2006; Bragadin & Kahkonen, 2013). For each activity of the Work Breakdown Structure (WBS) a quality weight is identified: the weight (Qwi) of the activity to represent the importance and contribution of the quality of the single activity to the overall quality of the construction project (El-Rayes & Kandil, 2005; Kandil and El-Rayes, 2006; Bragadin & Kahkonen, 2013). The Qwi indicator can be calculated with the following Eq. (14.1):

$$\mathrm{QWi} = \mathrm{Cib}\,/\,C \tag{14.1}$$

where C = overall cost of construction project; Cib = baseline cost of the activity (i); Qwi = weight of quality of activity (i) compared to other activity in the project. Then the quality index weighted (Qi) is calculated with the following Eq. (14.2):

$$\mathrm{Qi} = \mathrm{Qei} \times \mathrm{Qwi} \tag{14.2}$$

where Qi = Quality index weighted, of the activity (*i*); Qwi = weight of the quality of the activity I on the overall quality of the project Q; Qei = Quality estimate of the activity (*i*).

The construction project quality KPI, representing the overall quality of the Construction Project, is estimated as the sum of the quality of each activity of the project weighted to represent the importance and contribution of the quality of every activity to the overall quality of the project. The relative weights of the activities of the sample project can be found in Fig. 14.3 (El-Rayes & Kandil, 2005; Kandil and El-Rayes, 2006; Bragadin & Kahkonen, 2013).

The time indicator of each activity of the project is its duration. The duration of the activity can be computed based on labor hours and crew members of each activity (Bragadin et al., 2018).

$$\mathrm{Di} = \mathrm{MHi} / \mathrm{nm} \tag{14.3}$$

where Di = duration of the activity (*i*) in hours; MHi = total labor estimate of the activity (*i*) in manhours; nm = number of members of the working crew of the activity (*i*).

The cost indicator Ci for each activity (*i*) is the building price as detected from an official price list for public works (Regione Lombardia, 2011). The different design alternatives for each activity entail different initial products and different building procedures as indicated by the official price list. All design alternatives are suitable solutions for the final building products, meaning that the product alternatives generate activity alternatives consistent with building design and processes (Bragadin et al., 2018).

14.3.2 Computation Procedure

An initial random selection of options for each activity is performed and the corresponding objective function is computed. Next, GA uses genetic operators such as crossover, which divides two initial solutions exchanging their chromosomes to generate new solutions; mutation, which simulates the effect of random errors; and elitism, which maintains the best individual in the next generation or substitutes the son with the parent if it gives a better performance. The new solution is computed again, and the results of the objective function are compared with the previous ones. The best solutions are selected to improve the fitness function. Each solution has a fitness value different from the others and the best solutions are selected for future generations while worse solutions are set aside (Mitchell, 1996) (Fig. 14.2). The goal of the proposed procedure is to develop a multi-objective model that will help the decision-makers to select the alternatives for each project activity that will minimize construction cost (prices) and construction total duration while maximizing project quality (TCQT). Therefore, the proposed fitness function depends on the three WP parameters (time, cost, and quality) weighted (Bragadin et al., 2018). The following Eq. (14.4) is proposed:

$$\text{Fitness} = \max\left(\frac{\text{Cj} \times \text{Kc} + \text{Qj} \times \text{Kq} + \text{Ti} \times \text{Kt}}{3}\right) \tag{14.4}$$

where Cj is defined by the following:

$$\text{Cj} = 1 - \frac{\Sigma\text{Ci} - \text{Cmin}}{\text{Cmax} - \text{Cmin}} \tag{14.5}$$

where ΣCi is the total cost of each *i* work package of the project (*i* = 1, 2, 3, ….*n*) of each *j* generation (*j* = 1, 2, 3, ……*m*) and *m* is the number of generations.

Qj can be found by the following equation:

$$\text{Qj} = \frac{\Sigma\text{Qi}}{n} \tag{14.6}$$

where ΣQi is the total sum of quality indexes of each *i* work package of the project (*i* = 1, 2, 3, ….*n*) for the generation *j* and *n* the total number of work packages of the project.

Tj is the time parameter found for the *j* generation, defined by the following:

$$\text{Tj} = 1 - \text{NTdj} \tag{14.7}$$

where NTd is the normalized total duration:

$$\text{NTdj} = \frac{\text{TDj} - \text{Tmin}}{\text{Tmax} - \text{Tmin}} \tag{14.8}$$

where TDj is the total project duration found by network diagramming and critical path computation for the generation *j*. TDj is the maximum duration found by critical path analysis comparing each total duration TDjk of a single path k of the generation *j* composed by the work packages ik belonging to the k network path:

$$\text{TDj} = \max \text{TDjk} \tag{14.9}$$

The weighting parameters k_c, k_q, and k_t can range from 0 to 1 for cost, quality, and time, respectively. Aiming at balancing the three parameters, the following values have been set: $k_c = 1$, $k_q = 1$, and $k_t = 1$. The final evaluation of the found solutions can be performed by comparison with the maximum and minimum set limits of the three parameters, termed C_{max}, C_{min}, T_{max}, T_{min}, Q_{max}, and Q_{min} (Table 14.1).

Table 14.1 Limit values of total project alternatives

Total project values	Maximum limit	Minimum limit
Total project duration	Tmax = 1067 (h)	Tmin = 714 (h)
Total project cost	Cmax = € 113.309,99	Cmin = € 64.668,68
Total project quality index	Qmax = 109.1%	Qmin = 94.5%

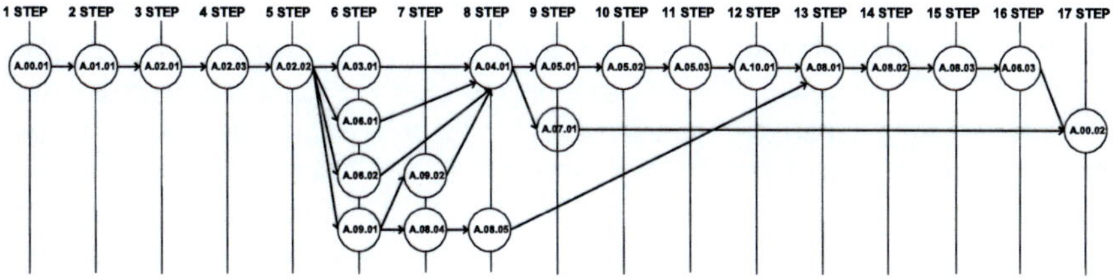

Fig. 14.4 Network diagramming of the pilot study project

14.4 Pilot Study Application

A GA-based algorithm has been implemented with Solver®, an add-in of MS Excel® (FrontlineSolvers, 2022). This application can explore the solution space very quickly and is able to identify a set of optimal solutions. The purpose of the pilot study is to test the proposed GA-based procedure. The pilot study consists of a small building renovation project that has been used also in a previous paper by the authors but processed with another computer application (Bragadin et al., 2018). For each work package of the pilot study, three different commercial product options have been considered and the corresponding activity durations, costs, and quality performances have been detected from a public works price list (Regione Lombardia, 2011). Quality indexes have been evaluated straightforwardly as product quality and its suitability for use (Fig. 14.3). Therefore, the proposed time–cost–quality trade-off procedure has been implemented using Solver®-based genetic algorithms, with the aim of finding a set of optimal solutions for the building construction project. Found data for each work package are presented in the following text.

The pilot study has 21 Work Packages (WP), and each has three possible alternatives of time, cost, and quality for project activities. The possible combinations of these alternatives create a large space of search, where each solution in this space can be a possible option for project delivery. A project model has been implemented with network diagramming. Therefore, critical path analysis can be performed, and total project duration can be found for each project alternative of the pilot project. Nevertheless, the search space is not 3^{21} because of different subpopulations that constitute the structure of base data of the problem, that is, different paths of the network (Fig. 14.4). Therefore, each possible path from project start to project finish constitutes a subpopulation, termed species. Within species, that is, single path, permutations of different WP alternatives are possible only after satisfaction of precedence constraints between succeeding WP.

No alternative permutations are possible between different species because of the structure of chromosomes, that is, the number of WP of each network path. The chromosome of a species is created by time, cost, and quality data of each chosen WP alternative belonging to a network path. The limit values of total project alternatives can be found by time-based computation of the critical path method (Fig. 14.4) and total sum of the cost and quality data of the project (Fig. 14.3). Minimum and maximum total values of the three project parameters, time, cost, and quality, can be found by manual computation of the corresponding alternatives of each WP (Table 14.1). Please note that the six limit values found in Table 14.1 belong to six different project alternatives. The aim of setting the min/max limits of each indicator is to assess the boundaries that define the min/max performance of project alternatives indicated by the outputs computed by the GA-based procedure.

Therefore, the Solver® application has been set for the specific problem using the evolutionary algorithm. The network diagram and the working options of the pilot study have been formalized in Microsoft Excel. The maximum number of generations has been set to 100 and the maximum time without improvements to 100 seconds. The best-found optimized result in the case of balanced weights (Kq = 1, Kc = 1, and Kt = 1) is the following: fitness value = 0,34,853, total project duration = 720 (h), total cost = € 64.668,68, and total quality index = 97,38%. Further testing of the developed model will be needed to assess its effectiveness in case of more complex projects with multiple environmental constraints.

14.5 Discussion and Conclusions

The well-known iron triangle of main project objectives, namely time, cost, and quality is of capital importance for project managers in construction, but balancing these three parameters for actual projects can be difficult because of the unknown or complex function linking all of these three parameters. With the aim of demonstrating that AI applications can address the TCQT problem, an innovative GA optimization has been developed and implemented with Solver®. Actual data of a pilot study concerning the expected duration of each work package WP, its quality index, and its cost have been detected, and three possible performing alternatives of the WP were developed. Therefore, the overall performance of the whole construction project, with the processing of all the work packages was simulated taking into account the possible different alternatives of activity duration, cost, and quality. The time estimate was developed by means of a CPM- based activity network, while the total cost estimate was the sum of the cost of all work packages, and the overall project quality index was estimated as the normalized sum of quality indexes of all work packages. The aim of the optimization is to find automatically (or semi-automatically) a balance between the time, cost and quality project objectives via genetic algorithm computation. When developed and implemented, the Evolutionary Algorithm extracts randomly one work package alternative for all the activities, thus creating a chromosome for each species (i.e., a path of the network model) of the project for each generation, and computes its suitability by fitness equation. New generations are then created and the found solutions in terms of total project duration, total cost, and total project quality are compared with the previous ones by fitness function computation. The fittest generations are maintained and developed, and the others are set aside from the evolutionary process. As a critical point, the developed implementation of a genetic algorithm needs a new and complex approach in project modeling to reach the result in terms of fitness of the final generation. Actually, the GA approach is of complex implementation and in comparison with the practical approach of construction managers need surely a greater effort in terms of model creation and computing. But the possibility of development of an AI computer application with this purpose creates a new perspective. Another key point of the proposed approach is the definition of the fitness function. The fitness function plays a major role in selecting the developed new generations, and therefore directly affects the procedure outputs. Actual data for a pilot study simulation of a building renovation project of two residential apartments have been used to demonstrate the possibility of implementing a GA-based optimization of project objectives, and the found results are consistent with the initial assumptions in terms of ranges of time, cost, and quality values. The limits of the research work are the single application to a small case study. Surely, the development of large actual project, with hundreds (or thousands) of activities is almost impossible with the presented computer application. Nevertheless, the use of AI applications and of genetic algorithms was presented as a possibility for understanding and mastering in an advanced manner the time-cost-quality trade-off (TCQT) in construction projects. The developed prototype and gained results represent proof-of-concept research based on which we are able to estimate the viability of this technology for the described purpose. Therefore, the application of the proposed approach can help construction managers to improve project quality, its timely implementation, and compliance with budget limits. Building sustainability can be understood as an aspect of quality, and sustainability itself is a fundamental requirement of construction projects as addressed by SDG goal no.9 – Building resilient infrastructure, this work can help to improve inclusive and sustainable industrialization and foster innovation (United Nations, 2015). Therefore, the application of the proposed approach can improve the quality and sustainability of constructed facilities by optimizing the time, cost, and quality trade-off. Future research work will be aimed at testing further the developed procedure with the imposed constraints and more complex projects.

References

Anderson, S. D., & Russel J. S. (2001). *Guidelines for Warranty, Multi-Parameter, and Best Value Contracting*. NCHRP report 451. Transportation Research Board – National Research Council, U.S.

Atkinson, R. (1999). Project Management: Cost, time and quality, two best guesses and a phenomenon, its time to accept other success criteria. *International Journal of Project Management, 17*(6), 337–342.

Babu, A. J. G., & Suresh, N. (1996). Project management with time, cost and quality considerations. *European Journal of Operational Research, 88*, 320–327.
Bragadin, M. A., & Kahkonen, K. (2013). Quality evaluation of construction activities for project control. *Journal of Frontiers in Construction Engineering*. - ISSN:2306-7713, *2*(1), 17–24.
Bragadin, M., Kähkönen, K., & Ballabeni, A. (2018). Using genetic algorithms for optimizing and modelling time, cost and quality trade-offs of construction projects. IN_BO - ISSN:2036-1602 vol. 2018, *9*(13), 200–207.
Demirkesen, S., & Ozorhon, B. (2017). Impact of integration management on construction project management performance. *International Journal of Project Management, 35*(2017), 1639–1654.
Darko, A., Chan, A. P. C., Adabre, M. A., Edwards, D. J., Hosseini, M. R., & Ameway, E. E. (2020). Artificial Intelligence in the AEC industry: Scientometric analysis and visualization of research activities. *Automation in Construction, 112*(2020), 103081.
Dong, N., Ge, D., Fischer, M., & Haddad, Z. (2012). A genetic algorithm-based method for look-ahead scheduling in the finishing phase of construction projects. *Advanced Engineering Informatics, 26*(2012), 737–748.
El-Rayes, K., & Kandil, A. (2005). Time-cost-quality trade-off analysis for highway construction. *Journal of Construction Engineering and Management, 131*(4), 477–486.
El Razek, R. H., Diab, A. M., Hafez, S. M., & Aziz, R. F. (2010). Time–cost-quality trade-off software by using simplified genetic algorithm for typical-repetitive construction projects. *World Academy of Science, Engineering and Technology, 37*, 312–321.
Faghihi, V., Reinschmidt, K. F., & Kang, J. H. (2014). Construction scheduling using genetic algorithm based on building information model. *Expert Systems with Application, 41*(16), 7565–7578.
FrontlineSolvers. (2022). *Genetic algorithms and evolutionary algorithms – introduction*. www.solver.com. Last access march 2022.
Goldberg, D. E. (1989). *Genetic algorithms in search optimization and machine learning*. Addison-Wesley Publishing Company Inc.
Goldberg, D. E., & Holland, J. H. (1988). Genetic algorithms and machine learning. *Machine Learning, 3*, 95–99. Kluwer Academic Publishers.
Hiun, H., Yoon, I., Hyun-Soo, L., Park, M., & Lee, J. (2021). Multiobjective optimization for modular unit production lines focusing on crew allocation and production performance. *Automation in Construction, 125*(2021), 103581.
Holland, J. H. (1975). *Adaption in natural and artificial systems*. The MIT Press.
ISO. (2010). ISO 10845-1:2010 Construction procurement – part 1: Processes, methods and procedures (ISO).
Kandil, A., & El-Rayes, K. (2006). MACROS: Multiobjective automated construction resource optimization system. *Journal of Management in Engineering, ASCE, 22*(3), 126.
Kyriklidis, C., & Dounias, G. (2014). *Application of evolutionary algorithms in project management*. 10th IFIP international conference on artificial intelligence applications and innovations. (AIAI), Sep 2014, Rhodes, Greece. pp. 335–343.
Khang, D. B., & Myint, Y. M. (1999). Time, cost and quality trade-off in project management: A case study. *International Journal of Project Management, 17*(4), 249–256.
Liu, J., Liu, Y., Shi, Y., & Li, J. (2019). Solving resource-constrained project scheduling problem via genetic algorithm. *Journal of Computing in Civil Engineering, 34*(2), 04019055.
Mitchell, M. (1996). *An introduction to genetic algorithms*. MIT Press.
Mishra, D., & Mahanty, B. (2016). A study of software development project cost, schedule and quality by outsourcing to low cost destination. *Journal of Enterprise Information Management, 29*(3), 454–478.
Monghasemi, S., Nikoo, M. R., & Fasaee, M. A. K. (2015). A novel multi criteria decision making model for optimizing time-cost-quality trade-off problems in construction projects. *Expert System with Applications, 42*(2015), 3089–3104.

Pan, Y., & Zhang, L. (2021). Roles of artificial intelligence in construction engineering and management: A critical review and future trends. *Automation in Construction, 122*, 103517.

Razali N. M., & Geraghty J. (2011). *Genetic Algorithm Performance with different selection strategies in solving TSP*. Proceedings of the world congress on engineering 2011 vol. II WCE, July 6–8, 2011 London, UK.

Regione Lombardia. (2011). *Prezzario Opere Pubbliche 2011*. Regione Lombardia.

San Cristóbal, J. R. (2009). Time, cost, and quality in a road building project. *Journal of Construction Engineering and Management © Asce, 135*, 1271–1274.

Soman, R. K., & Molina-Solana, M. (2022). Automating look-ahead schedule generation for construction using linked-data based constraint checking and reinforcement learning. *Automation in Construction, 134*(2022), 104069.

Sorrentino, M. (2013). *Genetic algorithms for construction time-cost-quality trade-off: A road project case study, Ricerche e progetti per il territorio, la città e l'architettura, Construction Management*, ISSN 2036 1602 | pp. 163–176.

Tiene, S., Bragadin, M. A., & Ballabeni, A. (2018). A genetic algorithm - based approach for project management and developed design of construction. TECHNE - ISSN:2239-0243, *16/2018*, 131–141. https://doi.org/10.13128/Techne-23027

United Nations. (2015). *Transforming our world: the 2030 Agenda for Sustainable Development*. Resolution adopted by the General Assembly on 25 September 2015.

Zhang, H., & Xing, F. (2010). Fuzzy-multi-objective particle swarm optimization for time-cost-quality tradeoff in construction. *Automation in Construction, 19*, 1067–1075.

Chapter 15
Conceptual Study and Literature Review of Integration of Lean Manufacturing and 3D Printing in Construction to Support Sustainability

Erik Gran, Javier Alonso Madrid, Guillermo Sotorrío Ortega, José Antonio Tenorio Rios, and Nils O. E. Olsson

Abstract This chapter presents a literature review and conceptual study of 3D printing of concrete in lean and sustainability perspectives. Three research questions are addressed. (1) What are expected challenges in practical applications of 3D printing of concrete; (2) To what extent can 3D printing of concrete support lean construction principles; (3) What are possible sustainability implications from applications of 3D printing of concrete in the construction sector? On the first question, we find that there are still a number of challenges that need to be addressed to move 3D printing from test sites to integrated use in construction projects. The integration of a 3D printer in the building process is then investigated using scenarios; façade elements and self-insulating wall elements. On the second question, we find that lean techniques provide for coordination and good information flow. If the information flow and planning process is taken care of in this manner, 3D printing can be successfully integrated into most larger construction processes and sup-

E. Gran
SINTEF Digital, Oslo, Norway
e-mail: erik.gran@sintef.no

J. A. Madrid
ATANGA Arquitectura e Ingeniería, Madrid, Spain
e-mail: javier@alonsomadrid.com

G. S. Ortega · J. A. T. Rios
Eduardo Torroja Institute for Construction Science, Spanish National Research Council (CSIC), Madrid, Spain
e-mail: sotorrio@ietcc.csic; tenorio@ietcc.csic.es

N. O. E. Olsson (✉)
Department of Mechanical and Industrial Engineering, Norwegian University of Science and Technology (NTNU), Trondheim, Norway
e-mail: nils.olsson@ntnu.no

G. Lindahl, S. C. Gottlieb (eds.), *SDGs in Construction Economics and Organization*, Springer Proceedings in Business and Economics,
https://doi.org/10.1007/978-3-031-25498-7_15

port sustainability in the construction sector. Regarding the third question, 3D printing can contribute to achieving several UN Sustainable Development Goals (SDGs), for example through the reduction of waste and more efficient use of natural resources, which is also related to the objective of lean construction. This means that lean appears to be a suitable approach for sustainable construction, and this paper discusses how 3D can be used as a technology to support the ambition of moving the construction industry in a more sustainable direction.

Keywords Additive manufacturing · 3D-printing · Lean · Construction

15.1 Introduction

This paper studies the use of three-dimensional (3D) printing in construction projects from a lean perspective. In the construction industry, additive manufacturing has emerged as an interesting option that can change how projects are done. As a consequence of the current movement toward digitization and automation, the industry is under pressure to re-think and renew its use of technology (Olsson et al., 2021).

3D printing is a type of additive manufacturing. Additive manufacturing relates to the process of joining materials to make objects from 3D model data, usually layer upon layer (Dietrich et al., 2019). Synonyms include additive fabrication, additive processes, additive techniques, additive layer manufacturing, layer manufacturing, and free-form fabrication (Astm f2792).

Since 2015, the UN has sought a more desirable future in 17 areas of concern, called Sustainable Development Goals (SDG). Additive manufacturing aims to contribute to several of these. 3D printing is actively contributing to SDG 9 (industry, innovation, and infrastructure), which aims to upgrade infrastructure and retrofit industries to make them sustainable, with increased resource-use efficiency and greater adoption of clean and environmentally sound technologies and industrial processes, with all countries taking action in accordance with their respective capabilities. 3D printing is relevant also for SDG 12 (Responsible consumption and production) which aims to achieve sustainable management and efficient use of natural resources. In both cases the contribution of additive manufacturing is through a reduction of waste and more efficient use of natural resources, achieving a transition to more sustainable construction methods. There is a growing trend in the use of more sustainable building materials and systems, so that in the construction sector, additive manufacturing may be displacing other construction systems with similar technical characteristics but with worse environmental values. As pointed out by Francis and Thomas (2020), lean and sustainability have common goals of promoting resource efficiency and reducing waste.

The purpose of this paper is to analyze 3D printing in a lean perspective and to discuss practical applications of 3D printing of concrete in the construction sector. The three research questions (RQs) of the study are:

RQ1. What are the expected challenges in the practical applications of 3D printing of concrete?

RQ2. To what extent can 3D printing of concrete support lean construction principles?

RQ3. What are the possible sustainability implications from the applications of 3D printing of concrete in the construction sector?

15.2 Applied Method

This study was performed as a part of the EU-funded research project HINDCON. To begin with, an extensive literature review was done on practical experiences from 3D printing of concrete for applications in the construction sector. This search included sustainability aspects. Lean construction and lean principles in general were known to the authors form before, but the literature search included a special focus on lean and 3D printing.

The paper uses two scenarios to illustrate how the construction sector can be affected by the implication of such technology: façade elements and self-insulating wall elements. The building elements used in the scenarios were all selected based on a workshop in 2016 concerning the applicability of 3D printing techniques. The discussion concerned building elements where the use of 3D printing techniques could realize elements with additional or better properties. The workshop resulted in a list of building elements that may be produced with the HINDCON printer and where this technology is likely to find usage. The paper also utilizes work in the HINDCON project especially targeted on sustainability considerations of the applied 3D printing technology.

15.3 3D Printing of Concrete

15.3.1 Additive Manufacturing with Concrete

Guaman-Rivera et al. (2022) found ongoing experimentation with 3D printing in construction and challenges related to the readiness of the technology. Currently, there are three large-scale additive manufacturing processes targeted at construction and architecture in the public domain, namely Contour Crafting, D-Shape, and Concrete Printing (Lim et al., 2011). All three have proven the successful in the manufacture of components of significant size and are suitable for construction and/or architectural applications. Contour crafting and concrete printing are both extrusion techniques while D-Shape is more like normal printing where a binding agent (water) is printed/sprayed onto a bed of compacted cement mix powder.

Concrete printing is based on the extrusion of cement mortar, and the process has been developed to retain three-dimensional freedom and has a smaller resolution of deposition, which allows for control of internal and external geometries.

15.3.2 Printing Structural Elements in Concrete

The setting or hardening of concrete is a chemical process that takes time. Maximum strength of concrete is usually considered to be the strength it has attained after 28 days of curing, but how fast the curing process is depends on the type of concrete and additives. For 3D printing purposes, it is important that the concrete retains shape after extrusion and cure fast enough to carry subsequent layers without deformation. On the other hand, the time between printing subsequent layers may decrease the strength of the bond between layers significantly, due to curing of the surface of the deposited layer. There is a time window for depositing a new layer depending on the stability of the previous layer due to rheological behavior and the fast concrete cure, enough to bear the weight of the new layer, and the time until the previous layer becomes less chemically active and thus result in a weaker bond. Experimental testing by the Eindhoven University of Technology (Salet et al., 2017) suggests a framework of three components (printable concrete, 3D printer, and print geometry) for describing a 3D concrete printing (3DCP) system where each of these components constitutes a range of parameters and variables.

Bos et al. illustrate the interdependencies of the components in a 3DCP system (Bos et al., 2016). They describe that the buildability of layers is limited as new layers add weight on top of the structure and thus depend on the stiffness and strength of the printed green layers. Fast-setting concrete mixes can eliminate this problem.

A simple straight stacking of layers is complicated. Cantilevering layers represent another type of problem with 3DCP. The contact surface between extruded strings (Gosselin et al., 2016) is another issue. In spite of this, 3D printing is interesting.

In 3D printing, the design, material, process, and product properties are recognized as interdependent. This interdependency is more pronounced when 3D printing in concrete, due to the slow setting reaction of concrete (Bos et al., 2016).

15.3.3 Reinforcing 3D Printed Elements

Concrete as a material has high compressive strength and low tensile or ductile strength. For most types of structural elements concrete needs reinforcing with other materials. Traditionally this is done by building a frame in rebar steel and casting concrete around it. Adding glass, steel, or carbon fibers to the concrete mix as well as some other additives may reduce the need for reinforcements, but these additives increase the cost of the concrete, and fibers increase the difficulty of pumping. There are also reported trials of embedding steel wire in the printed concrete layers to increase ductile strength.

With today's technology, there seem to be three different options for making structural elements in 3DCP:

- Limit the use of 3DCP to compression-loaded elements such as domes, arcs, and straight decorative columns.
- Use 3DCP to print an outer shell for the structure to be used as a mold/lost formwork. The shell will then be filled with traditional reinforcement and concrete and the shell is integrated into the structure.
- Print the structure with reinforcement integrated into the element on the *x*/*y* plane (horizontal at printing) and leave holes to be filled with tensioned steel wire or a rebar/concrete mix along the *z* axis (vertical).

15.3.4 Production of 3D-Printed Elements

The cycle time of the printer will in most cases be less than the curing time of the concrete. Effective use of the 3D printer then dictates that the printed structure must be moved after finishing printing (or the printer must be moved). This implies that the printed element must have cured adequately to withstand vibrations and small shocks in the movement without deformation.

At the building site, the finished 3D-printed structural elements will usually need to be lifted in place. The use of lifting equipment will introduce stresses to the elements other than the forces that will be working on it in the final assembled position. Unless the elements are left to be fully cured before assembly, it is thus necessary to evaluate the strength of the structure while it's not fully cured and with respect to other forces than it will be subjected to in the final position. The alternative is to set aside time to allow the elements to fully cure before assembly.

15.3.5 3D Printing and Sustainability

There are several potential advantages for 3D printing in a sustainability perspective. Environmental impact can be reduced by minimization of material waste, which is of particular interest, given the high carbon footprint of concrete. 3D printing can produce less waste and use less materials (Rael and San Fratello, 2011; Berman, 2012; Achillas et al., 2015). Additive construction such as 3D printing can also reduce emissions related to transport (Achillas et al., 2015; Strauss, 2013).

Muñoz et al. (2021) did a life cycle assessment (LCA) on a concrete 3D printing system. Their study included a complete supply chain of the 3D printing equipment, operation, and end-of-life, based on real data from a demonstration plant installed in Spain. The results showed that 3D printing can have a lower environmental impact than traditional construction for small production series. If the same structure needs to be produced in larger volumes, their results indicated that 3D printing and traditional production have a similar environmental impact.

15.4 Lean Construction

Lean construction (LC) is the continuous process of eliminating waste, meeting or exceeding all customer requirements, focusing on the entire value stream, and pursuing perfection in the execution of a constructed project (Diekmann et al., 2004). LC principles represent the implementation of major concepts to drive changes, such as increasing transparency and defining the value stream. The literature contains a number of attempts to define lean principles for construction (Koskela, 1992, 2000; Koskela & Howell, 2002; Höök & Stehn, 2008; Towill, 2008; Jørgensen & Emmitt, 2009).

The synthesis of these studies results in the following set of lean construction principles: (1) define and deliver the value for customer systematically; (2) reduce complexity and variability; (3) eliminate non-value-adding activities; (4) achieve continuous process flow; (5) increase the output flexibility; (6) continuous improvement and knowledge building; and (7) leadership for lean implementation.

15.4.1 Lean Construction Processes

Construction projects consist of several phases, including project design, product design, process design, procurement, and on-site installation (Ballard, 2008).

The first phase, project design and planning, includes alignment of customer and stakeholder purposes, definition of the project, concept design, and planning of the activities. Lean projects are structured and managed as a value-generating process for customer, with minimized number of steps. Activities are synchronized through all steps, focusing on flow as opposed to productivity.

The second phase, lean product and process design deal with aligning product and process design, considering the alignment of values, concepts, and criteria. Lean design actively involves the client, user, stakeholders, and suppliers. Value and waste consequences of decisions are made transparent at the design stage, exploring alternative solutions with maximum design space flexibility.

The third phase, Lean procurement and supply chain management, consists of detailed engineering, fabrication, and delivery of materials. Lean procurement includes principles such as establishing a supply network at the best value for the customer, reducing purchasing and supply lead times, simplifying material management processes, and pulling materials from the supply network as needed from the workplace.

Finally, lean construction/ assembly/on-site installation begins with the delivery of the materials and information to the site and deals with the on-site installation of materials, completing the construction, and commissioning. Lean assembly aims to standardize the work and quality of the outcomes on site.

15.4.2 Lean Construction Tools

Tools allow the implementation of principles, such as the use of information-sharing indicators and safety instructions. In this section, the term tools also incorporate methods, practices, and techniques. Different tools address different principles at different phases of the projects. Chief Engineer System (Morgan & Liker, 2006) facilitates the lean principle of defining and delivering the desired value for customers, by front-loading and involving the customer at the product development phase.

Last planner system (LPS; Ballard, 2000) aims to eliminate the non-value-adding activities, by introducing a pull-based production control system at the construction site. A big room (or Obeya room) that gathers all the important information from a project and supports knowledge building and sharing (Oppenheim, 2004) is an example of tool that supports the principle of continuous improvement and knowledge building. Takt time control (Duggan, 2013) in make-to-order industry a bottleneck activity is often chosen (or even made) to control the output from a production line. The Takt is simply defined as the desired output from the activity divided by the time to make one unit. By placing this bottleneck late in the value chain two things are achieved: a predictable rate of output from the production line and a focal point for the coordination of inputs.

According to Carvajal-Arango et al. (2019), the implementation of lean construction practices can have positive effects on all three dimensions of sustainability; economic, social, and environmental. They point out that several of the established lean construction practices have documented positive effects in the three dimensions of sustainability. Related to 3D printing, those of particular interest include modular construction, BIM, and Concurrent Engineering. El Sakka and Hamzeh (2017) applied the lean construction tool value stream maps to compare 3D printing with traditional production and found a 60% reduction in lead time, along with reductions in cost, reduction of waste, and improved quality.

15.5 Case Illustration of Manufacturing and Logistics Flow for 3D-Printed Structural Elements

3D printing offers great advantages to traditional place-built formwork and prefab elements when it comes to relatively cheap and effective customization of building elements. Compared to place-built one-of-a-kind structures this new technology offers customization at a lower cost by either printing the formwork in support materials or by integrating the formwork in the structure. On the other hand, the lead time of a 3D printed structural element that first must be printed as a shell, cured, filled with reinforced concrete, and cured again, will be substantial. The cost and properties of the elements will depend on what materials are used but can generally be thought of as more expensive. With the relatively long lead time from manufacturing the element to assembly, lean production governed by pull mechanisms is also not easily achieved. The following is a discussion of the industrial application of 3D printing to four different building elements that are used for illustration of the applicability of 3D printing.

15.5.1 Façade Elements

The ability to make freeform creative elements without costly formwork makes 3D printing in concrete an obvious choice for façade elements. 3D printing also offers new opportunities compared to prefab elements in that these façade elements may be unique without adding additional costs. However, facade elements often have double functions in that they in addition to being aesthetically pleasing also is part of the building load-bearing structure and also provide the building with thermal and acoustic insulation and the necessary protection against external weather conditions, and also the transmission of other loads such as wind. Façade elements thus often need to be reinforced to have the necessary tensile/ductile strength. The necessary strength may be achieved by leaving holes in the printed element and filling these holes later with either tensioned wire or reinforced concrete.

The cycle time of the printer for printing large building elements will for practical purposes be shorter than the curing time of the same elements. This implies that the printed elements need to be moved from the printer before they are fully cured to free up the printer for other tasks. On the other hand, moving "green" elements represent the risk of deformation due to shocks and/or vibrations. To reduce this risk, the elements should be printed on pallets that may be moved by shock-free transport equipment. Heavy lift equipment of this type is available and in use in shipyards.

Another consideration concerning the strength of the façade elements is stresses during lifting into the final position. These are different from the forces the elements will be subjected to as part of the building. It's also likely that lifting may take place while the concrete is not fully cured due to variations in curing time due to external factors like weather, temperature, and humidity. The standard way of designing with all stresses accommodated by the reinforcement will also work with 3D printed elements but may limit the artistic freedom in designing the elements as the openings in the form must be available for reinforcement after printing.

All issues described above will cause long lead time, which makes the production process not very suited for lean thinking. Just-in-time delivery of the element to the building site must be initiated long before delivery and cannot accommodate late changes in the design. Some time may be saved in the process if the design of the element can use tensioned wires as reinforcement.

On the other hand, the production process for façade elements represents an efficient way of creating unique building features that would take costly and time-consuming formwork to produce in any other way. Precast elements are usually standardized for obvious reasons and place-built forms/scaffolding are both costly to erect and sometimes dangerous to fill. Some reports that 50% of the building time and cost of new building is formwork and reinforcement. Using 3D printed items as lost formwork for a one-of-a-kind precast façade element will thereby be a more efficient way with respect to today's technology.

15.5.2 Self-Insulating Wall Elements

The wall elements discussed here will not be load-bearing. Load-bearing walls imply reinforcement by rebar or other high tensile/ductile materials. Walls may be printed as lost formwork and also for load-bearing walls, but such elements will not be self-insulating as they will be filled later with reinforced concrete.

The idea behind a self-insulating wall is to leave enclosed air in the structure to provide insulation. Insulation may be achieved by leaving empty space between an inner and outer wall or fill the space with other insulating material. In both cases, the inner and outer walls need to be connected in a way that does not leave thermal bridges but holds the wall elements together. One way of achieving this is to print a third surface inside the outer layers that connect to the inner and outer walls at alternate points. This way the thermal distance between the inner and outer walls will be greater than the distance between the surfaces of the element. Gosselin et al. (2016) describe how a slight change in designing how inner and outer walls are separated can reduce heat transfer by 58% in their example.

The advantage of printing wall elements rather than using pre-cast elements lies in the possibility of curved surfaces and other integrated design elements in the structure of the wall. Precast elements are usually made lying flat on tables that are flipped over when the first surface is cured before the next surface is cast. The connection between the two surfaces is usually made by rebar steel which is also employed for lifting purposes.

The technique of 3D printing wall elements in concrete opens new possibilities in architectural design. Using traditional methods curved surfaces are costly and take a long time to make. Decorations and ornaments integrated into the walls also imply place-built molds which similarly add time and costs. With 3D printing, the process of making a curved and ornamented wall is the same as a straight wall. As an added benefit, the wall can be made self-insulating in the same process.

The production process, however, must include time for curing. The wall elements must be made sometime before they are assembled. Changes or other events that may happen while the element is curing will thus lead to rework. All these issues again make it difficult to implement lean principles.

15.6 Discussion

The ability to produce building elements directly from 3D model is possibly the most attractive feature of 3D printing in concrete. This opens up for almost zero-material waste in the production. This is well aligned with the philosophy of lean construction, which is based on strategies to reduce all types of waste in production, time, and effort. The basis for 3D printing and lean construction to support sustainability in construction is therefore in place.

There are, however, still two limitations to 3D printing of concrete that are not resolved by today's technology that inhibit working directly from BIM models. The first obstacle is the inability to reinforce the elements in all directions. As has been pointed out earlier, only reinforcement along the printed concrete strings is possible today whether this is done by adding fibers to the concrete or embedding wire in the concrete strings. For reinforcement in the vertical plane (when printing) holes must be left in the structure to be filled by either reinforced concrete or post-tensioned steel wires.

Post-tensioned wires need to be mounted in brackets supported by concrete that have achieved high compression strength. The throughput time of a structure printed this way will thus include time for the curing of the concrete. The curing time for the printed concrete will vary depending on the type of concrete and additives.

The other obstacle is that holes or channels in the horizontal plane need to be closed at the top. Unless the diameter of the hole or channel is small compared to the width of the printed string, this is difficult to achieve without support. A possible solution may be to print clay in the hole or channel that may be washed out when the concrete is set, but that implies a second extruder with a change of extruder for each layer to be printed. Another solution could be to insert supports when the channel or hole is closed. The hole or channel would then be square at the top.

The best way is probably to print technical wall elements with most channels or holes in the vertical plane. Some wall elements may then be printed with layers that are vertical in the final position. Unless the wall element is supposed to take up some tensile/ductile forces, this will probably not be a problem. The maximum ductile/tensile strength of the element is limited anyhow both from the limited possibilities of including reinforcement along the layer and because of the possibility of layer separation with no reinforcement between layers. Technical wall elements that need tensile/ductile strength are likely to be produced by using the 3D print as lost formwork that is filled with reinforced concrete after curing.

Efficient use of the 3D printer presupposes moving elements out of the printer before they are cured. As wall and wall elements are relatively thin and high and thereby unstable, moving "green" elements is an added risk. A longer curing time in the printer, thus increasing the cycle time of the printer, and/or extra transport supports might be needed. The lead time from starting printing the element to assembly will anyhow include time for curing as the element should be cured before assembly. Filling the structure with reinforced concrete to achieve tensile/ ductile strength will add time as the printed form need to cure before filling. Unfortunately. a long lead time from start production to finished element means that the process does not lend itself easily to lean production.

Carvajal-Arango et al. (2019) found only few papers that directly related lean construction to sustainability. Adaloudis and Roca (2021) identified some challenges from a sustainability perspective because their findings indicated that incentives to invest in 3D printing were not mainly related to the environmental benefits, but rather to the technology's potential to increase automation and thereby moderate a shortage of skilled labor in the construction sector.

15.7 Conclusions

In the following, we address the defined research questions of the study.

15.7.1 Expected Challenges in Practical Applications of 3D Printing of Concrete?

Our first research question addressed what challenges that can be expected in practical applications of 3D printing of concrete. While 3D printing and other types of digitalization have the potential to change the construction project process, there are still a number of challenges that need to be addressed to move 3D printing from test sites to integrated use in construction projects. This paper has studied such a change of projects in a lean perspective.

The main attraction of 3DCP is the possibility to produce almost any shape without first building a mold or form for the concrete. Formwork and reinforcement in a modern building are reported to account for up to 50% of the building time and proportionally high costs. Being able to "disregard" the costs of nonstandard shapes will in addition give architects greater freedom in how they design new building artistically.

However, as described by Salet et al. (2017), 3D printing is a process between the 3D printer, the printable concrete, and the architectural design. At present, there exists some knowledge of what can be printed but how the design may limit or extend these printing capabilities is still unknown. By integrating fibers in the concrete mix some ductile strength may be achieved along the direction of the extruded concrete. At present this limits in most cases the use of 3D printing applications to create "lost formwork" or molds to be filled with reinforced concrete.

Printing time and to some extent also curing time depend on the size of the element. The production time on any element will depend on what is being made and, as pointed out by Bos et al. (2016), there exists an interdependency between the material, printing process, and design that introduce uncertainty to estimates.

15.7.2 3D Printing and Lean Construction Principles

Our second research question relates to how and to what extent 3D printing of concrete can support lean construction principles. Lean production focus on flow and just-in-time delivery. As printed concrete needs to mature before use, it is not well suited for lean thinking. There will always be a delay between producing the element and assembling the structure This means that the design must be finalized before starting a print and will not accommodate changes without rework.

The complexity of integrating the production with a 3D printer in the building process demands robust and versatile planning and control. The principles and tools of Lean construction (Moujib, 2007) may support this role. The hierarchical structure of the Last planner tool for long-, mid-, and short-term planning combined with the pull-based start of production renders integration with the current situation in the building process possible. Lean construction techniques can provide for coordination and good information flow. Provided that the information flow and planning process is taken care of, 3D printing may be integrated into most larger building processes, which probably will be needed to deliver the potential sustainability benefits.

A 3D printer is typically aimed at producing unique elements, so buffer stocks of standard elements are not possible. In fact, a 3D printer is close to the lean ideal with little or no changeover between different products. Emphasis is thus needed on production planning and control. The recommendations are to make use of the concurrent engineering methodology combined with the lean practices, a broader scope on each of the phases in the construction process, and information hubs and big rooms (Aasland & Blankenburg, 2012).

15.7.3 Sustainability Implications from Applications of 3D Printing of Concrete in the Construction Sector

Finally, the third research question addresses possible sustainability implications from applications of 3D printing of concrete in the construction sector. The reduction of waste is a key sustainability argument in favor of 3D printing. It is documented in previous studies that a 3D-printed object can have a lower environmental footprint, compared to traditional construction processes. 3D printing can contribute to achieving several UN Sustainable Development Goals (SDGs), including SDGs 9 and 12. One key contribution that aligns well with lean philosophy and supports sustainability is a reduction of waste and more efficient use of natural resources. However, this and previous studies indicate that the 3D printing technology is still relatively novel, and it may still take some time until it makes significant contributions to a more sustainable construction industry. But we need to work to get there.

Acknowledgments This research has been partially funded by the European Commission under the Horizon 2020 Framework Programme. Grant Agreement No 723611.

References

Aasland, K., & Blankenburg, D. (2012). *An analysis of the uses and properties of the Obeya*. 18th international ICE conference on engineering, technology and innovation (pp. 1–10). IEEE.

Achillas, C., Aidonis, D., Iakovou, E., Thymianidis, M., & Tzetzis, D. (2015). A methodological framework for the inclusion of modern additive manufacturing into the production portfolio of a focused factory. *Journal of Manufacturing Systems, 37*(1), 328–339.

Adaloudis, M., & Roca, J. B. (2021). Sustainability tradeoffs in the adoption of 3D Concrete Printing in the construction industry. *Journal of Cleaner Production, 307*, 127201.

Astm f2792. https://www.astm.org/Standards/F2792.htm. Accessed: 2019-03-08.

Ballard, G. (2000). *Positive vs negative iteration in design*. Brighton, Proceedings eighth annual conference of the international group for Lean Construction.

Ballard, G. (2008). The Lean Project delivery system: An update. *Lean Construction Journal*, 1–19.

Berman, B. (2012). 3-D printing: The new industrial revolution. *Business Horizons, 55*(2), 155–162.

Bos, F., Wolfs, R., Ahmed, Z., & Salet, T. (2016). Additive manufacturing of concrete in construction: Potentials and challenges of 3D concrete printing. *Virtual and Physical Prototyping, 11*(3), 209–225.

Carvajal-Arango, D., Bahamón-Jaramillo, S., Aristizábal-Monsalve, P., Vásquez-Hernández, A., & Botero, L. F. B. (2019). Relationships between lean and sustainable construction: Positive impacts of lean practices over sustainability during construction phase. *Journal of Cleaner Production, 234*, 1322–1337.

Diekmann, J. E., Krewedl, M., Balonick, J., Stewart, T., & Won, S. (2004). *Application of lean manufacturing principles to construction* (p. 191). Construction Industry Institute.

Dietrich, D. M., Kenworthy, M., & Cudney, E. A. (2019). *Additive manufacturing change management: Best practices*. CRC Press.

Duggan, K. J. (2013). *Creating mixed model value streams*. CRC Press, Taylor & Francis.

El Sakka, F., & Hamzeh, F. (2017). *3D concrete printing in the service of lean construction*. 25th Annual Conference of the International Group for Lean Construction (IGLC), 9–12 July 2017, Heraklion, Greece.

Francis, A., & Thomas, A. (2020). Exploring the relationship between lean construction and environmental sustainability: A review of existing literature to decipher broader dimensions. *Journal of Cleaner Production, 252*, 119913.

Gosselin, C., Duballeta, R., Roux, P., Gaudillière, N., Dirrenberger, J., & Morel, P. (2016). Large-scale 3D printing of ultra-high performance concrete – A new processing route for architects and builders. *Materials and Design, 100*, 102–109.

Guaman-Rivera, R., Martinez-Rocamora, A., Garcia-Alvarado, R., Munoz-Sanguinetti, C., Gonzalez-Böhme, L. F., & Auat-Cheein, F. (2022). Recent developments and challenges of 3D-printed construction: A review of research fronts. *Buildings, 12*, 229.

Höök, M., & Stehn, L. (2008). Applicability of lean principles and practices in industrialized housing production. *Construction Management and Economics, 26*(10), 1091–1100.

Jørgensen, B., & Emmitt, S. (2009). Investigating the integration of design and construction from a "lean" perspective. *Construction Innovation, 9*(2), 225–240.

Koskela, L. (1992). *Application of the new production philosophy to construction*. Stanford University.

Koskela, L. (2000). *An exploration towards a production theory and its application to construction*, VTT Technical Research Centre of Finland.

Koskela L., & Howell G. (2002). *The underlying theory of project management is obsolete.* Proceedings of the PMI Research Conference, PMI.

Lim, S., Buswell, R. A., Le, T. T., Wackrow, R. Austin, S. A., Gibb, A., & Thorpe A. (2011). *Development of a viable concrete printing process*. Proceedings of the 28th international symposium on automation and robotics in construction, ISARC.

Morgan, J. M., & Liker, J. K. (2006). *The Toyota product development system*. Productivity Press.

Moujib, A. (2007). *Lean project management*. Paper presented at PMI® Global Congress 2007—EMEA, Budapest, Hungary. Newtown Square, PA: Project Management Institute. https://www.pmi.org/learning/library/lean-project-management-7364.

Muñoz, I., Alonso-Madrid, J., Menéndez-Muñiz, M., Uhart, M., Canou, J., Martin, C., Fabritius, M., Calvo, L., Poudelet, L., Cardona, R., Lombois-Burger, H., Vlasopoulos, N., Bouyssou, C., Dirrenberger, J., Papacharalampopoulos, A., & Stavropoulos, P. (2021). Life cycle assessment of integrated additive–subtractive concrete. 3D printing. *The International Journal of Advanced Manufacturing Technology*, *112*, 2149–2159.

Olsson, N. O. E, Arica, E., Woods, R., Alonso-Madrid, J. (2021). Project Leadership and Society 2 (2021) Industry 4.0 in a project context: Introducing 3D printing in construction projects.

Oppenheim, B. W. (2004). Lean product development flow. *Systems Engineering, 7*(4), 352–376.

Rael, R., & San Fratello, V. (2011). *Developing concrete polymer building components for 3D printing*. Proceedings of the ACADIA 31st Annual Conference of the Association for Computer Aided Design in Architecture, Banff, BA, Canada

Salet, T. A. M., Bos, F. P., Wolfs, R. J. M., & Ahmed, Z. Y. (2017). 3D concrete printing – A structural engineering perspective. In D. A. Hordijk & M. Luković (Eds.), *High tech concrete: Where technology and engineering meet: Proceedings of the 2017 fib symposium, held in Maastricht, The Netherlands, June 12–14, 2017* (pp. xliii–lvii). Springer.

Strauss, H. (2013). *Am envelope: the potential of additive manufacturing for facade constructions* (Vol. 1, No. 2013). TU Delft.

Towill, D. R. (2008). Construction supply chain and the time compression paradigm. In W. J. O'Brien, C. T. Formoso, R. Vrijhoef, & K. A. London (Eds.), *Construction supply chain management handbook*. CRC Press.

Chapter 16
BIM Adoption Processes: Findings from a Systematic Literature Review

Kaleem Ullah, Emlyn Witt, and Irene Lill

Abstract Significant improvements in the performance of the construction industry have been expected from Building Information Modeling (BIM) and this has led to widespread attempts at its adoption. In parallel, there has been growing interest among researchers to examine BIM adoption processes together with the constraints they face, and this suggests that there is now a need for an up-to-date, state-of-the-art overview of BIM adoption research. The purpose of this study is to review and analyze existing BIM adoption research in order to synthesize their findings and derive an overall understanding of BIM adoption processes. To achieve this purpose a systematic review methodology was followed. The scope of the review is limited to academic articles written in English that are focused on BIM adoption processes and the factors affecting BIM adoption. A total of 410 relevant articles comprising mainly exploratory surveys and case studies on BIM adoption were identified and reviewed. Content analysis of the articles resulted in the classification of BIM adoption literature into project, organization, and industry levels and classification of factors affecting the BIM adoption process. This research has implications for practice and research that the classification of factors that affect the BIM adoption process can be used to help analyze BIM adoption in different organizations.

Keywords Building information modeling · BIM adoption · Construction industry · Literature review · Content analysis

16.1 Introduction

BIM as an innovative technology offers various potential benefits, and as a result, BIM adoption in the AEC/FM industry is on the rise, particularly in recent years (Ullah et al., 2019). It is believed that BIM implementation can also assist in

K. Ullah (✉) · E. Witt · I. Lill
Department of Civil Engineering and Architecture, Tallinn University of Technology, Tallinn, Estonia
e-mail: kaleem.ullah@taltech.ee; emlyn.witt@taltech.ee; irene.lill@taltech.ee

G. Lindahl, S. C. Gottlieb (eds.), *SDGs in Construction Economics and Organization*, Springer Proceedings in Business and Economics,
https://doi.org/10.1007/978-3-031-25498-7_16

achieving the UN's Sustainable Development Goals (Umar, 2021) particularly SGG9 (Industry, Innovation, and Infrastructure) and SDG 11 (Sustainable Cities & Communities) through its applications in visualization, energy simulations, life cycle assessments, and collaborative working environment.

The term adoption is often used interchangeably with implementation and is defined by Rogers (1983) as "a decision to make full use of an innovation as the best course of action available". Based on Roger's Diffusion of Innovation theory, Hochscheid and Halin (2018) described the BIM adoption process as a five-stage process, that is, awareness about BIM followed up by possible intention toward it, decision to adopt BIM, implementation, and confirmation to continue the use of BIM. In recent years academic research in the area of BIM has increased, including studies investigating the BIM adoption process. Research related to BIM adoption has mainly focused on motivation or readiness towards BIM (Juan et al., 2017), practical case studies of BIM implementation (Shibeika & Harty, 2015), implementation frameworks (Kouch, 2018), maturity models and adoption rate (Succar & Kassem, 2015), and factors affecting BIM adoption (Ahuja et al., 2020). These studies provide useful insights about various aspects of BIM adoption; however, there is still a scarcity of research that holistically review BIM adoption processes in project, organization, and industry wise and factors (drivers and impediments) affecting it since given scattered in these studies. Recently, many review studies have been carried out on BIM research for instance, Gao and Pishdad-Bozorgi (2019); however, most of the reviews were concentrated on certain BIM applications such as energy analysis, knowledge management, BIM for facility management, BIM for sustainable construction, and BIM for existing buildings. Consequently, this research is aimed to systematically review studies that were performed particularly on BIM adoption to state the current state of the BIM adoption research and factors affecting it.

The remainder of the paper is structured as follows: the methodology used for the systematic literature review is described in the "Methodology" section. The findings of the study are reported in the "Findings" section, and this is followed by a discussion and conclusion.

16.2 Methodology

Systematic literature review is a widely used methodology in academic research aimed at generating robust knowledge from already existing or published literature (Kraus et al., 2020). This study adopted a systematic literature review methodology following the guidelines recommended by Bearman et al. (2012).

16.2.1 Review Question

What is the current state of the research on the BIM adoption process and what are the factors (drivers and impediments) influencing the BIM adoption process?

16.2.2 Inclusion and Exclusion Criteria

As BIM is a wide topic, papers from peer-reviewed journals and conference proceedings of high quality focused on the BIM adoption process and factors affecting it were considered relevant. Only papers published in the English language were included. No restriction on the years of publications was imposed.

16.2.3 Search Strategy

A desktop study was performed using the search engines provided by ASCE Library, EBSCOhost, Emerald Insight, Science Direct, Scopus, and Web of Science. The main search strings and keywords were ((“Building Information Mode?l*” OR “BIM”) AND (“Adoption” OR “Implementation”). The keywords (Process) or (Diffusion) or (Factors) or (Drivers) or (Challenges) Or (Impediments) were also tried alongside main search strings and keywords. This, however, did not result in any additional articles being found compared to the initial main search strings and keywords.

16.2.4 Screening and Search Results

Papers relevant to the review question were screened from the database search results, looking into their titles and abstracts if necessary. The relevant papers were exported through the Mendeley reference manager and duplicates were removed. Search results are given in the “Findings” section.

16.2.5 Data Extraction and Quality Evaluation

The data relevant to the review question was extracted using NVivo (V.12) software by qualitative content analysis of reviewed papers returned from the selected databases.

16.2.6 Synthesis and Reporting Findings

Each article was read, and all content related to the review question was coded using NVivo (V12). The findings of the systematic literature review are given below.

16.3 Findings

16.3.1 Current State of the Research on BIM Adoption Process

16.3.1.1 Database Search Results

From the databases search, a total of 410 papers were found relevant to the review question. Out of 410 papers, only a few papers were practical cases of BIM implementation. However, the rest of the papers were found relevant because of their focus on themes directly relating to BIM adoption, e.g., awareness, readiness, frameworks/models, and strategies. Tables 16.1 and 16.2 show results from data searches and main sources of articles, respectively.

16.3.1.2 Theme of the Reviewed Papers

A number of themes related to BIM adoption process were found and these include:

- BIM awareness and readiness (Juan et al., 2017; Eadie et al., 2015)
- BIM acceptance models (Kim et al., 2016; (Acquah et al., 2018)
- BIM implementation frameworks (Almuntaser et al., 2018; Kouch, 2018)
- Factors affecting BIM adoption (Ahuja et al., 2020; Hong et al., 2016)
- Drivers/Motivations for BIM adoption (Liao & Teo, 2019),

Table 16.1 Database research results

Databases	ASCE library	EBSCOhost	Emerald insight	Science direct	Scopus	Web of science	Overall, after removing duplicates
Number of relevant articles after screening	62	119	57	40	218	185	410

Table 16.2 Top sources of relevant articles

Publication source	Number of articles
Automation in Construction	21
Journal of Management in Engineering	19
Engineering, Construction and Architectural Management	18
Journal of Construction Engineering and Management	15
Construction Innovation	13
Architectural Engineering and Design Management	9
Journal of Information Technology in Construction	9
International Journal of Construction Management	7
Sustainability	7
Journal of Civil Engineering & Management	6

- Challenges to BIM adoption (Vass & Gustavsson, 2017)
- BIM maturity model (Succar & Kassem, 2015)

16.3.1.3 Theoretical Lenses

From the literature review, it was observed that themes related to BIM adoption were studied using different theoretical lenses as shown in Table 16.3. These theoretical lenses, particularly innovation diffusion theories are based on well-established bodies of knowledge from sociology, psychology, and communication (Kale & Arditi, 2005) and are useful for understanding all aspects of innovation adoption or diffusion (Hosseini et al., 2015). A large number of papers did not use any theoretical lens, however, an increasing trend of using theories to explain aspects of BIM adoption can be observed in recent papers.

Table 16.3 Theoretical lenses

Type of theory	Explanation	Used by studies
Diffusion of innovation theory (DOI) (Rogers, 1983)	DOI explains "innovation diffusion" is a process through which innovation is communicated through certain channels over time among the member of social system.	Gledson and Greenwood (2017) and Shibeika and Harty (2015)
Technology Acceptance Model (TAM) (Davis, 1989)	TAM based on two attributes "perceived usefulness" and "perceived ease of use" aims to predict information technology acceptance by users and explains the behavior of users in acceptance of information technology.	Hong et al. (2016) and Kim et al. (2016)
Technology, Organizational & Environment (TEO) framework (Tornatzky et al., 1990)	TEO framework describes the influences of technological, the organizational, and the environmental context on innovation adoption.	Ahuja et al. (2020)
Theory of Reasoned Action (TRA) Fishbein and Ajzen (1975)	To explain acceptance or action behavior, TRA proposed that the behavioral intention (BI) of a person is influenced by his/her attitude and subjective norms.	Ding et al. (2015)
Theory of Planned Behaviour (TPB) Ajzen (1991)	TRA adds Perceived BehaviouralBehavioral Control (refers to people's perception of the ease or difficulty of performing the behavior of interest) to the exist theory of TRA	Nnaji et al. (2019)
Institutional Theory DiMaggio and Powell (1983),	Institutional Theory suggest that isomorphic pressures (coercive, mimetic, and normative) motivates structural and behavioral changes in organizations while gaining social legitimacy	Cao et al. (2017)
Task-Technology Fit Theory (TTF) Goodhue and Thompson (1995)	TTF describes that degree of suitability of technology to the user's task affects the use of information technology.	Gurevich et al. (2017)

16.3.1.4 Overview of Reviewed Papers on BIM Adoption

The papers covered in the systematic literature review on the various aspects of BIM adoption are categorized into three types of studies: studies focused on BIM adoption at the level of industry, organization, and project are shown in Tables 16.4, 16.5, and 16.6, respectively.

Table 16.4 BIM adoption studies at the industry level

Highlight of focus	Methodology	Lens	Country	Studies
BIM implementation challenges	Quantitative	None	Poland	Lesniak et al. (2021)
Key actors' perspective on BIM adoption	Qualitative	None	New Zealand	Doan et al. (2021)
BIM adoption in SMEs in UK AEC sector	Mixed method	None	UK	Vidalakis et al. (2020)
Factors affecting BIM adoption in AEC industry	Quantitative	None	China	Ma et al. (2019)
Organizational change framework for BIM adoption	Quantitative	None	Singapore	Liao and Teo (2019)
Management strategies for BIM adoption	Quantitative	Psychological climate theory	Hong Kong	Chan et al. (2018)
Level of BIM acceptance	Quantitative	TAM	Ghana	Acquah et al. (2018)
BIM in UK construction industry	Quantitative	DOI	UK	Gledson and Greenwood (2017)
Factors influencing BIM adoption	Quantitative	TAM	Australia	Hong et al. (2016)
Factors affecting BIM acceptance	Quantitative	TAM	Korea	Kim et al. (2016)
Conceptual structure on Macro BIM adoption	Literature review	None	Australia	Succar and Kassem (2015)
BIM adoption process on basis of hierarchy of needs	Mixed methods	Maslow's motivational theory & DOI	Australia & Finland	Singh and Holmstrom (2015)
BIM readiness and changes required for BIM adoption	Quantitative	None	UK	Eadie et al. (2015)
BIM acceptance in construction industry	Quantitative	TAM	Korea & USA	Lee et al. (2015)
Roadmap for BIM implementation	Mix method	None	Finland & UK	Khosrowshahi and Arayici (2012)
Collaborative BIM decision framework for BIM adoption	Qualitative	None	Australia	Gu and London (2010)

Table 16.5 BIM adoption studies at the organization level

Highlights/focus	Methodology	Lens	Organization type	Country	Reference
Readiness framework for BIM implementation	Qualitative	None	Design firms	Vietnam	Tong and Phung (2021)
Clients' role in BIM implementation	Case study	None	Public client	Sweden	Lindblad and Guerrero (2020)
BIM implementation framework	Case study	PMI framework	Architecture	KSA	Almuntaser et al. (2018)
BIM adoption for asset management	Quantitative	None	Public organization	Canada	Brunet et al. (2019)
BIM implementation framework for SME	Literature review	None	Contractors	Finland	Kouch (2018)
BIM adoption actions of public facility agencies	Case study	TAM, TTF, BAM	Public facility agencies	UK	Gurevich et al. (2017)
Motivations for BIM implementation	Quantitative	Institutional theory	Design & contractor	China	Cao et al. (2017)
Organizational challenges to BIM implementation	Case study	IT business value model	Public client	Sweden	Vass and Gustavsson (2017)
BIM acceptance and readiness	Quantitative	TAM & Artificial Neural Network	Architecture	Taiwan	Juan et al. (2017)
Factors affecting BIM adoption behaviors	Quantitative	None	Architecture & Engineering firms	Thailand	Ngowtanasawan (2017)
Factors influencing BIM adoption	Quantitative	TOE frame- work	Architecture	India	Ahuja et al. (2020)
BIM implementation in FM	Case study	None	Public client	Canada	Cavka et al. (2015)
Factors affecting BIM adoption	Quantitative	TRA	Architecture	China	Ding et al. (2015)
Contractors' transformation strategies for BIM adoption	Case study	None	Contractor	USA	Ahn et al. (2016)
Perceptions, challenges, and drivers for BIM adoption	Mixed method	None	Engineering firms	Malaysia	Rogers et al. (2015)

(continued)

Table 16.5 (continued)

Highlights/focus	Methodology	Lens	Organization type	Country	Reference
Diffusion process of BIM	Case study	DOI	Engineering firms	UK	Shibeika and Harty (2015)
Factors affecting behavioral intentions to adopt BIM	Quantitative	TAM	Architecture	Korea	Son et al. (2014)
Factors affecting BIM adoption	Quantitative	None	Constructer	Nigeria	Abubakar et al. (2014)
BIM implementation planning procedures	Case study	None	Facility owners	USA	Chunduri et al. (2013)
BIM adoption process	Case study	None	Architecture	UK	Arayici et al. (2011)

Table 16.6 BIM adoption studies at the project level

#	Description	Methodology	References	Country
1	BIM adoption process in a residential project	Case study	Whitlock and Abanda (2020)	UK
2	BIM implementation in the design processes	Case study	Pruskova and Kaiser (2019)	Czech Republic
3	BIM implementation in a renovation building project	Case study	Roorda and Liu (2008)	Canada
4	BIM Adoption process in a real-time refurbishment project	Case study	Okakpu et al. (2020)	New Zealand
5	BIM implementation in tunnel project	Case study	Daller et al. (2016)	Austria and Slovenia
6	BIM adoption process and its impacts on the design and construction phase	Case study	Rowlinson et al. (2010)	Hong Kong
7	BIM implementation process of a building project	Case study	Harty et al. (2010)	UK

The above tables show that the studies covered in the systematic literature review were mainly focused on examining the BIM adoption process and the factors affecting it. Table 16.5 indicates that BIM adoption studies on an organizational level were mostly concentrated on architecture firms. It shows that architecture firms are some of the most active organizations in BIM adoption. The studies focused on investigating BIM adoption on the project level were practical cases of BIM adoption.

Considering the BIM adoption process as a social phenomenon, various authors have tried to explain it rationally. To understand the BIM adoption process, Roger's Diffusion of Innovation (DOI) Theory is one of the widely used theoretical lenses. Based on the DOI theory, Hochscheid and Halin (2018) developed a model showing the BIM adoption process, which starts from *awareness* (exposure to BIM), *intention* (evaluating the possibility of using BIM), *decision* (decision of adopting/ Rejecting BIM), and, if the decision is made to adopt BIM, then *implementation* (the actual start of using BIM) and *confirmation* (BIM already in use, reached some level of mastery and willingness to continue).

In the literature review, Gu and London (2010), Cao et al. (2017), Singh and Holmstrom (2015) studies were found related to the early stages of the BIM adoption process. Gu and London (2010) developed a collaborative BIM decision framework for industry stakeholders who are likely to adopt BIM. Based on institutional theory, Cao et al. (2017) studied the motivations of contraction professionals to implement BIM and found that image motives and cross-project economic motives were the main reasons to adopt BIM. Similarly, Singh and Holmstrom (2015) explored the motivation of actors in BIM adoption decisions using Maslow's motivational theory & DOI. Studies by Almuntaser et al. (2018) and Arayici et al. (2011) were practical cases of BIM adoption, reporting various actions/techniques undertaken by organizations to implement BIM.

16.3.2 Factors Influencing BIM Adoption Process

16.3.2.1 Drivers for BIM Adoption

The literature review reveals various drivers motivating stakeholders toward BIM adoption. According to Khosrowshahi and Arayici (2012), the main driver for full BIM implementation is the project participants' understanding of its potential benefits over the traditional project delivery. External forces (BIM mandate, client demands, etc.) can also lead to organizations taking the BIM adoption decision. Table 16.7 shows the drivers for BIM adoption in the technology, organization, and environmental contexts.

16.3.2.2 Impediments to the BIM Adoption Process

BIM adoption is a complex process and can face a number of impediments. The impediments to the BIM adoption process identified from the literature review are listed in Table 16.8.

16.4 Discussion

With the potential BIM applications throughout the building life cycle enabling sustainable developments, the topic of BIM has attracted the attention of researchers, particularly in the last decade and a number of studies have been carried out. This current paper reviewed the research on the BIM adoption processes. The results

Table 16.7 Drivers for BIM adoption

Context	Drivers	References
Technology	1. Perceived usefulness of BIM 2. Perceived ease of use 3. Observability of BIM benefits	Gurevich et al. (2017) Acquah et al. (2018) Ding et al. (2015)
Organization	1. Top management support 2. Organizational readiness 3. Finical resources 4. Supportive organization culture 5. Leadership leading the adoption process	Liao and Teo (2019) Juan et al. (2017) Chan et al. (2018) Gu and London (2010) Liao and Teo (2019)
Environment	1. Client demands 2. BIM mandate 3. Partner's influence 4. Competitor's pressure	Chen et al., (2019) Rogers et al. (2015) Ahuja et al. (2020) Eadie et al. (2015)

Table 16.8 Impediments to *the* BIM adoption process

Context	Impediments	References
Technology	1. Complexity of BIM tools 2. Interoperability issues 3. Large size of BIM files	Ahuja et al. (2020) Ma et al. (2019) Ngowtanasawan (2017)
Organization	1. Insufficient senior management support 2. Less capability of BIM team 3. Cost constraints 4. Inadequate BIM training system 5. Lack of collaboration among project participants	Siebelink et al. (2018) Siebelink et al. (2018) Chan et al. (2018) Ahn et al. (2016) Ahuja et al. (2020)
Environment	1. Insufficient government support 2. Legal issues 3. No BIM use from the other/partners organizations	Ngowtanasawan (2017) Gu and London (2010) Chan et al. (2018)

found that a limited number of studies were carried through theoretical lenses, which reflects the findings of Akintola et al. (2020) stating that only 64 out of 1040 reviewed papers on BIM published from 2005 to 2016 were found to have used a theoretical lens.

The classification of papers into industry, organization, and project levels showed that many studies were centered on architecture firms, this can be related to Ahuja et al. (2020) reporting architecture firms among early adopters of BIM. Similarly, the results revealed that a majority of the publications were from developed countries, arguably due to high BIM adoption and BIM mandates in these countries compared to developing countries. Further, quantitative approaches were observed in the majority of the reviewed papers which were mostly focused on factors affecting BIM adoption and BIM acceptance models. While studies on real-time BIM implementation in AEC organizations dominantly used qualitative methods.

The findings also revealed the drivers and impediments to the BIM adoption process, which were grouped into technology, organization, and environmental contexts. This shows that apart from technology, other contexts (organization and environment) are also important for successful BIM adoption and these too require due attention.

16.5 Conclusion

This research has systematically reviewed existing literature on BIM adoption processes, and classified the studies concentrating on BIM adoption into industry, organization, and project levels. Theoretical lenses used to understand the BIM adoption phenomenon in the construction industry have also been identified and summarised. Most of the current BIM adoption research has focused on themes such as readiness and acceptance models, BIM adoption frameworks, factors affecting adoption, drives, and barriers to adoption. However, there are limited studies on actual cases of BIM implementation in AEC organizations describing the whole adoption process. Real-time case studies on BIM implementation are useful for successful BIM implementation as these provide lessons and detailed information on strategies and techniques followed during the BIM adoption process in a specific organization.

In the systematic literature review, a scarcity of studies on BIM adoption in public organizations or agencies was observed, future research could focus on this area. Further, future works can be carried out to minimize the impediments to BIM adoption in AEC/FM industry. Meanwhile, future studies should explore BIM implementation for building permits in the building control authorities. One limitation of this paper is that some relevant studies may have been missed due to the keyword search strategy selected. Lastly, this study contributes to the existing body of knowledge on BIM adoption and offers useful insights for researchers and industry practitioners. The results of this study can be useful for successful BIM implementation, which provides an opportunity to achieve sustainable construction and thus towards the fulfillment of SDG 9 (Industry, Innovation, and Infrastructure) and SDG 11 (Sustainable Cities and Communities).

Acknowledgments This research was supported by the BIM-enabled Learning Environment for Digital Construction (BENEDICT) project (grant number: 2020-1-EE01-KA203-077993), Minimizing the influence of coronavirus in a built environment (MICROBE) project (grant number: 2020-1-LT01-KA203-078100), Strengthening University-Enterprise Collaboration for Resilient Communities in Asia (SECRA) project (grant number: 619022-EPP-1-2020-1-SE-EPPKA2-CBHE-JP) and the Integrating Education with Consumer Behaviour relevant to Energy Efficiency and Climate Change at the Universities of Russia, Sri Lanka, and Bangladesh (BECK) project (grant number: 598746-EPP-1-2018-1-LT-EPPKA2- CBHE-JP) all co-funded by the Erasmus+ Programme of the European Union. The European Commission's support to produce this publication does not constitute an endorsement of the contents which reflect the views only of the authors, and the Commission cannot be held responsible for any use which may be made of the information contained therein.

References

Abubakar, M., Ibrahim, Y. M., Kado, D., & Bala, K. (2014). Contractors' perception of the factors affecting Building Information Modelling (BIM) adoption in the Nigerian Construction Industry. In Computing in civil and building engineering (2014) (pp. 167–178).

Acquah, R., Eyiah, A. K., & Oteng, D. (2018). Acceptance of building information modelling: A survey of professionals in the construction industry in Ghana. *Journal of Information Technology in Construction, 23*, 75–91.

Ahuja, R., Sawhney, A., Jain, M., Arif, M., & Rakshit, S. (2020). Factors influencing BIM adoption in emerging markets – The case of India. *International Journal of Construction Management, 20*(1), 65–76.

Akintola, A., Venkatachalam, S., & Root, D. (2020). Understanding BIM's impact on professional work practices using activity theory. *Construction Management and Economics, 38*(5), 447–467.

Ajzen, I. (1991). The theory of planned behavior. *Organizational Behavior and Human Decision Processes, 50*(2), 179–211.

Almuntaser, T., Sanni-Anibire, M. O., & Hassanain, M. A. (2018). Adoption and implementation of BIM – Case study of a Saudi Arabian AEC firm. *International Journal of Managing Projects in Business, 11*(3), 608–624.

Ahn, Y. H., Kwak, Y. H., & Suk, S. J. (2016). Contractors' transformation strategies for adopting building information modeling. *Journal of Management in Engineering, 32*(1), 05015005.

Arayici, Y., Coates, P., Koskela, L., Kagioglou, M., Usher, C., & O'Reilly, K. (2011). BIM adoption and implementation for architectural practices. *Structural Survey, 29*(1), 7–25.

Bearman, M., Smith, C. D., Carbone, A., Slade, S., Baik, C., Hughes-Warrington, M., & Neumann, D. L. (2012). Systematic review methodology in higher education. *Higher Education Research & Development, 31*(5), 625–640.

Brunet, M., Motamedi, A., Guénette, L.-M., & Forgues, D. (2019). Analysis of BIM use for asset management in three public organizations in Québec, Canada article information: BIM in facilities management applications: A case study of a large university. *Built Environment Project and Asset Management, 9*(1), 261–277.

Cao, D., Li, H., Wang, G., & Huang, T. (2017). Identifying and contextualising the motivations for BIM implementation in construction projects: An empirical study in China. *International Journal of Project Management, 35*(4), 658–669.

Cavka, H. B., Staub-French, S., & Pottinger, R. (2015). Evaluating the alignment of organizational and project contexts for BIM adoption: a case study of a large owner organization. *Buildings, 5*(4), 1265–1300.

Chan, I. Y. S., Liu, A. M. M., & Chen, B. (2018). Management strategies for 5D-BIM adoption in Hong Kong. In *Proceedings of the 21st international symposium on advancement of construction management and real estate* (pp. 1023–1039).

Chen, Y., Yin, Y., Browne, G. J., & Li, D. (2019). Adoption of building information modeling in Chinese construction industry: The technology-organization-environment framework. Engineering, construction and architectural management.

Chunduri, S., Kreider, R., & Messner, J. I. (2013). A case study implementation of the BIM planning procedures for facility owners. In *AEI 2013: Building solutions for architectural engineering* (pp. 691–701).

Daller, J., Žibert, M., Exinger, C., & Lah, M. (2016). Implementation of BIM in the tunnel design–Engineering consultant's aspect. *Geomechanics and Tunnelling, 9*(6), 674–683.

Davis, F. D. (1989). Perceived usefulness, perceived ease of use, and user acceptance of information. *Management Information Systems Research Center, University of Minnesota, 13*(3), 319–340.

DiMaggio, P. J., & Powell, W. W. (1983). The iron cage revisited: Institutional isomorphism and collective rationality in organizational fields. *American Sociological Review, 48*, 147–160.

Ding, Z., Zuo, J., Wu, J., & Wang, J. Y. (2015). Key factors for the BIM adoption by architects: A China study. *Engineering, Construction & Architectural Management, 22*(6), 732–748.

Doan, D. T., Ghaffarian Hoseini, A., Naismith, N., Ghaffarianhoseini, A., Zhang, T., & Tookey, J. (2021). Examining critical perspectives on building information modelling (BIM) adoption in New Zealand. *Smart and Sustainable Built Environment, 10*(4), 594–615.

Eadie, R., Browne, M., Odeyinka, H., Mckeown, C., & Mcniff, S. (2015). A survey of current status of and perceived changes required for BIM adoption in the UK. *Built Environment Project and Asset Management, 5*(1), 4.

Fishbein, M., & Ajzen, I. (1975). *Belief, attitude, intention, and behavior: An introduction to theory and research*. Addison-Wesley.

Gao, X., & Pishdad-Bozorgi, P. (2019). BIM-enabled facilities operation and maintenance: A review. *Advanced Engineering Informatics, 39*, 227–247.

Gledson, B. J., & Greenwood, D. (2017). The adoption of 4D BIM in the UK construction industry: An innovation diffusion approach. *Engineering, Construction and Architectural Management, 24*(6), 950–967.

Goodhue, D. L., & Thompson, R. L. (1995). Task-technology fit and individual performance. *MIS Quarterly, 19*(2), 213–236.

Gu, N., & London, K. (2010). Understanding and facilitating BIM adoption in the AEC industry. *Automation in Construction, 19*(8), 988–999.

Gurevich, U., Sacks, R., & Shrestha, P. (2017). BIM adoption by public facility agencies: Impacts on occupant value. *Building Research and Information, 45*(6), 610–630.

Harty, C., Throssell, D., Jeffrey, H., & Stagg, M. (2010). Implementing building information modeling: a case study of the Barts and the London hospitals. In *Proceedings of the International Conference on Computing in Civil and Building Engineering*. Nottingham (UK): ICCBE.

Hochscheid, E., & Halin, G. (2018, September). BIM implementation in architecture firms interviews, case studies and action research used to build a method that facilitates implementation of BIM processes and tools. In *Proceedings of the 36th eCAADe annual conference*, Lodz, Poland.

Hong, Y., Sepasgozar, S. M. E., Ahmadian, A. F. F., & Akbarnezhad, A. (2016). Factors influencing BIM adoption in small and medium sized construction organizations. In *33rd international symposium on automation and robotics in construction* (pp. 452–461).

Hosseini, M. R., Chileshe, N., Zuo, J., & Baroudi, B. (2015). Adopting global virtual engineering teams in AEC projects: A qualitative meta-analysis of innovation diffusion studies. *Construction Innovation, 15*(2), 151–179.

Juan, Y.-K., Lai, W.-Y., & Shih, S.-G. (2017). Building information modeling acceptance and readiness assessment in Taiwanese architectural firms. *Journal of Civil Engineering and Management, 23*(3), 356–367.

Kale, S., & Arditi, D. (2005). Diffusion of computer aided design technology in architectural design practice. *Journal of Construction Engineering and Management, 131*(10), 1135–1141.

Khosrowshahi, F., & Arayici, Y. (2012). Roadmap for implementation of BIM in the UK construction industry. *Engineering, Construction and Architectural Management, 19*(6), 610–635.

Kim, S., Park, C. H., & Chin, S. (2016). Assessment of BIM acceptance degree of Korean AEC participants. *KSCE Journal of Civil Engineering, 20*(4), 1163–1177.

Kouch, A. M. (2018). A three-step BIM implementation framework for the SME contractors. In *IFIP international conference on product lifecycle management* (pp. 15–24). Springer.

Kraus, S., Breier, M., & Dasí-Rodríguez, S. (2020). The art of crafting a systematic literature review in entrepreneurship research. *International Entrepreneurship and Management Journal, 16*(3), 1023–1042.

Lee, S., Yu, J., & Jeong, D. (2015). BIM acceptance model in construction organizations. *Journal of Management in Engineering, 31*(3), 04014048.

Lesniak, A., Gorka, M., & Skrzypczak, I. (2021). Barriers to BIM implementation in architecture, construction, and engineering projects-the polish study. *Energies, 14*(8), 2090.

Liao, L., & Teo, E. A. L. (2019). Managing critical drivers for building information modelling implementation in the Singapore construction industry: An organizational change perspective. *International Journal of Construction Management, 9*(3), 240–256.

Lindblad, H., & Guerrero, J. R. (2020). Client's role in promoting BIM implementation and innovation in construction. *Construction Management and Economics, 38*(5), 468–482.

Ma, G., Jia, J., Ding, J., Shang, S., & Jiang, S. (2019). Interpretive structural model-based factor analysis of BIM adoption in Chinese construction organizations. *Sustainability, 11*, 1982.

Ngowtanasawan, G. (2017). A causal model of BIM adoption in the Thai architectural and engineering design industry. *Procedia Engineering, 180*, 793–803.

Nnaji, C., Okpala, I., & Kim, S. (2019). A simulation framework for technology adoption decision making in construction management: A composite model. In *Computing in civil engineering 2019*.

Okakpu, A., Ghaffarian Hoseini, A., Tookey, J., Haar, J., & Ghaffarianhoseini, A. (2020). Exploring the environmental influence on BIM adoption for refurbishment project using structural equation modelling. *Architectural Engineering and Design Management, 16*(1), 41–57.

Pruskova, K., & Kaiser, J. (2019). Implementation of BIM technology into the design process using the scheme of BIM execution plan. *IOP Conference Series: Materials Science and Engineering, 471*(2), 022019.

Rogers, E. M. (1983). Diffusion of innovations. In *Free Press* (3rd ed.).

Rogers, J., Chong, H.-Y. Y., & Preece, C. (2015). Adoption of building information modelling technology (BIM): Perspectives from Malaysian engineering consulting services firms. *Engineering, Construction and Architectural Management, 22*(4), 424–445.

Roorda, D., & Liu, M. K. (2008). Implementation of building information modeling (BIM) on the renovation of the art gallery of Alberta in Edmonton, Alberta. In *Structures congress 2008: 18th analysis and computation specialty conference* (pp. 1–15).

Rowlinson, S., Collins, R., Tuuli, M. M., Jia, Y., Lu, J. W. Z., Leung, A. Y. T., & Mok, K. M. (2010). Implementation of building information modeling (BIM) in construction: A comparative case study. *AIP Conference Proceedings, 1233*(1), 572–577.

Shibeika, A., & Harty, C. (2015). Diffusion of digital innovation in construction: a case study of a UK engineering firm. *Construction Management and Economics, 33*(5-6), 453–466.

Siebelink, S., Voordijk, J. T., & Adriaanse, A. (2018). Developing and testing a tool to evaluate BIM maturity: Sectoral analysis in the Dutch construction industry. *Journal of Construction Engineering and Management, 144*(8), 05018007.

Singh, V., & Holmstrom, J. (2015). Needs and technology adoption: Observation from BIM experience. *Engineering, Construction & Architectural Management, 22*(2), 128–150.

Son, H., Lee, S., Hwang, N., & Kim, C. (2014). The adoption of building information modeling in the design organization: An empirical study of architects in Korean design firms. In ISARC. Proceedings of the International Symposium on Automation and Robotics in Construction (Vol. 31, p. 1). IAARC Publications.

Succar, B., & Kassem, M. (2015). Macro-BIM adoption: Conceptual structures. *Automation in Construction, 57*, 64–79.

Tong, N., & Phung, Q. (2021). Developing an organizational readiness framework for BIM implementation in large design companies. *International Journal of Sustainable Construction Engineering and Technology, 12*(3), 57–67.

Tornatzky, L. G., Fleischer, M., & Chakrabarti, A. K. (1990). *The process of technological innovation*. Lexington Books.

Ullah, K., Lill, I., & Witt, E. (2019). An overview of BIM adoption in the construction industry: Benefits and barriers. In *10th Nordic conference on construction economics and organization*. Emerald Publishing Limited.

Umar, T. (2021). *Challenges of BIM implementation in GCC construction industry*. Engineering, Construction and Architectural Management.

Vass, S., & Gustavsson, T. K. (2017). Challenges when implementing BIM for industry change. *Construction Management & Economics, 35*(10), 597–610. https://doi.org/10.1080/01446193.2017.1314519

Vidalakis, C., Abanda, F. H., & Oti, A. H. (2020). BIM adoption and implementation: Focusing on SMEs. *Construction Innovation, 20*(1), 128–147.

Whitlock, K., & Abanda, H. F. (2020). Making a business case for BIM adoption. In *Lecture notes in networks and systems* (Vol. 72, pp. 231–246).

Chapter 17
Defining Requirements for a BIM-Enabled Learning Environment

Emlyn Witt, Kalle Kähkönen, and Marco Alvise Bragadin

Abstract Digitalization is transforming the real estate and construction (REC) sector and a key feature of this transformation is Building Information Modelling (BIM). BIM offers opportunities for improving education and training through data-rich virtual environments in which project-based learning experiences can be designed and delivered. This could lead to considerable changes and improvements to the education and training of different professionals in the REC sector from managers to site workers. Researchers at Tallinn University of Technology, Tampere University, and the University of Bologna are currently developing a BIM-enabled Learning Environment (BLE) with the intention of providing more realistic, immersive, and integrated learning experiences. A desk study identifying existing initiatives, mapping the educational-technological systems in the partner universities, and analyzing existing BIM-enabled learning experiences at each partner university revealed an initial list of functional requirements for a BLE. This list was elaborated through 31 interviews of REC stakeholders in the three partner countries. A validation workshop was then held to confirm the relevance of the identified functional requirements, rank them in terms of their relative importance and identify any missing requirements. The resulting, validated functional requirements were then presented to technical experts at each of the universities for recommendations on how these functions could be technically delivered in the context of a BLE. Analysis of their feedback suggests a system design comprising a host platform based on a

E. Witt (✉)
Department of Civil Engineering and Architecture, Tallinn University of Technology, Tallinn, Estonia
e-mail: emlyn.witt@taltech.ee

K. Kähkönen
Faculty of Built Environment, Tampere University, Tampere, Finland
e-mail: kalle.kahkonen@tuni.fi

M. A. Bragadin
Department of Architecture, University of Bologna, Bologna, Italy
e-mail: marcoalvise.bragadin@unibo.it

G. Lindahl, S. C. Gottlieb (eds.), *SDGs in Construction Economics and Organization*, Springer Proceedings in Business and Economics,
https://doi.org/10.1007/978-3-031-25498-7_17

Learning Management System (LMS) installation with additional collaboration tools and further integrations with (open) BIM solutions to enable course-specific BIM functionalities. This research relates to Sustainable Development Goals (SDGs) 4-Quality Education, as the BLE is intended to enhance education and training possibilities in the REC sector and 11-Sustainable Cities and Communities, as the result of improved REC education and training will lead to a more sustainable Built Environment.

Keywords Building Information Modelling (BIM) · BIM-enabled learning · Virtual learning environment · Real estate and construction education

17.1 Introduction

17.1.1 Background to the BIM-Enabled Learning Environment

Digitalization is transforming the real estate and construction sector and a key feature of this transformation is building information modeling (BIM) – the virtual representation of all building-related information. By enabling the creation of digital data equivalents (digital twins) of real buildings, BIM generates opportunities to do many things in new and better ways including education and training. Addressing the current challenge of ensuring that graduates' competencies are properly aligned with the emerging, digitalized REC industry (Du et al., 2017; Hwang & Safa, 2017; Tranquillo et al., 2018) is one such opportunity as is the expected improvement in students' motivation, satisfaction, and academic and professional performance when their learning is mediated through technological innovations such as BIM (Ferrandiz et al., 2018).

Interest in BIM-based education has been growing over the last decade or so with most study programs incorporating BIM awareness creation and educating students and trainees on how to use different BIM-related software packages (e.g., Revit, ArchiCAD, etc.) for model viewing, creation, viewing, simulating, data sharing or coordination (e.g., Ren & Zhang, 2014; Comiskey et al., 2017). Courses typically highlight the benefits and barriers of BIM and the reasons for BIM adoption in the industry (e.g., Ahn et al., 2013; Puolitaival & Forsythe, 2016), and progress to BIM knowledge and authoring/manipulation skills (e.g., Ghosh et al., 2015). Beyond developing BIM software skills, BIM technology has also been used to impart other learning such as coordination, collaboration, communication, and interpersonal relationships among students, etc. (e.g., see Barham et al., 2011; Bozoglu, 2016; El Zomor et al., 2018; Zhao et al., 2015; Ghosh et al., 2013).

Underwood et al. (2013) categorized the evolution of BIM education into three progressive stages:

1. BIM-aware – where graduates are made aware of the uses and exigencies of BIM relating to its implications for both digital and cultural transformation of the construction industry.
2. BIM-focused – involving graduates' abilities to use and manipulate BIM software in performing specific tasks such as modeling, clash detection, simulation, etc.
3. BIM-enabled – where education takes place in a BIM-mediated virtual environment and BIM acts as a platform for learning (Underwood et al., 2013).

Both BIM-aware and BIM-focused education have become widespread while BIM-enabled education is still in an early stage of development. Accounts of extant BIM education cases have been documented in Abdirad and Dossick (2016) and, more recently, in Olowa et al. (2020).

17.1.2 Existing BIM-Enabled Learning Environments

BIM-enabled learning environments have been experimented with over the past decade. Ku and Mahabaleshwarkar (2011) proposed a BIM interactive model (BiM) – a platform combining a virtual environment (Second Life) with BIM for learning purposes and proposed a theoretical web-based virtual world for engaging construction stakeholders in real-time social interaction. Their intention is to supplement construction education with intelligent 3D BIM models in order to overcome the limitation of location-based learning and make it accessible to anyone with an internet connection. Shen et al. (2012) used the 3D-UNITY game engine to create a web-based training environment for HVAC rehabilitation and improvement using a BIM model. In contrast to Ku and Mahabaleshwarkar (2011) BiM proposal, Shen et al. (2012) argued that game engines have been sufficiently developed for BIM interoperability thereby making game creation cheaper and easier and demonstrating how BIM could be leveraged for teaching. More recently, Zamora-Polo et al. (2019) have argued that, for the education of engineers, BIM can be considered a Virtual Learning Environment (VLE).

17.1.3 Purpose of This Research

Researchers at Tallinn University of Technology, Tampere University, and the University of Bologna are currently developing a prototype BIM-enabled Learning Environment (BLE) with the intention of providing more realistic, immersive, and integrated learning experiences. In addition to the BLE platform itself, pilot learning modules are being created to demonstrate the potential for this approach and, to determine their effectiveness, evaluation tools for these modules need to be designed. This research investigates existing, applicable evaluation models and derives an evaluation model and tools specifically adapted for the immersive project-based learning experiences provided through the BLE. The BLE is intended to be:

- Open – supports participation in BIM-enabled learning regardless of software tools being used
- Compatible – BLE must be compatible with different Learning Management Systems and with other systems, approaches, and requirements for wide-spread use
- Accessible – both in the sense of:
  - Enabling public access including the possibility of hosting BIM-enabled MOOCs
  - User-friendly and easy to use so that BIM-enabled learning can be made available to a wide range of learners and teachers
- Replicable – fully documented, available, complete, a detailed specification which can be used by others to replicate the BLE or develop alternatives to it
- Inclusive – addressing the learning needs of all professions in the REC sector
- Scalable – for sector-wide use in the REC sector

The primary target of this research is to provide a new BIM-based stepping-stone, that is, the BLE solution for the education of various professionals in the REC sector. For this purpose, the main attention is on the use of digital models rather than in their creation. Digital models with their data have already developed and become part of all REC professionals' working infrastructure. Whether you are project manager, cost engineer, site inspector, or installer (just to mention examples of several professions), the skills for taking advantage of digital models in your work are of prime importance. Education and training need to be transformed accordingly.

In the following section of the paper, the research methodology is described in terms of the data collection and analysis process steps and their logic. The findings at each of these steps are then presented and discussed before a proposed BLE system concept is outlined. Conclusions are then drawn and related to the Sustainable Development Goals (SDGs).

17.2 Methodology

A desk study of related initiatives and applicable standards relevant to the proposed BLE was first undertaken. An exercise to identify and compare the existing platforms and systems in use at the three partner universities was then carried out to ascertain the education-technological systems contexts into which the BLE should integrate.

An initial list of functional requirements was compiled by analyzing existing courses or learning experiences which had elements of BIM-enabled learning (one such case study being analyzed from each partner university). The functional requirements were further elaborated through a series of 31 interviews of

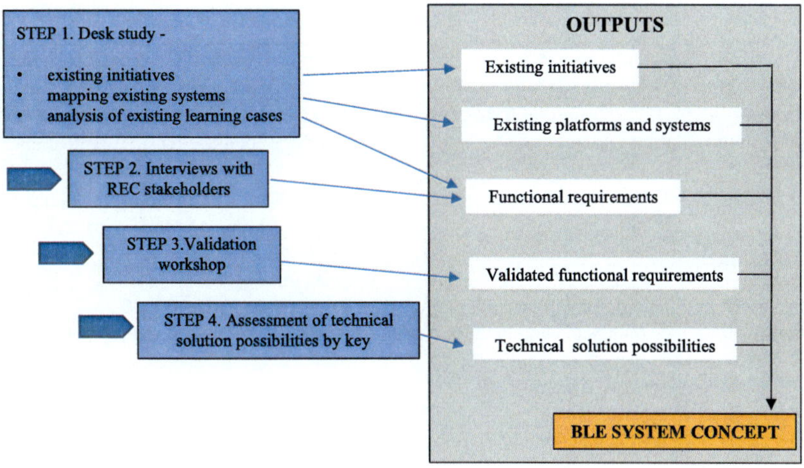

Fig. 17.1 Methodology for defining BIM-enabled learning environment requirements

stakeholders connected to education and training throughout the REC sector and in three countries (Estonia, Finland, and Italy).

The identified functional requirements were then subjected to a validation exercise in the form of an online workshop where the functional requirements were presented and participants were asked to rank them in terms of their relative importance and to add any further, missing requirements.

The list of validated functional requirements was then assessed by key informants in each country who commented on how these functional requirements could be technically delivered. The feedback from these key informants was compiled and an emergent technical solution in the form of a system concept design for the BLE was proposed.

These process steps, their logic and outputs are shown in Fig. 17.1.

17.3 Findings

17.3.1 Existing Educational-Technological Systems

A mapping exercise was conducted to identify the current educational-technological systems in use within the three partner universities. The intention was to establish commonalities and differences between institutional system arrangements and it was found that all three universities were using Moodle as a learning management system in addition to other systems with varying levels of integration between them. Table 17.1 summarizes the systems that were identified and investigated.

Notably, Moodle installations are relied on to a significant degree and some integrations exist between these Moodle systems and other, complementary systems in certain cases (e.g., with the BigBlueButton plugin in TalTech Moodle and with the

Table 17.1 Existing educational-technological systems in the partner universities

System type	Tallinn University of Technology (TalTech)	Tampere University (TAU)	University of Bologna (UNIBO)
Study information systems	OIS2 (custom-built study information system)	SISU (custom-built study information system)	SOL – Studenti Online (custom-built study information system)
LMS/VLE	TalTech Moodle (customized Moodle installation)	TAU Moodle (customized Moodle installation)	Virtuale (customized Moodle installation)
Other educational systems	KnowledgeHUB (project-specific Moodle installation – this project is led by the University of Huddersfield)	EXAM (Electronic Exam Platform)	BOOK (uniBO Open Knowledge) MOOC platform e-CLA (customized Moodle installation) AlmaEsami (Electronic exam platform – only for management purposes) EOL – Esami On-Line (customized Moodle Installation for online exams)
Video conferencing	MS Teams Big Blue Button (integrated with TalTech Moodle) WORKSUP (virtual conference platform developed at TalTech) Zoom	MS Teams Zoom	MS Teams Zoom
Other systems	–	Panopto (video sharing) –integrated (plugin) with TAU Moodle Kahoot! (game-based learning platform)	-Panopto (video sharing) – integrated with Virtuale

Panopto plugin in TAU and Unibo Moodle). In addition, custom, project-specific Moodle installations (e.g., KnowledgeHUB) have been set up in order to facilitate inter-university collaborations which enable broader usership beyond universities' own students and staff.

17.3.2 Initial Functional Requirements

The three partner universities each selected an existing learning experience that was either a direct example of BIM-enabled learning or involved closely related learning activities. The learning activities selected were all aimed at Masters students within Architecture, Engineering, and Construction (AEC) study programs as follows.

The BIM-enabled Cash Flow Exercise is a learning experience in which students develop, analyze, and optimize company cash flows in the context of a construction project scenario. This exercise had been piloted for 2 years within a Construction Investments course at Tallinn University of Technology (TalTech).

The Simulation of the Building Design Process course has been implemented since 2006 at Tampere University. For this course, the participating students from different disciplines form a design group that consists of students from architecture, structural engineering, earth and foundation structures, HVAC, engineering, electrical engineering, and construction management and economics. Students' task is designing a building for the predefined need. Each student has a specific role in meeting his/her major subject of study.

The BIM-enabled Construction Job Site Layout Exercise requires students to develop, analyze, and optimize a building job site layout within a construction project context. This exercise had been piloted for 1 year within the Building Sites and Production course at the University of Bologna (UNIBO).

From the analysis of the teaching and learning activities involved in each of the three cases above, an initial list of 19 capabilities that would be required from a BLE to host and enable these activities was derived and their relative importance to performing the desired activities was estimated as shown in Table 17.2.

17.3.3 Validated Functional Requirements

Thirty-one semi-structured online interviews were carried out (12 in Italy, 10 in Estonia, and 9 in Finland) in the spring of 2021 among REC stakeholders involved in education and/or training. The interviewees' affiliations were:

- 15 affiliated with Higher Education Institutions
- 14 affiliated with the Construction (AEC) Industry
- 2 affiliated with the Real Estate/Facilities Management Industry

In addition to describing their current teaching/training activities, interviewees were played a short video explaining the general concept of a BIM-enabled Learning Environment and asked to identify functions/capabilities that they would expect such a learning environment to have. A thematic analysis of their responses, together with the initial functionalities identified from the study of three existing BIM-enabled learning activities (described above), led to the compilation of a list of 30 functional requirements. This list was subsequently the focus of a (virtual) validation workshop held in June 2021 at which BIM educators and others interested in BIM-enabled education and training from five-countries (Estonia, Finland, Italy, Nigeria, and Sri Lanka) participated. Workshop participants were asked to validate each of the proposed BLE functional requirements by rating each on their level of importance with a five-point Likert-type scale (1: not important, 2: slightly important, 3: moderately important, 4: very important, 5: critically important). They were also asked to identify any gaps (or missing functions) in the list. A detailed analysis

Table 17.2 BLE functional requirements – initial list

#	Requirement description	Importance
1	The upload, storage, download, sharing, and viewing of typical file types – documents, spreadsheets, etc.	Critical
2	Collaborative (in groups) viewing and editing of typical file types – documents, spreadsheets, presentation slides, etc.	Essential
3	BIM model viewing, sharing, editing, and data extraction	Critical
4	Collaborative (in groups) viewing and editing of BIM models	Essential
5	A Common Data Environment for project data (for project-based learning scenarios)	Essential
6	Repository for BIM models and associated project data for learning	Critical
7	Student group collaboration functionality – group formation, group communications, collaborative viewing and editing of files, whiteboards, etc. Instructor access to groups for advising, monitoring and assessing, instructor options to make changes to groups	Essential
8	Between groups collaboration functionality – communications between groups, collaborative viewing and editing of files, whiteboards, etc.	Recommended
9	Questionnaire creation, completion, submission, and analysis	Recommended
10	Student feedback functionality	Essential
11	Assessment/grading functions – grade entering for individuals/groups, grade book	Critical
12	Collaborative BIM model viewing and sharing	Critical
13	Collaborative BIM model checking	Critical
14	Collaborative BIM model editing	Essential
15	BIM model information take off	Recommended
16	Project team collaboration functionality – communications with project team and collaborative viewing of files	Recommended

of the interviews, validation workshop, and emergent functional requirements are reported in a journal article that is currently in press: Olowa et al. (2021). The emerging list of validated functional requirements is shown in Table 17.3.

The average ratings received indicate that even the lowest ranked functional requirement ("Recording of group sessions and lessons") was accorded an average rating higher than "3: moderately important" and this attests to the general importance and, therefore, validity of the identified functional requirements.

17.3.4 Assessment of Technical Solution Possibilities

Key informants (technical experts in BIM and/or educational technology systems) from each of the partner universities were presented with the list of validated functional requirements and asked to give recommendations of how these functionalities could be delivered through a BLE. Their recommendations are summarized in Table 17.4.

No technical solutions were offered for:

Table 17.3 BLE functional requirements – validated and importance ranked list

Importance		
Rank	Ave. rating	Functional requirement
1	4.6	Student feedback
2	4.5	BIM model data extraction
2	4.5	Simulation of the project development process (realistic BIM workflow, key stakeholder roles, etc.)
2	4.5	Collaboration in groups
2	4.5	Links between courses (to build on previous courses' results and to track impacts on/inputs to future courses)
2	4.5	Integration of platform with external systems/business
7	4.4	BIM model viewing
7	4.4	BIM model sharing
7	4.4	Common Data Environment (CDE) for project data
7	4.4	Data security/password protection
11	4.3	BIM model collaborative viewing and editing
11	4.3	File upload, storage, download, sharing, editing
13	4.2	Instructor access and monitoring of groups and group work
13	4.2	Hosting of different courses
15	4.1	BIM model checking
16	4.0	Collaborative viewing and editing of documents and spreadsheets
16	4.0	Questionnaire creation, completion, submission, and analysis
16	4.0	Video playback
19	3.9	Repository of example BIM models
20	3.8	BIM model editing
20	3.8	Collaboration between groups
20	3.8	Live interactions between users
20	3.8	Gamification support functions
24	3.7	Group formation
24	3.7	Registration of users (learners/instructors)
26	3.6	BIM model creating
27	3.5	Extended reality (XR) functions: AR/MR/VR
27	3.5	Assessment/grading functions – grade entering for individuals/groups, grade book
29	3.4	BIM model version management
30	3.3	Recording of group sessions and lessons
[a]		BIM object creating and editing
[a]		Linking to extra learning materials
[a]		Individual learners' storage for learning materials

[a]Suggestions of additional functions received at the validation workshop

Table 17.4 Technical solutions possibilities

Technical solution type	Functional requirements which could be delivered using this technical solution [Square brackets indicate that more than one type of technical solution was suggested]
Learning Management Systems (LMS), e.g., Moodle, with integrated learning tools (using the LTI standard)	Hosting of different courses
	Registration of users (learners/instructors)[a]
	Data security/password protection[a]
	Linking to extra learning materials
	Individual learners' storage for learning materials
	Questionnaire creation, completion, submission, and analysis
	Assessment/grading functions – grade entering for individuals/groups, grade book
	Student feedback
	Group formation
	Video playback
	Gamification support functions
	[File upload, storage, download, sharing, editing]
	[Instructor access and monitoring of groups and group work]
	[Collaboration in groups]
	[Collaboration between groups]
	[Live interactions between users]
	[Recording of group sessions and lessons]
	[BIM model sharing]
	[Repository of example BIM models]
Collaboration platforms with integrated applications (e.g., Office 365, Google for Education)	[File upload, storage, download, sharing, editing]
	[Instructor access and monitoring of groups and group work]
	[Collaboration in groups]
	[Collaboration between groups]
	[Live interactions between users]
	[Recording of group sessions and lessons]
	Collaborative viewing and editing of documents and spreadsheets
	[Common Data Environment (CDE) for project data]
Integrations with specific BIM solutions	[BIM model sharing]
	[Repository of example BIM models]
	[Common Data Environment (CDE) for project data]
	BIM model viewing
	BIM model data extraction
	BIM model version management
	BIM model checking
	BIM model editing
	BIM model collaborative viewing and editing
	BIM model creating
	BIM object creation and editing
	Extended reality (XR) functions: Augmented Reality (AR)/Mixed Reality (MR)/Virtual Reality (VR)

[a]eduGAIN offers a solution to allowing widespread access to the BLE for the global research and education community

- Simulation of the project development process (realistic BIM workflow, key stakeholder roles, etc.)
- Links between courses (to build on previous courses' results and to track impacts on/inputs to future courses)
- Integration of platform with external systems/business

These were acknowledged as potentially useful features but, in all cases, potential technical solutions for them would need to be considered once they were more clearly defined.

17.4 Discussion

The identified functional requirements can be separated into three general categories:

1. Learning Management System (LMS) functions
2. Collaboration functions
3. BIM functions

These are clearly reflected in the potential technical solutions offered which reference the use of LMS platforms, collaboration platforms, and BIM solutions. Some functional requirements (e.g., "Collaboration in groups"), depending on how, precisely, they are defined, could be delivered using more than one of these possibilities.

These findings suggest that the proposed BLE could be achieved on the basis of a host platform that would allow:

- User registrations to be authenticated and managed
- Hosting of course materials
- Assessment, grading, and feedback
- Collaboration between learners, between learners and instructors, and between groups

Depending on individual course needs, further integrations between this platform and specific BIM solutions would extend the capabilities available to the hosted courses. In addition, for a wider, global usership, the possibility of utilizing the eduGAIN system for authenticating and managing users is available. Figure 17.2 illustrates this conceptualization of a BLE system design.

17.5 Conclusions

The need for BIM-enabled education and training in the REC sector has been established and, to facilitate this, researchers at Tallinn University of Technology, Tampere University, and the University of Bologna are currently developing a BIM-Enabled Learning Environment (BLE) to provide more realistic, immersive, and

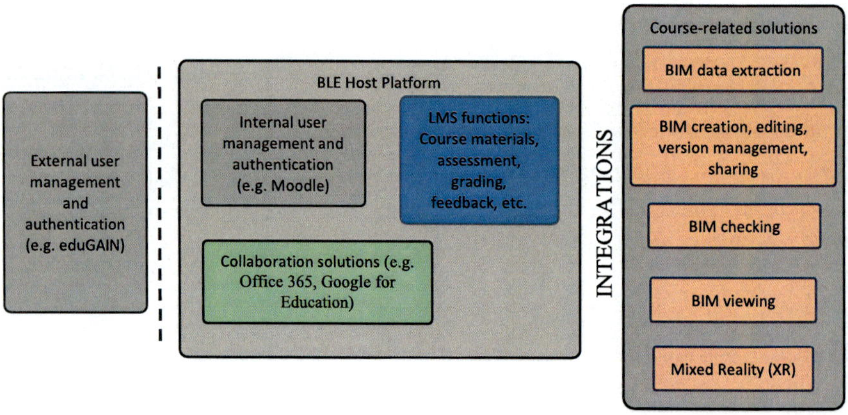

Fig. 17.2 BIM-enabled learning environment system design concept

integrated learning experiences. This research has reported the derivation of a validated list of the functional requirements of the proposed BLE and, based on these, an initial concept of technical solutions and a system design concept to deliver them.

The system design comprises a host platform based on an LMS installation (e.g., Moodle) with additional collaboration tools and further integrations with (open) BIM solutions to enable course-specific BIM functionalities. Within the framework of the current research project, a demonstration version of the BLE is currently being developed with the intention of enabling proof of concept, and pilot courses are being developed for this purpose. The extent of the BIM functionalities that can successfully be integrated with the BLE is, at this stage, unknown but the intention is to achieve some key integrations (enabling BIM viewing, data extraction from BIM models, and more) in order to demonstrate the value of the BLE as an education and training platform. All information and documentation will be made available so that other research teams are able to replicate and take forward these developments.

This research relates to Sustainable Development Goals (SDGs) 4-Quality Education, as the BLE is intended to enhance education and training possibilities in the REC sector and 11-Sustainable Cities and Communities, as the result of improved REC education and training will lead to a more sustainable Built Environment.

Acknowledgments This research was supported by the BIM-enabled Learning Environment for Digital Construction (BENEDICT) project (grant number: 2020-1-EE01-KA203-077993) co-funded by the Erasmus+ Programme of the European Union. The European Commission's support to produce this publication does not constitute an endorsement of the contents which reflect the views only of the authors, and the Commission cannot be held responsible for any use which may be made of the information contained therein.

References

Abdirad, H., & Dossick, C. S. (2016). BIM curriculum design in architecture, engineering, and construction education: A systematic review. *Journal of Information Technology in Construction, 21*, 250–271. http://www.itcon.org/2016/17

Ahn, Y. H., Cho, C.-S., & Lee, N. (2013). Building information modeling: Systematic course development for undergraduate construction students. *Journal of Professional Issues in Engineering Education and Practice, 139*(4), 290–300. https://doi.org/10.1061/(ASCE)EI.1943-5541.0000164

Barham, W., Meadati, P., & Irizarry, J. (2011). Enhancing student learning in structures courses with building information modeling. In *Congress on Computing in Civil Engineering: Proceedings* (pp. 850–857). American Society of Civil Engineers. http://ascelibrary.org/doi/10.1061/41182%28416%29105. Accessed February 8, 2019

Bozoglu, J. (2016). Collaboration and coordination learning modules for BIM education. *Journal of Information Technology in Construction, 21*, 152–163.

Comiskey, D., McKane, M., Jaffrey, A., Wilson, P., & Mordue, S. (2017). An analysis of data sharing platforms in multidisciplinary education. *Architectural Engineering and Design Management, 13*(4), 244–261. https://www.tandfonline.com/doi/full/10.1080/17452007.2017.1306483. Accessed February 15, 2019

Du, J., Zou, Z., Shi, Y., & Zhao, D. (2017). Simultaneous data exchange between BIM and VR for collaborative decision making. In *Computing in Civil Engineering 2017* (pp. 1–8). American Society of Civil Engineers. http://ascelibrary.org/doi/10.1061/9780784480830.001. Accessed February 13, 2019

El Zomor, M., Mann, C., Doten-Snitker, K., Parrish, K., & Chester, M. (2018). Leveraging vertically integrated courses and problem-based learning to improve students' performance and skills. *Journal of Professional Issues in Engineering Education and Practice, 144*(4), 04018009. http://ascelibrary.org/doi/10.1061/%28ASCE%29EI.1943-5541.0000379. Accessed February 8, 2019

Ferrandiz, J., Banawi, A., & Peña, E. (2018). Evaluating the benefits of introducing "BIM" based on Revit in construction courses, without changing the course schedule. *Universal Access in the Information Society, 17*(3), 491–501. http://link.springer.com/10.1007/s10209-017-0558-4

Ghosh, A., Parrish, K., & Chasey, A. D. (2013). From BIM to collaboration: A proposed integrated construction curriculum. In *ASEE annual conference and exposition, conference proceedings*. https://www.scopus.com/record/display.uri?eid=2-s2.0-84884339291&origin=resultslist&sort=r-f&src=s&nlo=&nlr=&nls=&sid=cd7563182192ef5ebc090f14d762cb3e&sot=a&sdt=cl&cluster=scolang%252c%2522English%2522%252ct%252bscopubyr%252c%25222018%2522%252ct%252c%2522

Ghosh, A., Parrish, K., & Chasey, A. D. (2015). Implementing a vertically integrated BIM curriculum in an undergraduate construction management program. *International Journal of Construction Education and Research, 11*(2), 121–139.

Hwang, S., & Safa, M. (2017). Learning advanced decision-making techniques and technologies through a collaborative project. In K.-Y. Lin, N. M. El-Gohary, & P. Tang (Eds.), *Computing in Civil Engineering 2017* (pp. 35–42). American Society of Civil Engineers. https://ascelibrary.org/doi/pdf/10.1061/9780784480830.005

Ku, K., & Mahabaleshwarkar, P. S. (2011). Building interactive modeling for construction education in virtual worlds. *Journal of Information Technology in Construction, 16*, 189–208. https://www.scopus.com/record/display.uri?eid=2-s2.0-79954551580&origin=resultslist&sort=r-f&src=s&st1=%28%22Education%22+OR+%22Teaching%22+OR+%22Training%22%29+AND+%28%22Building+Information+Modeling%22+OR+%22Building+Information+Modelling%22+OR+%22Virtu

Olowa, T., Witt, E., & Lill, I. (2020). Conceptualising building information modelling for construction education. *Journal of Civil Engineering and Management, 26*(6), 551–563. https://doi.org/10.3846/jcem.2020.12918. Accessed November 7, 2020.

Olowa, T., Witt, E., & Lill, I. (2021). Building information modelling (BIM) – Enabled construction education: Teaching project cash flow concepts. *International Journal of Construction Management*. https://doi.org/10.1080/15623599.2021.1979300

Puolitaival, T., & Forsythe, P. (2016). Practical challenges of BIM education. *Structural Survey, 34*(4/5), 351–366. http://www.emeraldinsight.com/doi/10.1108/SS-12-2015-0053. Accessed February 7, 2019

Ren, S., & Zhang, W. (2014). Application of BIM software in construction design education. *World Transactions on Engineering and Technology Education, 12*(3), 432–436. https://www.scopus.com/record/display.uri?eid=2-s2.0-84916237291&origin=resultslist&sort=r-f&src=s&nlo=&nlr=&nls=&sid=cd7563182192ef5ebc090f14d762cb3e&sot=a&sdt=cl&cluster=scolang%252c%2522English%2522%252ct%252bscopubyr%252c%25222018%2522%252ct%252c%2522

Shen, Z., Jiang, L., Grosskopf, K., & Berryman, C. (2012). Creating 3D web-based game environment using BIM models for virtual on-site visiting of building HVAC systems. In *ASCE Construction Research Congress 2012*, May 21–23, West Lafayette, IN, USA.

Tranquillo, J., Kline, W. A., & Hixson, C. (2018). Student-created canvases as a way to inform decision-making in a capstone design sequence. ASEE Annual Conference and Exposition, 2018- June.

Underwood, J., Khosrowshahi, F., Pittard, S., Greenwood, D., & Platts, T. (2013). *Embedding Building Information Modelling (BIM) within the taught curriculum: Supporting BIM implementation and adoption through the development of learning outcomes within the UK academic context for built environment programmes*. Higher Education Academy. https://www.heacademy.ac.uk/system/files/bim_june2013.pdf

Zamora-Polo, F., Luque-Sendra, A., Aguayo-Gonzalez, F., & Sanchez-Martın, J. (2019). Conceptual framework for the use of building information modeling in engineering education. *Journal of International Engineering Education, 35*(3), 744–755.

Zhao, D., McCoy, A. P., Bulbul, T., Fiori, C., & Nikkhoo, P. (2015). Building collaborative construction skills through BIM-integrated learning environment. *Journal of International Construction Education Research, 11*, 97–120.

Part V
Policy Support for Sustainability and the Circular Economy

Chapter 18
Circularity in the Built Environment: A Goal or a Means?

Tom B. J. Coenen, Klaasjan Visscher, and Leentje Volker

Abstract To reach a circular built environment, the changes and solutions of such mission need to align with the challenges a Circular Economy (CE) aims to solve. Despite the rather uniform policies on the goals and challenges of circular construction, the understanding of CE among practitioners appears divergent. Using the concept of problem space and solution space in relation to missions, the various perceptions, interpretations and framings of the CE mission in the Dutch construction industry were studied by means of 20 semi-structured interviews. Results indicate that the perceptions of the underlying challenges vary from mere resource scarcity to wide societal reforms, including social equity and climate neutrality. Also, the relation with other concepts seems contested, particularly regarding sustainability. The problems CE aims to address and the solutions to reach such CE turned out to interact and even intertwine in the conceptualizations of CE. Mission achievement hence calls for convergence of both the problem and the solution space acknowledging the mission's co-evolving nature. Given the resulting positioning of CE as both a means for underlying challenges and a goal in itself, we propose to understand CE as a mediation concept that couples directed solutions to a wide set of societal challenges. This implies that rather than aiming for a uniform definition, action should be aimed at dynamically guided solutions towards addressing the evolving societal challenges. Further research in other sectoral and geographical contexts is required to explore the validity and implications of CE as a mediation concept.

T. B. J. Coenen (✉) · L. Volker
Civil Engineering & Management, University of Twente, Enschede, The Netherlands
e-mail: t.b.j.coenen@utwente.nl; l.volker@utwente.nl

K. Visscher
Department of Technology, Policy & Society, University of Twente, Enschede, The Netherlands
e-mail: k.visscher@utwente.nl

G. Lindahl, S. C. Gottlieb (eds.), *SDGs in Construction Economics and Organization*, Springer Proceedings in Business and Economics,
https://doi.org/10.1007/978-3-031-25498-7_18

Keywords Circular economy · Mission-oriented innovation policy · Sustainability transition · Framing · System perspective

18.1 Introduction

Circular Economy (CE) has become one of the major themes in the construction industry, especially due to the large use of virgin resources and waste in the sector (Benachio et al., 2020). Addressing various challenges, particularly in line with SDGs 11 (sustainable cities and communities) and 12 (responsible consumption and production), the CE concept has been operationalized through the articulation of a mission. Such missions can be understood as narratives for challenge-oriented change (Janssen et al., 2021). In this respect, missions have increasingly been connected to directing system transformation in line with the societal challenges (Schot & Steinmueller, 2018). To achieve such missions, the direction of the underlying challenges (problem space) and required changes and innovations (solution space) need to be aligned (Wanzenböck et al., 2020). While CE has been introduced as a mission to address several underlying goals and challenges in the environmental sustainability domain, scholars have been defining CE in various ways, most notably considering resource flows, either or not including economic aspects (Kirchherr et al., 2017). This includes various ways for circular futures to take shape (Bauwens et al., 2020). However, CE has been widely regarded as an essentially contested concept (Korhonen et al., 2018b). Next to the conceptual critiques on feasibility and effectivity (Corvellec et al., 2021), its variety of discourses and myriad of framings adds to the concept's elusiveness (Calisto Friant et al., 2020), which impedes an effective mission attainment (Janssen et al., 2021).

While CE has experienced huge advancements in academia in terms of operationalization, conceptualization and adoption (Goyal et al., 2021), most research in the CE construction context addresses (technological) solutions (Benachio et al., 2020). Charef and Lu (2021), nonetheless, showed that, next to changing the thinking of professionals and novel technologies, a change of policy and regulation for adoption and implementation of CE is needed. For example, standards for circular construction need to be established (Anastasiades et al., 2021). Many framings and interpretations exist on the meaning of CE throughout the sector (Hossain et al., 2020), which is a major barrier to the transition to a circular construction sector. At the same time, implementation depends on the perceptions, interpretations and framings in practice. To determine and eventually steer the directionality of the solutions, it is of critical importance that the understandings of the CE concept are aligned and convergent in construction practice.

Considering the contradiction between the aim of alignment and convergence of missions and the essential contestation of the CE concept, we study the potential of CE to provide direction of change in line with societal challenges in the construction context. To that end, we analyse the framings and interpretations of the problem space, solution space and interactions as perceived by professionals in the Dutch construction sector in light of the formal mission policies. This approach reveals, for example, how CE is perceived in the construction context and to what extent CE is seen as a means or as an end. Next to insights into the framings and alignment of CE in the Dutch context to support CE policy, we aim to contribute to the ways of understanding and positioning the CE concept in academia and how future research in circular construction should deal with the equivocality of the concept.

18.2 Theoretical Framework

Societal missions have become increasingly important in innovation policy to guide change and innovation into societally desired directions. These missions are not only considered *wicked* but also constitute a wide heterogeneity of problem structures and specific designs of solutions with respect to these challenges (Wanzenböck et al., 2020). Therefore, these should be seen as "open-ended, [...] concerning the socio-economic system as a whole, even inducing (or requiring) system transformation" (Kuhlmann & Rip, 2018, p. 450). To guide and steer system change in the direction of missions, the concept of mission-oriented innovation policies (MIP) has been introduced. According to Wanzenböck et al. (2020, p. 475), "[a] MIP should be seen as a policy that provides directionality in supporting the process towards converging problem-–solution constellations [and] aims at advancing problem–solution constellations which become sufficiently stable to serve as common frame of reference for actors." Hence, the convergence of problem–solutions space guides the directionality and attainability of missions.

To conceptualize the underlying problem–solution structure of missions, Wanzenböck et al. (2020) presented a framework with two axes comprising the level of wickedness of both the problem and the solution space (Fig. 18.1). By studying the perceptions of the problem and solution spaces separately as well as by

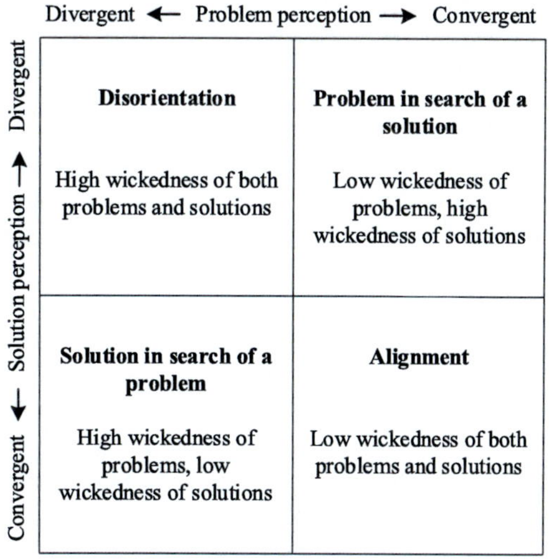

Fig. 18.1 Four types of problem–solution spaces. (Adapted from Wanzenböck et al. (2020))

analysing their interactions, the problem–solution perception in the current stage of transition towards circular construction can be investigated. Here, the problem space is determined by the level of normativity, complexity and available knowledge, while the solution space is characterized by contestation of the feasibility and prioritization of solutions, the alignment of solutions with the existing system and fragmentation of the various solutions. The resulting quadrants are: (1) disorientation in which both the problems and solutions are highly wicked such as social justice; (2) problems in search of solutions such as bullying; (3) solutions in search of problems such as spaceships to Mars; and (4) alignment between problems and solutions such as water retention. Depending on the blurriness between the problem and the solution space, mission alignment can either or both be reached through problem-led directionality and solution-led directionality. This implies a variation in pathways to increase the convergence and alignment of missions (Wanzenböck et al., 2020). Given that MIPs aim to converge towards an aligned framing (lower right quadrant), we use the framework to analyse the CE as a mission in the Dutch construction industry using the problem–solution axes.

18.3 Research Approach

To analyse the consequences of the variation in framings of the CE concept, we studied the Dutch context, where CE has since 2016 been positioned as one of the central pillars of environmental sustainability policy. We have focused the research on the perceptions of CE in construction practice to study the alignment and convergence of the CE mission interpretation in the Dutch construction industry. Following Wanzenböck et al. (2020), we distinguished the challenges addressed by the mission from the priorities of and interactions between the solutions for achieving the mission. Together, these interacting dynamics take place in the problem–solution space as addressed in Fig. 18.1. Accordingly, we took the following research steps.

First, we have mapped the development of CE policy in the Dutch construction context by means of an analysis of grey literature. This enabled us to reflect upon the relation between the formal policies and interpretations and framings in practice. Next, we identified the various perceptions and interpretations of the societal problems CE aims to address in the sector and the potential solution pathways connected to it by means of semi-structured interviews. Using purposive sampling (Campbell et al., 2020), individuals were selected such that both the various subsectors and the actor types were covered, which resulted in a diverse representation of the sector. To reach this actor variety, we used the innovation system categories presented in Kuhlmann and Arnold (2001). This resulted in interviews with 20 individuals with varying levels of experience with CE, consisting of two market parties, three policy makers, four public clients, two scholars, two network managers, four engineering/consultancy firms and three individuals from boundary organizations such as financers and legal specialists (see Appendix). Although the interviews were conducted as open conversations, several indicative questions were used to start and to guide the conversations (Table 18.1).

Table 18.1 List of indicative interview questions

Problem space	What does CE mean to you? And what societal challenges does it address?
	Who determines and affects what CE means?
	How does CE relate to other missions? And is there overlap with other missions or societal challenges?
	Is there, in your view, consensus on the meaning of CE?
	How is this problem interpretation related to the formal strategy?
Solution space	What are the ultimate solutions for circular construction?
	Who affects the solutions? And who determines what solutions become dominant?
	What are the roles of government and market in the solution space?
	To what extent are solutions introduced and adopted successfully?
	Are the current developments sufficient to achieve the mission goals?
Problem–solution interaction	To what extent do solutions address the mission goals?
	Should the CE mission be addressed integrally or separately with respect to other missions and challenges?
	How do the problem and solution spaces interact?
	How do the solutions feed into the problem space?

The transcripts of the 20 interviews were qualitatively coded in the Atlas.ti software tool. The interviewees' quotations were linked to the three separate categories presented in Table 18.1, being problem space, solution space and problem–solution interaction. The former two focus on both the level of convergence of the framings and the latter addresses how those spaces affect each other. Next, we compared the quotations per theme to find the framings of the CE mission, as well as their diversity. The final step was to link those overviews to the formal strategies to analyse the alignment of practice with the formal strategies. Altogether, this led to the results presented in the next section. First, we introduce the mission as formally stated in the Dutch national policies and strategies. Second, we present the interpretations and framings in practice based on the interviews, distinguishing the problem space, solution space and problem–solution interaction. Quotations were translated into English by the authors.

18.4 Results and Analysis

18.4.1 Mission Context: Circular Construction by 2050

The Dutch government was one of the first to launch a national mission for CE in Europe (Giorgi et al., 2022). The mission was formally introduced in 2016 and aimed for the Netherlands to be circular in 2050, meaning: (1) high-grade use of existing resource and waste flows; (2) replacement of fossil and non-sustainably produced resources by renewable and widely available alternatives; and (3) redesign of production methods, products and domains and rethinking consumption (IenW and EZK, 2016). Although building upon existing waste policy, the CE mission was introduced rather radically. CE was in the first strategy report already presented as an imaginary to mobilize action for directed change: "The idea of a Circular Economy as a fully closed system is a mobilizing ideal image. The use of primary resources and the generation of waste flows can probably never be fully prevented" (IenW and EZK, 2016, p. 15). Transition agendas were developed for five priority domains, including construction. Each priority sector was considered a separate pathway and Transition Teams were established to direct and lead the transition towards a circular system by formulating transition agendas and policies. Four strategic priorities have been identified in the transition agenda for construction, being market development, measurement of CE, policy and legislation, and knowledge and awareness.

In parallel to this formal strategy, in 2017, a large, industry-wide platform (CB'23) was launched by two governmental organizations and a normalization institute. This platform aimed at wide sectoral participation to uniformize the understanding and operationalization of circularity in the Dutch construction industry, which also affected the formal policies. Gradually, the goals and definitions found in the policies, strategies and networks shifted in focus. While the initial goals strongly approached CE as a means to address material scarcity and resource supply risks, the formal goals for the construction industry became increasingly integral, widening the formal CE mission to issues such as the pollution (particularly CO_2 reduction), biodiversity and environmental impact.

18.4.2 Problem Space

The previous section indicates that CE is positioned in the Dutch construction context as a "promise" rather than a clear description of challenges or goals. However, the challenges and goals addressed by the mission were perceived remarkably divergent. The interviews revealed different takes on how this imaginary is substantiated in practice, not only through the formal strategies but also through the practitioners' personal experiences and framings.

Several interviewees explicitly stated the need to distinguish CE from circular construction. According to them, CE is about the transformation of society at large towards a future circular system, while circular construction should be directed at resource/material efficiency in the built environment, with other environmental benefits as indirect benefits. One consultant explained: "*[in contrast to CE, circular construction] is specifically about the inexistence of waste and that all resources are put in the resource to be reused at a high grade.*" However, most of the interviewees did not explicitly distinguish between CE and circular construction, but their interpretations resonated with the notion of the latter. This addresses primarily the reduction of primary materials as the focus of circularity and is in line with the definition provided by the Transition Team. Apart from the noted differences, this indicates a rather subconscious interpretation of the CE concept and its underlying challenges.

When digging deeper into the societal problems that are addressed by these reduction efforts, only two interviewees mentioned resource depletion or material supply risks as a primary challenge. All the others saw circularity as a means to reduce the wider environmental impact, including emissions, biodiversity and social equity, as well as economic impacts such as long-term cost reductions and labour efficiency. Moreover, CE was linked to the deteriorating state of infrastructure and the housing shortages that call for large-scale construction or replacement activities in the coming decade. Interviewees argued that linking this building, maintenance and replacement challenge with circularity principles could increase the feasibility and reduce the costs while addressing sustainability goals. For example, a public client noted: "*The focus is increasingly shifting towards replacement and renovation [of assets]. That means that asset management is becoming more prominent in the steering on [circular] measures.*" It turned out that the interpretations of CE of practitioners closer to the national CE policy in government organizations were strongly aligned with the formal strategies, while the more deviating interpretations were found in market organizations. In sum, there was a large variation of the specific societal challenges that were expected to be addressed by CE. Despite this apparent contestation, three interviewees surprisingly perceived consensus on the problems that CE addresses, which indicates an unknown struggle for convergence of the various framings.

The relation and integration to other missions and their challenges was perceived divergently too. Whereas most interviews agreed that CE should always be considered in relation to wider sustainability issues, because the underlying goal – reducing environmental impact – is similar, there was considerable disagreement on the extent to which CE must be considered either separately or integrally from a governance perspective. Some interviewees urged for considering CE as a "way of thinking" that is inherently sustainable, while other stressed to limit CE to the scope of circular construction, and, as such, to focus on resource flows. One consultant even considered CE to be the materials part of sustainability, indicated by: "*How can we do things circular in terms of materials, but also do it sustainable in terms of CO_2 emissions? In this way we weigh various sustainability aspects.*" Apart from the conceptual interpretations, several interviewees mentioned the dilemma between conceptual purity and pragmatism in terms of governability. To illustrate the coupling of CE to the wider sustainability topic in order to be eligible for funding, a policy officer noted: "*we didn't get funds for CE in 2017, but we coupled it to the climate goals, [because] if we start working circular, we will contribute to the CO_2 reduction targets. [...] Such coupling increases the urgency of and support for CE.*"

Altogether, the interpretations and framings of the problem space are divergent. Not only do the underlying challenges differ but also the scopes differ in extent and direction. It is hence unclear how CE relates to other missions and how the responsibilities should be allocated. This indicates a high wickedness of the problem space.

18.4.3 Solution Space

Construction works and innovations are, especially in the infrastructure domain, predominantly developed by market parties as a response to (public) tenders. However, it is in such tenders that the CE concept is operationalized into solutions. As indicated in the paragraphs below, the types of innovations, other social and processual changes and their priorities are contested too.

According to all interviewees, operationalization of CE is still fragmented and the solutions depend on the varying client specifications. As illustrated by a contract manager from a market party: "*despite the ambition to be [the Netherlands'] most circular contractor, we need to address client's specifications in tenders in order to win those.*" Nevertheless, clients and public organizations increasingly support and participate in initiatives that investigate specific innovations. Although there are many circular alternatives developed since the introduction of CE policy, according to a public manager, "*the strategies and design principles are not that much under debate anymore, but the degree to which these ought to contribute to solving the problems is still contested.*" According to a sector banker, the share of supply-driven circular solutions in building construction is higher than in infrastructure due to the sector's structure. In building construction, there are several upscaling developments in modular and industrial housing innovations. This indicates a low but increasing convergence of circular innovations.

On a more abstract level, solutions were mentioned by the interviewees in terms of, for example, reuse, avoiding materials and recycling quota without specifying the (technological) solutions to do so. The formal policies also include a hierarchy of these abstract strategies, which are operationalized in the R-ladder,[1] which was mentioned by three interviewees. For example, a public manager stated that "*we try to find solutions [in the market] as high as possible on the R-ladder in order to retain as much value as possible in our assets.*" Still, there remains a discussion on the extent to which such hierarchy should be used to rank or assess solutions or to direct the transition. While various interviewees stress the necessity to aim for as high as possible R strategies, others argue that this does not always lead to solutions that address underlying problems most effectively. Moreover, the interpretations of strategies, such as the difference between reuse and recycling appear equivocal too. Nevertheless, there seems to be consensus that on this abstract level, different types of solutions apply to different types of construction assets, particularly distinguishing bulk materials (e.g. asphalt, concrete) from other resources. There is hence no single CE strategy that applies to all construction assets.

When addressing the solution space, all interviewees mentioned changes and innovations that facilitate processual and social changes that allow for technological solutions. As a consultant mentioned: "*I think the biggest challenge is to look [at CE] as a system transformation rather than technological innovations.*" In the

[1] The R-ladder consists of a hierarchy of waste management strategies in which certain activities are considered more circular than others. Examples of such Rs are Reuse, Reduce, Refurbish and Recycle (Potting et al., 2017).

infrastructure context, many interviewees mentioned that reconsiderations in asset management are fundamental for becoming circular from a systemic perspective. To exemplify, a public client's manager remarked: "*If you treat stuff properly it lasts very long, but [...] asset managers do not realize that CE is an obvious element of asset management.*" Also, data management and data transparency were mentioned in both building construction and infrastructure to play a crucial role in making circular decisions, for example, "*[it can only be treated circular] if it remains digitally insightful what materials are used in buildings, what the quality is, [...] and which dimensions are used.*" Such data include as-built drawings, structural monitoring data, and changes to and renovations of the asset to allow for circular choices in later lifecycle stages of the asset. Another condition mentioned for operating circularly depends on client–contractor relationships. This includes procurement methods (i.e. offering the space for circular innovation), but, even more, building relationships and collaborations that align the incentives between actors for considering whole lifecycles of assets. A researcher argued: "*Circularity is partly innovation, but largely not-so-innovation; namely, effectively using stuff that is already there. [To achieve this,] the strictly contractual relationship between client and contractor needs to be dropped to go to more constructive [relationships] to put actors in their strengths.*" Despite the agreement on the need for conditional changes, the extent to which and by whom those need to be enforced are still under debate. However, both market parties and public clients express a leading role for government organizations.

Similar to the problem space, the perception of the solution space shows a divergent image with a high level of fragmentation throughout the sector in terms of dominant solutions for a successful transition towards a circular infrastructure sector. Although interviewees all have images on circular futures, those images differ. The interviewees show yet an increasing understanding of the systemic implication of circularity beyond technological "add-ons," particularly in larger organizations. However, the structural aspects of the sector pose major barriers to achieve such solutions on an integral and systemic level. Nevertheless, there are several promising initiatives, particularly launched by large public organizations and networks.

18.4.4 Problem–Solution Interaction

Problems and solutions interact in the context of a mission. After having established that the problem and solution spaces are both wicked, the interaction will have a major impact in the evolvement of the mission. Various types of interactions are discussed below.

The level to which CE was perceived as a goal rather than a means to address underlying challenges was remarkably high. As a CE researcher stated: "*I have never consciously thought about it before, but there seems to be a shift from CE as a means to CE as a goal in itself.*" This was also implicitly confirmed by many interviewees. For example, CE has, according to a policy maker increasingly become a strategic pillar next to, for example, sustainability rather than under it.

Still, it is often presented as a solution for various other (societal) challenges. Specifically in the operationalization of CE in projects or specific products, circularity is interpreted as a goal serving other challenges. As an asphalt expert illustrated: "*when acting circular by reusing more asphalt, the overall CO_2 emissions will also drop.*" Within the set of interviews, it was visible that professionals with more experience and expertise in CE see circularity as a means, while professionals newer to the subject tend to consider it as a stand-alone objective.

A striking observation in the interviews was that, rather than considering it from a problem perspective, several interviewees explicitly defined CE in terms of solutions, most notably in terms of modularity and reuse. For example, a sector banker stated that: "*in CE, you try [...] as much as possible to take the reuse and lifespan extension of materials into account.*" This illustrates a blurry line between the problem and the solution space. The same goes for the R strategies, which are by some presented as solutions to become circular and by others as targets to develop solutions for. It indicates a co-evolution between the spaces. This co-evolution was also acknowledged by a network manager: "*I think if we want to start seeing CE as the new normal, we need to collectively live through it. We just need to put it to work to start understanding what the implications are and to synchronize towards a common perception.*"

For these reasons, several interviewees argued that it is not fruitful to aim for a fixed and al-encompassing definition of CE or circular construction, nor to define fixed problems and solutions. As indicated by a network manager: "*I have never believed in such uniform definition. [...] The focus should instead be on turning it from an elitist sport to something for the masses and on moving people into the right direction.*" A researcher added: "*if we keep adapting the definition of what a 100% CE means, we will eventually get there, but if we stubbornly hold on to the current definition, we won't.*" This illustrates the evolving character of the CE concept, which can be at times converging, but also diverging. Nevertheless, about half the interviewees explicitly mentioned the necessity for a uniform definition, most notably because of the measurability of progress and the political legitimization.

In conclusion, some interpret CE as a challenge in itself, while the majority is increasingly seeing it as a means to address underlying challenges. Nevertheless, there is shared perception that CE as a concept is still developing, not just in terms of solutions but also conceptually.

18.5 Discussion

Results show that the problem and solution spaces of CE in construction are both wicked. Many interpretations of the problem space do not only vary among respondents but also when compared to literature on CE in construction (e.g. Hossain et al., 2020). The problem and the solution space interact in an unpredictable way. Applying this to the four quadrants (Fig. 18.1), the mission finds itself in a disoriented stage and evidence of a process towards an increasing alignment was not found.

Hence, an effective transition towards CE requires learning about the problem while experimenting with solutions, following a co-evolutionary logic (Wanzenböck et al., 2020). A practical example of this co-evolution was confirmed in another study considering the case of the pilot project on the development of a Circular Viaduct in the Netherlands (Coenen et al., 2021). In line with our findings, the viaduct case shows that while the pilot was designed according to the best knowledge of making it circular at the time, the outcome of this process – the design – also affected how others in the sector interpreted CE. As suggested in the Result section, the meaning of circularity in the construction context must be discovered and converged by experimenting and learning with solutions. Despite the exploratory nature of this venture, coordination of the solutions is necessary to achieve this alignment and to provide direction and share knowledge and insights from a policy perspective (Mazzucato et al., 2020).

Because this notion of CE fits the conceptualization of a super-wicked problem (Levin et al., 2012), CE is becoming a discursive resource to guide framings in addressing other challenges, such as climate change (Green & Sergeeva, 2020). This fits the purpose of the mission as a narrative for guiding and coordinating directed change (Janssen et al., 2021). Nevertheless, this overlap limits the power of CE as a mission because the CE does not simply offer a narrative for coupling challenges to solutions. Instead, it acts as an intermediary concept by offering an ideal view of a configuration of society (Kirchherr et al., 2017). However, results of our research indicate that, while acknowledging the CE as a societal transformation, circular construction is rather seen as a way of doing things more resource efficient – especially regarding the reduction of primary resources. Such activity-based view of CE has obvious limitations in terms of conceptual purity and possibly also for the transformation potential. In other words, recycling more, reusing more and building modularly, will not lead to a regenerative, closed-loop system (Korhonen et al., 2018a). Moreover, such framings are fundamentally different from the economic and political conceptualizations that challenge the current consumeristic system itself (e.g. Lowe & Genovese, 2022). Nonetheless, the solution-oriented interpretations offer action perspective to actors needed to explore the problem space.

Although the societal challenges mentioned (e.g. resource depletion and carbon emissions) underpin the various interpretations of CE, results indicate that the concept is treated as a goal in itself too. Considering it as a target helps in guiding change into a specific direction and hence to position it as a mission. Inspired by Jasanoff's socio-technical imaginaries (Konrad & Böhle, 2019), we hence propose to understand the CE as a mediation concept. This means that, as an ideal imaginary, CE has a mediating role between systemic change and societal challenges. As an illustration, this means that, when considering it as a mediation concept, a designer does not need to understand the full implications of the societal challenges that CE aims to address to contribute to addressing those challenges, while policy makers do not need to steer the actual solutions as long as the CE principles relate to the underlying challenges.

As a guide to change in practice, such mediation concept helps to tackle an interacting set of societal challenges without the need to understand the full complexity of them. This implies that the interpretation of the societal challenges is, through the

CE concept, affected by circular solutions, while the circular solutions, in turn, are operationalized through CE and substantiated by the underlying challenges. Such conceptualization contradicts the many definitions and operationalizations of CE in literature and practice, because these assume mostly a rather clear interaction between the problems it addresses and the solutions that aim to solve these, such as the closing and narrowing of the loops (e.g. Geissdoerfer et al., 2017; Kirchherr et al., 2017).

18.6 Conclusions

To reveal the potential of CE in construction to guide change in line with wider societal challenges, we studied the relation between the articulation of the CE mission in the Dutch construction policies and the perceptions of the problem and the solution space of this mission in construction practice. We found that while the policy documents provide generally clear terms and uniform conceptualizations, the framings in practice vary strongly, even among the more CE-experienced practitioners. This leads to a disoriented problem–solution space. Results indicate that the interpretations of the problems and solutions regarding CE co-evolve. This demands an experimentational yet coordinated approach to circular solutions to explore the conceptual implications of the problems addressed by CE. Overall, the understanding of CE in the Dutch construction industry is narrower than the interpretations of CE in literature and can be understood as circular construction rather than CE. In both policy and in practice, the focus is primarily on resource-reducing practices to address not only resource scarcity but also reduction of wider environmental impacts and loss of biodiversity (SDGs 11, 12, 13 and 15).

The inherent contestation of the CE concept, which was confirmed by the case results and becomes transparent when positioning CE as a mediation concept, has implications for mission-oriented policies. The CE concept contributes to aligning the various perspectives with those challenges and, consequently, narrowing the solution space by offering a long-term yet evolving ideal imaginary. Although being instrumental in addressing a wider set of underlying challenges, it is essential to define the conceptual boundaries of CE from a policy perspective. Entanglement of the problem and the solution space, subsector-specificity and inability to envision the eventually dominant solutions hinder the formulation of uniform objectives. Given that construction is a highly demand-oriented sector with a major public clientship, public policy should focus on the formulation and operationalization of long-term objectives. This requires, on the one hand, the inclusion of the market to safeguard feasibility of the objectives, and, on the other hand, the clients' preparedness to let go of actors that do not take the required steps.

This study was based on 20 interviews in a single context. Extending the sample, for example through surveys, would give more insights into the variation of interpretations and framings of circularity in construction. Furthermore, the results indicate a large sectoral and geographic element in the framings of CE. However, the current study is limited to the framings in the Dutch construction context, while the nature and diversity of framings might be different in other countries or contexts. Doing similar research in other sectors, regions and countries would help understanding the CE as a mediation concept, would increase generalizability of the results and would reveal insights in the opportunities for steering for circularity in order to address a wide set of societal challenges. Such research would also inform on the cross-sectoral alignment of the CE concept with respect to the addressed challenges and potential solutions.

Acknowledgements The authors wish to thank the interviewees from the Dutch construction sector for their time and enthusiastic responses. Also, the authors thank Rijkswaterstaat infrastructure agency for the funding.

Appendix: Anonymized Overview of Interviewees

No.	Interviewee type
Public client organizations	
1	Programme manager infrastructure at Dutch province
2	Programme manager infrastructure agency
3	Programme manager infrastructure agency
4	Policy coordinator water board association
Ministries	
5	Coordinator, Ministry of Interior and Kingdom Relations
6	Policy officer, Ministry of Infrastructure and Water Management
7	Coordinator, Ministry of Infrastructure and Water Management
Industry	
8	Consultant sustainable and circular construction
9	Consultant sustainable and circular construction
10	Consultant sustainable and circular construction
11	Sustainability coordinator contractor organization
12	Project coordinator contractor organization
13	Consultant sustainable and circular construction
Knowledge institutions	
14	Asphalt expert independent research organization
15	Circular infrastructure scholar
Network organizations, platforms, and associations	
16	Director circularity network organization
17	Sustainability manager industry association
Financial, legal and process experts	
18	Economic expert sustainable construction
19	Legal expert sustainable construction
20	Standardization expert

References

Anastasiades, K., Goffin, J., Rinke, M., Buyle, M., Audenaert, A., & Blom, J. (2021). Standardisation: An essential enabler for the circular reuse of construction components? A trajectory for a cleaner European construction industry. *Journal of Cleaner Production, 298*, 126864. https://doi.org/10.1016/j.jclepro.2021.126864

Bauwens, T., Hekkert, M., & Kirchherr, J. (2020). Circular futures: What will they look like? *Ecological Economics, 175*, 106703. https://doi.org/10.1016/j.ecolecon.2020.106703

Benachio, G. L. F., Freitas, M. d. C. D., & Tavares, S. F. (2020). Circular economy in the construction industry: A systematic literature review. *Journal of Cleaner Production, 260*, 121046. https://doi.org/10.1016/j.jclepro.2020.121046

Calisto Friant, M., Vermeulen, W. J. V., & Salomone, R. (2020). A typology of circular economy discourses: Navigating the diverse visions of a contested paradigm. *Resources, Conservation and Recycling, 161*, 104917. https://doi.org/10.1016/j.resconrec.2020.104917

Campbell, S., Greenwood, M., Prior, S., Shearer, T., Walkem, K., Young, S., Bywaters, D., & Walker, K. (2020). Purposive sampling: Complex or simple? Research case examples. *Journal of Research in Nursing, 25*, 652–661. https://doi.org/10.1177/1744987120927206

Charef, R., & Lu, W. (2021). Factor dynamics to facilitate circular economy adoption in construction. *Journal of Cleaner Production, 319*, 128639. https://doi.org/10.1016/j.jclepro.2021.128639

Coenen, T. B. J., Volker, L., & Visscher, K. (2021). Introducing circular innovation in the construction industry: The case of the circular viaduct. In *37th annual ARCOM conference* (pp. 624–633).

Corvellec, H., Stowell, A. F., & Johansson, N. (2021). Critiques of the circular economy. *Journal of Industrial Ecology, 1–12*, 421. https://doi.org/10.1111/jiec.13187

Geissdoerfer, M., Savaget, P., Bocken, N. M. P., & Hultink, E. J. (2017). The circular economy – A new sustainability paradigm? *Journal of Cleaner Production, 143*, 757–768. https://doi.org/10.1016/j.jclepro.2016.12.048

Giorgi, S., Lavagna, M., Wang, K., Osmani, M., Liu, G., & Campioli, A. (2022). Drivers and barriers towards circular economy in the building sector: Stakeholder interviews and analysis of five european countries policies and practices. *Journal of Cleaner Production, 336*, 130395. https://doi.org/10.1016/j.jclepro.2022.130395

Goyal, S., Chauhan, S., & Mishra, P. (2021). Circular economy research: A bibliometric analysis (2000–2019) and future research insights. *Journal of Cleaner Production, 287*, 125011. https://doi.org/10.1016/j.jclepro.2020.125011

Green, S. D., & Sergeeva, N. (2020). The contested privileging of zero carbon: Plausibility, persuasiveness and professionalism. *Buildings and Cities, 1*, 491–503. https://doi.org/10.5334/bc.49

Hossain, M. U., Ng, S. T., Antwi-Afari, P., & Amor, B. (2020). Circular economy and the construction industry: Existing trends, challenges and prospective framework for sustainable construction. *Renewable and Sustainable Energy Reviews, 130*, 109948. https://doi.org/10.1016/j.rser.2020.109948

IenW, EZK. (2016). *Nederland circulair in 2050*. Rijksoverheid.

Janssen, M. J., Torrens, J., Wesseling, J. H., & Wanzenböck, I. (2021). The promises and premises of mission-oriented innovation policy – A reflection and ways forward. *Science and Public Policy, 48*, 438–444. https://doi.org/10.1093/scipol/scaa072

Kirchherr, J., Reike, D., & Hekkert, M. (2017). Conceptualizing the circular economy: An analysis of 114 definitions. *Resources, Conservation and Recycling, 127*, 221–232. https://doi.org/10.1016/j.resconrec.2017.09.005

Konrad, K., & Böhle, K. (2019). Socio-technical futures and the governance of innovation processes—An introduction to the special issue. *Futures, 109*, 101–107. https://doi.org/10.1016/j.futures.2019.03.003

Korhonen, J., Honkasalo, A., & Seppälä, J. (2018a). Circular economy: The concept and its limitations. *Ecological Economics, 143*, 37–46. https://doi.org/10.1016/j.ecolecon.2017.06.041

Korhonen, J., Nuur, C., Feldmann, A., & Birkie, S. E. (2018b). Circular economy as an essentially contested concept. *Journal of Cleaner Production, 175*, 544–552. https://doi.org/10.1016/j.jclepro.2017.12.111

Kuhlmann, S., & Arnold, E. (2001). *RCN in the Norwegian research and innovation system* (Background Report No.12). Oslo.

Kuhlmann, S., & Rip, A. (2018). Next-generation innovation policy and grand challenges. *Science and Public Policy, 45*, 448–454. https://doi.org/10.1093/SCIPOL/SCY011

Levin, K., Cashore, B., Bernstein, S., & Auld, G. (2012). Overcoming the tragedy of super wicked problems: Constraining our future selves to ameliorate global climate change. *Policy Sciences, 45*, 123–152. https://doi.org/10.1007/s11077-012-9151-0

Lowe, B. H., & Genovese, A. (2022). What theories of value (could) underpin our circular futures? *Ecological Economics, 195*, 107382. https://doi.org/10.1016/j.ecolecon.2022.107382

Mazzucato, M., Kattel, R., & Ryan-Collins, J. (2020). Challenge-driven innovation policy: Towards a new policy toolkit. *Journal of Industry, Competition and Trade, 20*, 421–437. https://doi.org/10.1007/s10842-019-00329-w

Potting, J., Hekkert, M., Worrell, E., & Hanemaaijer, A. (2017). *Circular economy: Measuring innovation in the product chain* (Vol. 46). PBL Netherlands Environmental Assessment Agency.

Schot, J., & Steinmueller, W. E. (2018). New directions for innovation studies: Missions and transformations. *Research Policy, 47*, 1583–1584. https://doi.org/10.1016/j.respol.2018.08.014

Wanzenböck, I., Wesseling, J. H., Frenken, K., Hekkert, M. P., & Weber, K. M. (2020). A framework for mission-oriented innovation policy: Alternative pathways through the problem–solution space. *Science and Public Policy, 47*, 474–489. https://doi.org/10.1093/scipol/scaa027

Chapter 19
Circular Supplier Partnerships for Resource Economic Marketization in the Construction Industry

Lasse M. Fredslund, Stefan Christoffer Gottlieb, and Christian Koch

Abstract Radical changes in supply chains relations are critical in a world of environmental problems and limited resources. Presently, there is a partnership trend in the Danish construction industry, where attempts to reconfigure supply chains through strategic and collaborative arrangements are taking place. These partnerships are legitimized by their ability to capitalize on e.g. innovations and repetition effects. However, it has been difficult to involve the supply chain actors as strategic partners, which is problematic as a new resource economic marketization is dependent on transformations of supply and demand throughout the entire supply chain. Drawing on institutional theory, we analyze and discuss the role of concurrent organizational dynamics, market mechanisms, and supranational regulations in legitimizing what is here labeled circular supplier partnerships (CSP) in the construction industry. We illustrate the possible underpinnings of CSPs and analyze the barriers to the development and legitimation processes of CSPs. In doing so, the paper substantiates CSPs as a possible driver for sustainable development by the consolidation of the 17 UN Sustainable Development Goals.

Keywords Procurement · Partnerships · Circular economy · Marketization · Legitimacy

19.1 Introduction

Human development and corporate actions of the linear economy have led to increased societal prosperity and wealth (Lacy & Rutqvist, 2015). However, the resulting population growth and unsustainable harvesting of natural resources have led to environmental problems and resource scarcity. The global construction

L. M. Fredslund (✉) · S. C. Gottlieb · C. Koch
Department of the Built Environment, Aalborg University, Copenhagen, Denmark
e-mail: laf@build.aau.dk; scg@build.aau.dk; christiank@build.aau.dk

G. Lindahl, S. C. Gottlieb (eds.), *SDGs in Construction Economics and Organization*, Springer Proceedings in Business and Economics,
https://doi.org/10.1007/978-3-031-25498-7_19

industry is one of the main contributors to these problems, being responsible for almost 40% of all human-induced GHG emissions, and the construction and maintenance of buildings and infrastructure represent nearly 50% of the global material consumption (IPCC, 2021). Resource economic business models and marketization, also known as the circular economy (CE), is a conceptual remedy to these problems and has been promoted as business concepts and public policies, e.g. in the EU Circular Economy Action Plan and the EU taxonomy for sustainable activities (European Commission, 2021). However, the process of generating common understandings and gaining legitimacy for CE poses a particular challenge for actors who want to negotiate the current market conditions of the linear economy. Moreover, there are various interpretations of circular economy business models (CEBM) causing limited industry sensitivity (Lüdeke-Freund et al., 2019) and lack of knowledge on how actors legitimize (Demil et al., 2018) their business models according to relevant CE aspects (Centobelli et al., 2020). According to Geissdoerfer et al. (2018), CEBM and supply chains are critically linked in the realization of a resource economic marketization, which calls for new understandings of how circular supply chain collaborations can be organized in the complex and project-based construction industry, including the Danish.

The construction industry is characterized by opportunistic behavior and short-term individual transactions, which often leads to misaligned goals and misaligned incentives structures (Davies & Doherty, 2019). However, two concurrent tendencies, circular economy and strategic partnerships, encourage new types of collaboration. The merging of concepts and approaches from these two trends can be understood as a possible joined concept of "circular supplier partnerships" (CSPs), which in this paper, are seen as strategic and collaborative arrangements between a client and several supplier organizations based on CEBM. We argue that CSPs possibly have the capacity to overcome the structural issues of the construction industry. However, to achieve this necessitates the theorization and legitimization of CSPs as driver for a resource economic marketization. Combining an analytical lens of institutional theory stressing the role of legitimacy (Deephouse et al., 2017) with a mix-method case study of a strategic partnership, the purpose and aim is thus to discuss how CSPs can be legitimized and what role concurrent organizational dynamics, market mechanisms, and supranational regulations play in this process. In doing so, the paper addresses the 17th UN Sustainable Development Goals (UN SDG) "*Partnership for Sustainable Development*" (UN SDG, 2020).

19.2 Theory

The project adopts the theoretical aspects of theorization and legitimacy from organizational institutionalism (Díez Martín et al., 2021) to create an opportunity for analyzing barriers and drivers on multiple-levels, e.g. organizational dynamics, market mechanisms, and supranational regulations in relation to CSPs in the Danish construction industry. Theorization is vital because it connects to the centrality of institutional thinking, namely how new ideas become legitimated and made available in simplified form for wider adoption (Greenwood et al., 2002). Organizational legitimacy can be understood as: "*the perceived appropriateness of an organization to a social system in terms of rules, values, norms, and definitions*" (Deephouse et al., 2017, p. 32). The concept of legitimacy is a suitable analytical framework to understand what rules, values, norms, and definitions are essential to theorize in the understanding of CSPs in the transition toward a resource economic marketization. We moreover draw on the concept of legitimacy (Bitektine & Haack, 2015) to examine how moral, pragmatic, regulatory, and cognitive forms of legitimacy are challenged by altering current linear regulations, values, and exchange systems (Deephouse et al., 2017). Legitimacy is a corporate survival-enhancing phenomenon (Boxenbaum & Jonsson, 2017) derived from conforming to institutionalized myths in a field (Meyer & Rowan, 1991). It affects an organization's ability to attract both social and economic resources as well as its status, reputation, and market access, where an organization can eliminate competition if it appears more legitimate on the market (Deephouse & Carter, 2005). We draw on these insights to analyze the actions required to legitimize circular rules, norms, and definitions on multiple levels, which often compete with the actions and beliefs of the dominant linear economy of the neoliberal society (Lounsbury & Glynn, 2001).

19.3 Method

The paper aims to provide an understanding of different legitimacy dynamics and mechanisms of CSPs at multiple levels. The empirical basis is a case analysis of the strategic partnership TRUST that constitutes as a paradigmatic case (Flyvbjerg, 2010) on how a strategic partnership is organized and operates in the Danish construction industry. The TRUST partnership is focal in a CSP perspective because the internal members are practicing a more relational, long-term, and dependent approach through, e.g. risk sharing, knowledge development, and repetitive effects legitimizing the partnership in the Danish construction industry (Gottlieb et al., 2020a, 2020b).

The empirical material is collected through a mixed method approach involving interviews, a survey and relevant background literature. Ten semi-structured interviews have been conducted with employees and related suppliers in TRUST focusing on different perceptions across organizational levels e.g. suppliers, wholesalers, senior management, purchasers, process coordinators, project managers, and craftsmen. The interviews focused on barriers and drivers to better understand supply chain integration in a strategic partnership. Moreover, a survey with 130 respondents quantifying the current maturity level for supply chain collaboration in the Danish construction industry have been carried out addressing underlying aspects of four maturity levels e.g. prices competition, prices and quality competition, project partnerships, and strategic alliances. Thus, the case study is limited by the context of the TRUST partnership and an overall survey, which cannot paint the full picture of an entire industry. We can therefore not draw universal generalizations, but contextual generalizations that require insight into how this case study differ from other cases (Siggelkow, 2007).

19.3.1 Case Description

TRUST is a partnership formed in 2016 in response to the City of Copenhagen's decision to tender a substantial part of their municipal renovation and construction projects through strategic partnerships. This alternative approach is created based on several urgent challenges related to increased urbanization, the need to strengthen the welfare system, and decrease time overruns and construction costs, which were considerably higher in Copenhagen than in other comparable municipalities. One of the partnership tenders was won by the TRUST consortium, which created a strategic partnership of six key partners in a 4-year project portfolio worth four billion DKK. The strategic partnerships were believed to address the mentioned and pressing issues through a shared location involving long-term cooperation across a project portfolio targeting common benefits and objectives (Frederiksen, 2021). The initial vision was to extend the strategic partnership to both subcontractors and suppliers creating an opportunity to involve the competence and resources of the external actors, while also ensuring effective competition (Construction Copenhagen, 2016). However, it quickly became apparent that it was difficult to involve the supply chain on an equal basis with the other partners regardless of initial efforts. Despite this problem, the TRUST partnership is hailed as a prominent, effective, and innovative partnership, due to its focus on continuously optimizing productivity, while issues of both the supply chain and sustainability have had a less prominent role in the partnership. The case is nevertheless interesting in a CSP perspective because the internal members are practicing a more relational, long-term, and dependent approach through risk sharing, common objectives, knowledge development, and harnessing of repetition effects than usually found in the Danish construction industry (Gottlieb et al., 2020a, 2020b).

19.4 Analysis

The analysis takes a multi-level perspective to scrutinize how CSPs can be legitimized and the role of concurrent developments of organizational dynamics, market mechanisms, and supranational regulations in this process. Thus, the analysis provides a foundation for discussing how current supply chain initiatives together with new regulations can be a lever for development of new resource economic forms of marketization. First, we analyze organizational processes, strategies, and practices affecting the legitimizing of the supply chain as a driver for CE. Second, we analyze the role that CSPs play in the formation and realization of a resource economic marketization as an arena for a broader diffusion of CE. Finally, we analyze institutional and regulatory responsibilities to better understand structural change in relation to legitimizing CE and CSPs as a precondition for resource economic marketization.

19.4.1 Circular Economy in the Supply Chain

Geissdoerfer et al. (2018) argue that CE business models are highly related to the actors of the supply chain, where a circular supply chain is theorized by "*reverse*" or "*closed*" resource loops that can maximize value over the lifetime of a product by remanufacturing, renovation, or recycling. Various strategic concepts to consolidate and manage circular supply chains have been proposed across different industries, e.g. strategic alliances, partnerships, or networks involving a more relational, dependent, and long-term relationship (Schenkel et al., 2015). Lahane et al. (2020) have defined a new domain of "*Circular Supply Chain Management*," which focuses on resource regeneration to utilize and capture the value of resources in a better way. According to Geissdoerfer et al. (2018), this approach must be able to address several objectives to achieve economic, environmental, and societal objectives, including how different beliefs and competences can be reconciled in strategic relationships. CE business models, however, require a rethinking of the construction industry's local practices and structures in terms of how to create collaboration beyond the current organizational boundaries (Angelis et al., 2018).

The TRUST partnership has tried to involve the suppliers to a greater extend, which is expressed by the Director of TRUST: "*We created a local supplier forum, where there was a great interest from the suppliers creating a more open dialogue about products, functionality, and pricing.*" However, according to a TRUST supplier, the forums were motivated by the suppliers to increase their share of the project portfolio and were based on typical supply and demand negotiations. This reflects a general lack of strategic supplier involvement in the Danish construction industry, where it has been difficult to involve the actors of the supply chain in a more strategic effort (Fredslund, 2021). As such, new endeavors entail negotiations of different relevant legitimacy aspects, e.g. perceptions, performance, regulations, and values (Deephouse et al., 2017) integrating the suppliers in a strategic and prolonged relationship. This is also seen in the survey that illustrates a greater consideration of a narrow pricing and one-time collaborations, rather than a consideration of prolonged and relational conditions. Moreover, it has been problematic to create a more strategic involvement of the suppliers due to considerations of pricing. This issue also applies in the TRUST partnership, as argued by the Business Area Manager: "*Both parties [client and suppliers] need to see an advantage of the partnership, that's what you have to ensure. But, we focus on how much we have to pay when we buy it. In reality, we should measure the costs when it is assembled. We do not get that strategic dialogue with the supplier.*" In order words, it has been difficult to legitimize a more strategic involvement of the supplier's competences and resources due to a "*merchant culture*" focusing on narrow prizing instead of multi-criteria, which otherwise define a higher procurement maturity (Meng et al., 2011). Findings from the survey indicate that this paradox is a problem in the Danish construction industry, and, consequently, that there is a need to rethink the relations with the suppliers addressing the strategic concepts of CE (Górecki et al., 2019).

The understanding of the CE principles is thus based on a basic representation of a much more complex organizational system (Geissdoerfer et al., 2018), which includes involvement of the suppliers and differs from the conventional linear system and thinking. CE business models represent a set of strategic decisions designed to preserve the embedded environmental and economic value of a product or service in either biological or technological loops (Centobelli et al., 2020). CE business models imply new social interactions (Ranta et al., 2018) across the entire value chain and can be perceived as a very abstract concept based on many different understandings and without a greater degree of industry anchoring and sensitivity (Lüdeke-Freund et al., 2019). This calls for new approaches and processes (MacArthur, 2015) that erode the "*taken for grantedness*" or cognitive legitimacy of the linear economy by expanding the organizational boundaries of current strategic partnerships in the Danish construction industry. Circular supply chains have a crucial role in the efforts to reduce GHG emissions, resources, and risks while maintaining competitiveness through business models that combine production and consumption systems with the CE principles (Lacy & Rutqvist, 2015; Nußholz, 2017). This means that CE structures are highly linked to the legitimation of CSPs, where definitions and realization of a resource economic marketization can be seen as an arena for mobilizing CSPs.

19.4.2 Resource Economic Marketization

Resource economic marketization entails the challenge of valuing material that changes status from being "*waste*" in a linear model to being a resource that can be reintroduced into buildings in a "*circular*" model. This change of status implies a liberation of the material from the context in which it was considered as waste, to how it can enter a resource economic market relation (Callon, 2017). To be included in a new context, the material must undergo a valuation that presupposes a standardization and a normative and pragmatic legitimation of how the properties of the material have value and can perform under new market conditions. According to Koch and Polesie (2021), this valuation requires "*intermediaries*" across the supply chains, such as the many portals and platforms emerging these years, to enable exchanges and relations between buyers and sellers. This involves a mixed qualitative and quantitative (Callon, 2017) valuation of goods, which is not specific to a particular market (Kahneman & Tversky, 1972). Yet, the balance between the quantitative and the qualitative in the CE is shifted toward greater legitimation of resource economic considerations. At the same time, the formation of new roles means that structure-setting organizations change character (Brunsson & Jutterström, 2018), e.g. the suppliers' or contractors' role in handling CE goods on a resource economic market.

A mixture of qualitative and quantitative valuations, also known as multi-criteria pricing, involves considerations, e.g. sustainability, collaborative evaluations, or the relations that are part of a specific product life cycle. However, this form of pricing is challenged by the merchant culture and one-off collaborations in the Danish construction industry. Moreover, qualitative prizing is based on different soft aspects, which are extremely difficult to document as stated by the Strategic Purchaser of TRUST partnership: "*That is the challenge for TRUST. There are many of these soft values that should provide value, but they are really hard to measure. And the paradox is whether we should have fewer suppliers and close relationships or total competition among many suppliers.*" TRUST has reduced the suppliers within each product category, creating a better foundation for cooperation and more holistic pricing. Yet, a lack of transparency, mutual trust, and monopoly conditions has made it difficult to create strategic relations with the suppliers. It is especially the case for complex and cost-intensive products, e.g. concrete elements or facades, as demonstrated by the Director of the TRUST partnership: "*You can do nothing, you stand 'naked', and they are the ones who rule over you. And you just have to adjust to what they want, otherwise you will be put back in the supply queue.*" Hence, the realization of a resource economic marketization involving the mobilizing, legitimation, and diffusing of CSPs will presumably be problematic in the current institutional environment of the Danish construction industry, which is characterized by different relational problems, as well as low earnings and high risks affecting the leeway for innovations. However, CSPs could be enhanced by organizations' future need to comply to regulatory demands from supranational authorities (EU/UN), where CE procurement processes are an extension of EU's formal ambitions for a stronger sustainable development (Eurocities, 2020; IBMin, 2021).

19.4.3 Institutional and Regulatory Responsibility

According to Kirchherr et al. (2018) and Rizos et al. (2016) dominant actors, e.g. politicians and governments, have an institutional responsibility because they can be either "*drivers*" or "*barriers*" to more circular transitions, and thus a lever for a regulative legitimation of CSPs. These dominant actors can remove existing barriers and theorize the development of more dedicated CE-initiatives across different societal levels and dimensions. This is important since CE challenges the current practices and structures as well as the regulatory legitimacy of the linear economy in the Danish construction industry. Moreover, there is an absence of standards (Nußholz et al., 2019), as well as a general lack of common understandings of CE visions for the Danish construction industry (Danish Construction, 2019) that counteract the establishment of new market conditions for CE products and services as well as recycled materials. Thus, circular construction can be considered as a "*market in the making*" (Köhler et al., 2019) characterized by competing perceptions of conflicting linear and circular rationales, values, and meanings systems, creating high complexity, contradictions, and over time competition.

In this context, the EUs Circular Economy Action Plan and EUs taxonomy for sustainable activities (European Commission, 2021) can be considered as a centrally driven governance initiative, which may have a decisive impact on a more legitimated shift to a resource economic marketization. The regulatory legitimacy mechanisms intervene in the financial and business markets, which is crucial for the concept of CE among other mechanisms financing construction projects in the future. In 2021, EU posited accounting requirements for the largest companies, and, from 2022, banks and other financial institutions that invest in buildings will have to document a high percentage of the materials used are part of circular loops. It may then be likely that the major construction contractors and relative municipalities and other clients addressing this legislation will make similar demands on potential strategic partnerships, the supply chain, subcontractors, and the rest of the value chain. This means that the regulation may also affect smaller actors in the industry and create a momentum for both regulatory and normative legitimacy in the development of CSPs. However, the EU is not alone in taking regulatory action in relation to the CE. A number of nations are preparing similar interventions in construction processes and practices, and also more voluntary initiatives, based on a series of actors including cities that ally themselves internationally, are profiled in circular construction and legitimize the development through local demands. Accordingly, the "*green transition*" calls for new institutional preconditions and structures (Farla et al., 2012), which can create interactions between different perceptions, resources, and competencies. This can take the form of new types of organizations, such as CSPs, to integrate and combine CE structures and rationales between already established sectors, organizations, and practices.

19.5 Discussion

The development of new forms of organization is already taking place in the Danish construction industry, where strategic and collaborative constellations in the form of strategic partnerships and project portfolios have begun to gain a high degree of legitimacy. The Strategic Partnership TRUST has shown its worth by reducing risk, conflicts, and costs by establishing stronger relationships, repetition effects, and innovations (Gottlieb et al., 2020a, 2020b). In an ideal form, the current Danish development attempts to create a new relational culture without regard to the original organizational boundaries. But for the most part, this effort never reaches the strategic level across the actors of the supply chain, where collaboration is often characterized by short-term relationships that do not necessarily reach beyond the individual project. Another barrier to strategic supplier involvement is a lack of economic transparency, which is an ingredient for trust, cooperation, creativity, and optimizations, which is stated by the Strategic Purchaser of the TRUST partnership: "*You have to open up, and we are incredibly bad at that. We only look in our own boxes, and we are not honest. That is why we have so many conflicts and no common interests. However, transparency go both ways and it stops already with ourselves in the partnership.*" Thus, price and data transparency seem to be key rationales for building up trustful, dependent, and long-term relationship through CSPs. These aspects are stressed by the majority of the survey respondents as key-development areas for promoting relational forms of procurement practices in the Danish construction industry.

The TRUST partnership is legitimized by the ability to capitalize on innovations, repetition effects, and fewer legal disputes (Frederiksen, 2021; Gottlieb et al., 2020a, 2020b), where the internal partners are addressing some of the above key-rationales. These aspects constitute the TRUST partnership as interesting in relation to the development of CSPs both building on established legitimacy and complying with supranational regulations (EU/UN). Despite the lack of strategic supplier involvement, the Director of TRUST can see a major potential in expanding the current organizational boundaries: "*There is a huge potential if we dare to open up the entire value chain. This is where the true trust reveals and you solve things together openly and honestly, which gives better processes and economy. The effect is that you are two about solving the challenges instead of just one.*" It is therefore surprising that a more strategic and long-term approach has not yet been initiated across the value chain and especially given that many issues in the EU construction industry can be attributed to traditional procurement practices. However, the current partnership trend, coupled with supranational regulations and policies (EU/UN), may act as an important lever for legitimizing the development of CSPs and resource economics marketization through joint activities and new value orientations, rather than relying on activities that are mobilized by individual actors (Gray & Purdy, 2018; Villani et al., 2017).

Products and services from the supply chain and the subcontractors account for the majority of costs, work activities, and resource consumption in projects. Moreover, the diffusion and mobilization of CE business models is highly linked to the development of CSPs, which together reveals the supplier's key-role in the realization of a resource economic market that may answer the expectations of supranational authorities. As such, it is difficult for a single actor, such as the contractor, or the TRUST partnership to address these urgent issues without closer and more relational cooperation with the supply chain or the subcontractors. The assumption is that the competencies, resources, and innovation potentials that are embedded at the external actors must be legitimized through the development of CSPs for a resource economic market to be established. This requires a confrontation with the various aspects of regulatory, moral, and pragmatic legitimacy, which currently only reproduce the cognitive legitimacy or "*taken-for-grantedness*" of the linear practices, structures, and economy. In perspective, CSPs can be seen as a new approach to better align strategic activities with the external actors of the supply chain in a common effort of orchestrating a resource economic marketization as an arena for CSPs.

There is a development gap in how actors rethink their value network, the organizational structure, and the relationship with, e.g. clients, supply chain partners, and end-users (Centobelli et al., 2020) complying with future regulations, policies, and ambitions (EU/UN). Resource economic marketization requires organizational change (Brunsson & Jutterström, 2018) and actors creating legitimacy through CSPs processes. This implies working on legitimacy within an interorganizational system and accepting a common set of structures, values, and beliefs (Demil et al., 2018) stressing the cognitive dimension of CSPs (Martins et al., 2015). Yet, the process of creating legitimacy for a new market is often a challenge for the involved actors, especially in relation to competing and conflicting linear perceptions and practices (Hedberg & Lounsbury, 2021). However, CSPs may potentially create a necessary nexus in the progress of new CE norms, regulations, and practices overcoming current tensions between competing practices and beliefs in the realization of a resource economic marketization in the Danish construction industry. Criticizing the mechanisms of the linear economy is important since the overall construction industry is one of the major contributors to global warming, pollution, and resource scarcity. The EU is trying to accelerate the transition to a CE with measures aimed at reducing resources and waste through regulations, action plans, and taxonomy for sustainable investment. These initiatives stress the need for an ambitious sustainable development, which is one of the most prevailing societal challenges in modern times. This involves changes in supply, demand, and distribution of resources, which may lead to either conflict or cooperation in the society. CSPs may thus be considered as an important initiative and driving force for sustainable development through new forms of strategic partnerships with the suppliers, which address the 17 UN SDG "*Partnership for Sustainable Development*" (UN SDG, 2020).

19.6 Conclusion

We have analyzed and discussed the role of legitimacy in understanding how CSPs may act as a driving force for resource economic marketization in the Danish construction industry. The case study of the TRUST partnership indicated that relational aspects can create legitimacy on the market by decreasing the negative effects of the current structures and practices of the Danish construction industry, which often leads to, e.g. high risks, misaligned objectives, and conflicts. However, processes of CE business models and a resource economic marketization invite new forms of partnership organizations with the actors of the supply chain and should lead to reflections on the present difficulties expanding the boundaries of strategic partnerships. These difficulties are affected by an absence of trust, transparency, monopolization, and a merchant culture focusing on the lowest price instead of a more holistic pricing. As such, it would be instrumental in the current partnership boundaries that were extended involving the resources and knowledge that are nested in the supply chain through CSPs. It is thus interesting that the respondents and other research recognize a significant potential in answering the different CE challenges through a more integrated and strategic collaboration with the supply chain. This highlights the necessity for CSPs, where the strategic partnership trend in the Danish construction industry can provide an important empirical understanding on how CSPs can be formed and developed by expanding organizational boundaries, perceptions, and practices.

However, these challenges are difficult to handle in the Danish construction industry, where the fragmented and complex environment poses a barrier to common and long-term pursuits of CE objectives involving the legitimization of CSPs. In this pursuit, the EU has formalized a "*Circular Economy Action Plan*" that emphasizes a need to stimulate a new market for CE in the EU construction industry involving cross-sectoral structures, where private and public sectors create a common mission together. Based on this plan and the 17th UN SDG, the paper is contributing to new understandings of how a Danish partnership trend and supranational regulations (EU/UN) can be a lever for CSPs and over time a resource economic marketization in the Danish construction industry. Moreover, the paper is strengthening the research field of supply chain operations with a greater regard for a circular transition and resource economics understandings. In this perspective, further research should address a deeper empirical understanding of legitimation processes that can promote strategic partnerships in the supply chain targeting an increase of circular and relational principles in organizations work mode.

References

Angelis, R. D., Howard, M., & Miemczyk, J. (2018). Supply chain management and the circular economy: Towards the circular supply chain. *Production Planning & Control – The Management of Operations, 29(6)*, 425–437.

Bitektine, A., & Haack, P. (2015). The "macro" and the "micro" of legitimacy: Toward a multilevel theory of the legitimacy process. *The Academy of Management Review, 40(1)*, 49–75.

Boxenbaum, E., & Jonsson, S. (2017). Isomorphism, diffusion and decoupling: Concept evolution and theoretical challenges. In R. Greenwood, C. Oliver, T. B. Lawrence, & R. E. Meyer (Eds.), *Organizational institutionalism* (pp. 77–102). SAGE.

Brunsson, N., & Jutterström, M. (2018). *Organizing and reorganizing markets*. Oxford Scholarship.

Callon, M. (2017). *L'emprise des marchés: Comprendre leur fonctionnement pour pouvoir les*. La Découverte.

Centobelli, P., Cerchione, R., Chiaroni, D., Del Vecchio, P., & Urbinati, A. (2020). Designing business models in circular economy: A systematic literature review and research agenda. *Business Strategy and the Environment, 29(4)*, 1734–1749.

Construction Copenhagen. (2016). *Appendix 2 – Task description offer of strategic partnerships*. Construction Copenhagen: Finance Administration.

Danish Construction. (2019). *Bæredygtighedspolitik*. Dansk Byggeri.

Davies, I. A., & Doherty, B. (2019). Balancing a hybrid business model: The search for equilibrium at cafédirect. *Journal of Business Ethics, 157(4)*, 1043–1066.

Deephouse, D. L., & Carter, S. M. (2005). An examination of differences between organizational legitimacy and organizational reputation. *Journal of Management Studies, 42*(2), 329–360.

Deephouse, D. L., Bundy, J., Tost, L. P., & Suchman, M. C. (2017). Organizational legitimacy: Six key questions. In R. Greenwood, C. Oliver, T. B. Lawrence, & R. E. Meyer (Eds.), *The SAGE handbook of organizationel institutionalism* (pp. 28–54). SAGE.

Demil, B., Lecocq, X., & Warnier, V. (2018). "Business model thinking", business ecosystems and platforms: The new perspective on the environment of the organization. *Management, 21(4)*, 1213–1228.

Díez Martín, F., Blanco González, A., & Prado Román, C. (2021). The intellectual structure of organizational legitimacy research: A co-citation analysis in business journals. *Review of Managerial Science, 15*, 1007–1043.

Eurocities. (2020). *Public procurement of circular construction materials*. EU Big Buyers Initiative Lessons Learned Report.

European Commission. (2021). *EU taxonomy for sustainable activities*. Retrieved from https://ec.europa.eu/info/business-economy-euro/banking-and-finance/sustainable-finance/eu-taxonomy-sustainable-activities_en

Farla, J., Markard, J., Raven, R., & Coenen, L. (2012). Sustainability transitions in the making: A closer look at actors, strategies and resources. *Technological Forecasting & Social Change, 79*(*6*), 991–998.

Flyvbjerg, B. (2010). Fem misforståelser om casestudiet. In S. Brinkmann & L. Tanggaard (Eds.), *Kvalitative metoder* (pp. 463–487). Hans Reitzels Forlag.

Frederiksen, N. (2021). *Configuration, collaboration and complexity in strategic partnerships: An institutional inquiry conducted from the inside of a strategic partnership in Danish construction*. Aalborg University Press.

Fredslund, L. (2021). *Implementation of construction supply chain management: An institutional analysis of logics, decouplings and hybrid procurement practices*. Aalborg University Press.

Geissdoerfer, M., Morioka, S. N., Carvalho, M. M., & Evans, S. (2018). Business models and supply chains for the circular economy. *Journal of Cleaner Production, 190*, 712–721.

Górecki, J., Núñez-Cacho, P., Corpas-Iglesias, F. A., & Molina, V. (2019). How to convince players in construction market? Strategies for effective implementation of circular economy in construction sector. *Cogent Engineering, 6*(*1*), 1–22.

Gottlieb, S. C., Frederiksen, N., Koch, C., & Thuesen, C. (2020a). Hybrid organisations as trading zones: Responses to institutional complexity in the shaping of strategic partnerships. *Construction Management and Economics, 38*(*7*), 603–622.

Gottlieb, S. C., Thuesen, C., Frederiksen, N., & Berg, J. B. (2020b). *Strategiske partnerskaber i Københavns Kommune: Erfaringer og resultater 2017–2019* (BUILD Rapport Bind Nr. 26). Institut for Byggeri.

Gray, B., & Purdy, J. (2018). *Collaborating for our future: Multistakeholder partnerships for solving complex problems*. Oxford Scholarship Online.

Greenwood, R., Suddaby, R., & Hinings, C. R. (2002). Theorizing change: The role of professional associations in the transformation of institutionalized fields. *The Academy of Management Journal, 45*(*1*), 58–80.

Hedberg, L. M., & Lounsbury, M. (2021). Not just small potatoes: Cultural entrepreneurship in the moralizing of markets. *Organization Science, 2*(*2*), 433–454.

IBMin. (2021). *National strategi for bæredygtigt byggeri*. Indenrigs- og Boligministeriet (IBMin).

IPCC. (2021). *Climate change 2021: The physical science basis* (IPCC Sixth Assessment Report). IPCC.

Kahneman, D., & Tversky, A. (1972). Subjective probability: A judgment of representativeness. *Cognitive Psychology, 3*(*3*), 430–450.

Kirchherr, J., Piscicelli, L., Bour, R., Kostense-Smit, E., Muller, J., Huibrechtse-Truijens, A., & Hekkert, M. (2018). Barriers to the circular economy: Evidence from the European Union (EU). *Ecological Economics, 150*, 264–272.

Koch, C., & Polesie, T. (2021). Building a market for circular economy – A social construction of business models. In *Proceedings New Business Models conference 2021*. Halmstad.

Lacy, P., & Rutqvist, J. (2015). *Waste to wealth – The circular economy advantage*. Palgrave Macmillan.

Lahane, S., Kant, R., & Shankar, R. (2020). Circular supply chain management: A state-of-art review and future opportunities. *Journal of Cleaner Production, 258*, 1–16.

Lounsbury, M., & Glynn, M. A. (2001). Cultural entrepreneurship: Stories, legitimacy, and the acquisition of resources. *Strategic Management Journal, 22*(*6–7*), 545–564.

Lüdeke-Freund, F., Gold, S., & Bocken, N. M. (2019). A review and typology of circular economy business model patterns. *Journal of Industrial Ecology, 23(1)*, 36–61.

MacArthur. (2015). *Potential for Denmark as a circular economy – A case study from delivering the circular economy*. Ellen MacArthur Foundation.

Martins, L. L., Rindova, V. P., & Greenbaum, B. E. (2015). Unlocking the hidden value of concepts: A cognitive approach to business model innovation. *Strategic Entrepreneurship Journal, 9(1)*, 99–117.

Meng, X., Sun, M., & Jones, M. (2011). Maturity model for supply chain relationships in construction. *Journal of Management in Engineering, 27(2)*, 97–105.

Meyer, J. W., & Rowan, B. (1991). Institutionalized organisations: Formal structure as myth and ceremony. In W. W. Powell & P. J. DiMaggio (Eds.), *The new institutionalism in organizational analysis* (pp. 41–82). Chicago Press.

Nußholz, J. L. (2017). Circular business models: Defining a concept and framing an emerging research field. *Sustainability, 9(10)*, 2–16.

Nußholz, J. L., Nygaard Rasmussen, F., & Milios, L. (2019). Circular building materials: Carbon saving potential and the role of business model innovation and public policy. *Resources, Conservation and Recycling, 141*, 308–316.

Ranta, V., Aarikka-Stenroos, L., & Mäkinen, S. J. (2018). Creating value in the circular economy: A structured multiple-case analysis of business models. *Journal of Cleaner Production, 201*, 988–1000.

Rizos, V., Behrens, A., & van der Gaast, W. (2016). Implementation of circular economy business models by small and medium-sized enterprises (SMEs): Barriers and enablers. *Sustainability (Basel, Switzerland), 8(11)*, 2–18.

Schenkel, M., Caniëls, M. C., Krikkea, H., & van der Laan, E. (2015). Understanding value creation in closed loop supply chains – Past findings and future directions. *Journal of Manufacturing Systems, 37*, 729–745.

Siggelkow, N. (2007). Persuasion with case studies. *Academy of Management Journal, 50(1)*, 20–24.

UN SDG. (2020). *The SDG report*. United Nations Publications.

Villani, E., Greco, L., & Phillips, N. (2017). Understanding value creation in public-private partnerships: A comparative case study. *Journal of Management Studies, 54(6)*, 876–905.

Chapter 20
The Impact of the EU Taxonomy of Sustainable Finance on the Building Field

Christian Koch, Martine Buser, and Rickard Andersson

Abstract If the capitalist economy could be redirected to serve sustainability, it would bring hope for having a chance to mitigate the ongoing climate change. Following the Paris accord, the EU has undertaken several initiatives to regulate the private sector activities among which the EU taxonomy. The taxonomy aims at directing private finance investments by providing a classification system establishing a list of environmentally sustainable economic activities. The question this paper aims to address is what are the implications of regulation of finance and investments for companies in the building and real estate sector? Drawing on institutional theory, focusing on legitimacy, we follow the process of introducing the taxonomy within three companies – a contractor, a real estate company, and a consulting engineering company – to identify the changes the taxonomy can introduce in their business. Keeping in mind that establishing the regulation and the taxonomy is a compromise between politicians, industry lobbyists' interventions, and technical experts' advices, the proposed solutions may not be able to achieve the aforementioned goals. The cases were selected as possible forerunners of the impact of the taxonomy in Sweden. Two of the companies, a contractor and a real estate company, engaged recently in systematizing their measures, documenting their climate impact, and establishing strategic goals for reduction of energy and material consumption. The third, a consulting engineer company, has so far felt comfortable doing progressive statements. The extra effort of establishing documentation was carried out within 3 years during which the companies have had an unusually good economy. The analysis shows that the taxonomy is not demanding enough to provide the needed push for living up to the Paris accord and consequently risks becom-

C. Koch (✉)
Department of the Built Environment, Aalborg University, Copenhagen, Denmark
e-mail: christiank@build.aau.dk

M. Buser · R. Andersson
Department of Architecture and Civil Engineering, Chalmers University of Technology, Gothenburg, Sweden
e-mail: buser@chalmers.se; rickande@chalmers.se

G. Lindahl, S. C. Gottlieb (eds.), *SDGs in Construction Economics and Organization*, Springer Proceedings in Business and Economics,
https://doi.org/10.1007/978-3-031-25498-7_20

ing a publicly sanctioned greenwashing. The paper aims at reinforcing the development toward a sustainability of the Swedish construction sector. In doing so, it contributes to SDGs 11, 12, and 13.

Keywords Sustainable finance · Taxonomy · Sweden · Institutional theory · Legitimacy

20.1 Introduction

Upon the Paris climate summit, it has become increasingly clear that public efforts would not be able to reach the Paris agreement without the help of the private sector. The latter needs to be mobilized – and its dynamics of development accelerated – if a 2-°C goal is to be obtained. The European Union has seen a host of initiatives following the summit, such as EU's 2030 climate target plan (Hadfield & Coenen, 2022). At a national level, Sweden, for example, has launched initiatives more specifically directed to the construction sector such as the Swedish governmental law on climate declaration for new buildings, an obligatory light-weight life-cycle analysis (LCA).

The most ambitious and systematic of these initiatives is probably the coming EU regulation package of sustainable finance, the taxonomy (EU, 2022a, b, c). This set of regulations is partly accepted and partly still under development, and negotiation aims at defining and specifying a classification system of economic activities that can be considered environmentally sustainable. As part of the Green Deal, it is designed to help the EU meet the coming 2030 and 2050 climate goals. In order to achieve these goals, companies and investors must be clearly informed about what are the types of investments to make to avoid greenhouse gas emissions. The taxonomy therefore establishes criteria for around 80 economic activities, which the companies' projects, processes, or products must fulfill to be classified as sustainable. Providing this common standard, should create security for investors, prevent greenwashing, help companies to become more climate-friendly, mitigate market fragmentation, and help investors compare investments across the member states. This should help guide investments where they are most needed. Companies will be required to assess how their activities perform against the taxonomy criteria and publish these results annually. From January 1, 2023, all large companies will be required to report their taxonomy alignment, along with relevant information that may help investors assess their performance (EU, 2022a, b, c).

The regulation targets companies domiciled in EU with more than 500 employees and SMEs, listed on the stock exchange (EU, 2021). It is thought to help steering and regulating financial activity in a sustainable direction by setting demands to banks and other financial institutions. According to the EU taxonomy, an economic activity is sustainable if it "*makes a substantial contribution to at least one of six environmental goals, does not significant harm to the remaining five goals and complies with the minimum standards set out by the EU, UN and OECD (social and governance demands)*" (Ahmad, 2021). The demand element to do no significant harm is called the "Do No Significant Harm" (DNSH) principle.

The system has been required and contracted by EU, which has tasked several expert groups to develop the sustainability criteria (Schütze & Stede, 2021), influenced by industry lobbyists and in turn politicians of the EU commission and the European parliament that have negotiated and passed the adjacent legislation.

The construction and real estate sector is dependent on project financing and therefore firmly rooted in the financial market. This goes especially for clients in the private sector, which are a series of private equity foundations. Moreover, the contractors, consulting engineering and real estate companies are also in need of financing their own company development.

Building on neo-institutional theory that gives us the possibility to connect transformation of the regulative system and the reactions of organizations to these changes, the paper aims at identifying what are the implications of a public-science-driven regulation of finance and investments aiming at creating sustainability for companies in the construction and real estate sector.

The empirical material has been collected through a desk study of press releases, companies' documents, legislations, and academic productions related to the EU taxonomy. To illustrate the activities taking places in the sector, the case of three forerunner companies have been selected, i.e., companies prepared to anticipate and follow the direction taken by the new regulation.

20.2 Theoretical Framework

20.2.1 Institutional Theory

To examine the role of the taxonomy for companies operating in the building and real estate sector in Sweden, we build on the theoretical frame of institutional theory. This enables us to explore how the socially constructed forces shape organizational reality and guide organizational behavior. Scott (2014, p. 56) has defined institutions as comprised of "*regulative, normative, and cultural-cognitive elements that, together with associated activities and resources, provide stability and meaning to social life.*" These three institutional pillars mold organizational behavior through beliefs and values by providing both guidance and the necessary resources to generate certain behaviors, while also preventing others (Scott, 2014). This also includes that deviations from those institutionally prescribed behaviors should be associated with some form of cost (Lawrence et al., 2011). The regulatory pillar involves the process and ability to establish rules and compliance mechanisms in terms of different forms of rewards and sanctions that enforce conformity to them e.g., laws, rules, directives, regulations, and other formal structures of control (Scott, 2014). Organizations are therefore expected to respond to regulatory changes as the basis of legitimacy is what is legally sanctioned (Ranta et al., 2018).

Legitimacy has been described as a key component to institutional change (Suddaby & Greenwood, 2005; Suddaby et al., 2017) and has been defined by Suchman (1995, p. 574) as a "*generalized perception or assumption that the action of an entity are desirable, proper or appropriate within some socially constructed system of norms, values, beliefs and definitions.*" An underlying assumption is that organizations are actively seeking to either maintain or increase their legitimacy to gain acceptance within their institutional environment and thereby ensure their long-term survival (Mignerat & Rivard, 2009). Legitimacy can be described as a process of collective meaning-making that is actively negotiated through language, communication, and translation.

Organizations have been described to *strive* to gain greater legitimacy among the actors within their institutional environment as a means to thereby obtain better organizational results as well as improved access to resources (Díez-Martín et al., 2013). But even though the creation of a legislative framework may support the transition toward more sustainable practices, the lack of legitimacy for the practices proposed by it may result in dismissal among the actors (Ranta et al., 2018). Also, that organizations strategically carry decoupling activities where they ceremonially respond to formal rules to gain legitimacy from outside, but at the same time keep their internal practices untouched (Stål & Corvellec, 2018; Kern et al., 2017). This indicates that any given initiative is decided through a holistic combination of all institutional pillars where the legitimacy of the regulative pillar in itself is not capable to induce change within an institutional environment (Ranta et al., 2018). Although the regulative pillar is more easily manipulated and can be planned in a more strategic manner, changes in the regulative pillar still require support from elements in the normative and cultural-cognitive pillar to mitigate the risk of becoming superficial (Scott, 2010).

20.3 Method

This paper builds on an interpretative approach and a desk study to gather the empirical material. Following the emergence of an EU legislation is unusually complicated as it involves a heavy bureaucracy, European political institutions, a large number of private stakeholders, and multiple national arenas among others. To follow this development, we have followed a process theory perspective which gives particular attention to the chronological ordering of the contributory events as a way of capturing the key factors that explain the role of different actors in shaping policy and regulatory changes (Pierson, 2004). The desk study encompasses the regulative frame, the gathering of legislations, directives, consultations, analysis of selected responses to the hearing rounds, and other regulatory measures at the European and national level (namely Sweden), as well as public and professional media coverage and academic production on the topic of taxonomy. The collection of this material covers the period from April 2021 to February 2022 and is part of an EU-funded project since August 2021.

We have chosen to follow up three selected companies that can be described as forerunners. The three companies have communicated publicly about their process to comply with the EU taxonomy demands. All three are large shareholder companies are due to make their 2021 annual reports accordingly. Two companies, Wästbygg and Platzer have used the taxonomy to enter the Nasdaq Green Equity notation and the third COWI is a non-noticed shareholder company, which nevertheless have taken important steps in the direction pointed to by the EU taxonomy. To document the three case studies, we have collected information from companies' homepages, press coverages, annual reports, and external evaluations carried out by consulting companies (so-called "second opinion").

The frame of understanding is based on a selective literature review drawing papers related to sustainable development and institutional theory.

20.3.1 The Emergence of the Taxonomy

The emergence of the taxonomy frames the following three cases. An important milestone was the issue of an EU action plan related to the New Green Deal (European Commission, 2018). At this time, a technical expert committee was tasked to develop the needed instruments for the regulation of the financial market in a sustainable direction. The expert committee came up with a proposal of criteria for almost 80 climate change mitigation activities and almost 100 climate change adaptation activities, including do-no-significant-harm criteria across six environmental objectives. These are:

1. Mitigation criteria
2. Climate change adaptation
3. Sustainable use and protection of water and marine resources
4. Transition to a circular economy
5. Pollution prevention and control
6. Protection and restoration of biodiversity and ecosystems

An organization of panels of interest groups was established in the process. This included a member state expert group with ministry representatives from the member states; the technical expert group established in 2018, which contributed to the basic creation of the taxonomy and contained people with scientific background; and the sustainable finance platform established in 2020, which included a large number of industry interests (associations), but also NGOs such as Green Building Council and Ellen Macarthur Foundation among the main actors. We condense this contribution under what is called the "social dialogue panels." The legislation on non-financial disclosure regulating the listed companies was accepted in June 2020.

The commission established a second-generation experts' panel calling it the platform of sustainable finance. The appointment of members for this 50 members panel was done through an open call, meaning banks like SEB, Allianz, and BNP Paribas, became members. The Platform on Sustainable Finance is supposed to work on the missing parts of the taxonomy, i.e., protection of biodiversity, the transition to a circular economy, protection of water and marine resources, pollution prevention, social aspects, and environmentally harmful activities. The legislation of the first two aspects of the taxonomy, climate mitigation and adaption, was passed in June 2021. As a dispute between nations members over nuclear power and natural gas as being recognized as part of the transition, broke out, the technical expert's panel threatened to withdraw (November 2021). By January 2022, the taxonomy for climate mitigation covers seven business sectors: agriculture and forestry, manufacturing, energy, transport, buildings, water, waste and sewage, and information and communication technologies (ICT) (Schütze & Stede, 2021). More specific to the construction sector, the taxonomy identifies two technical screening to take into consideration: the first one for "substantial contribution to climate change mitigation" (EU, 2021) for construction of new building and includes the following:

1. The energy performance of the building should be at least 10% lower than the threshold set for the nearly zero-energy building requirements in national legislation.
2. Buildings larger than 5000 m^2 should be tested for air-tightness and thermal integrity, and any deviation should be disclosed to investors and clients.
3. As part of the transition to a circular economy, at least 70% (by weight) of the non-hazardous construction and demolition waste should be recycled to new buildings.
4. Building designs and construction techniques should support circularity and in particular demonstrate the ability for disassembly or adaptability of buildings, i.e., to be designed to be more resource efficient, adaptable, flexible, and dismantable to enable reuse and recycling.

The second for "substantial contribution to climate change adaption" (EU, 2021) for existing buildings and new buildings and includes the following:

1. The building is not dedicated to extraction, storage, transport, or manufacture of fossil fuels.
2. The energy performance of the building resulting from the construction does not exceed the threshold set for the nearly zero-energy building.
3. The specified water use for the following water appliances should comply with (a) wash hand basin taps, and kitchen taps have a maximum water flow of 6 L/min; (b) showers have a maximum water flow of 8 L/min; (c) WCs have a full flush volume of a maximum of 6 L and a maximum average flush volume of 3.5 L; (d) urinals use a maximum of 2 L/bowl/h. Flushing urinals have a maximum full flush volume of 1 L.
4. As part of the transition to a circular economy at least 70% (by weight) of the non-hazardous construction and demolition waste should be prepared for reuse, recycling, and other material recovery, including backfilling operations using waste to substitute other materials.
5. Building designs and construction techniques support circularity.

Schütze and Stede (2021) find that in the hearing, one of the most frequent critiques was that there was no absolute energy efficiency goal as part of the building renovation threshold as each country can adapt the measures. Besides, half of the stakeholders responding to the building renovation threshold proposed a tightening of the threshold. This is especially interesting because two-thirds of the respondents in this section are companies or industry associations, which underlines that there are also corporate stakeholders with an interest in higher efficiency goals. However, the Swedish real estate owners advocate a lowering of the demands with reference to the high level of the energy labels in Sweden.

20.3.1.1 Three Forerunners

Case 1: Wästbygg – A Forerunner of the Transformation the Taxonomy Is Triggering

Wästbygg, a west Swedish contractor, has grown significantly in the last 10 years from a 2012 turnover at 1.1–3.8 billion Swedish kroners in 2020 and 2021. Moreover, the company has restructured from a personally owned company, sold 2013, to becoming a noticed shareholder company in 2021. Its first independent sustainability report was issued in 2016. A goal of becoming fossil-free by 2030 was announced in 2017. By 2020, the company measure reports six key performance indicators for the sustainable performance of a contractor, including waste volumes, material transport (diesel consumptions and emission), machines on site and provisional site heating, electricity and district heating consumption and travel, using a greenhouse gas protocol methodology. The results of total emissions for 2020 was 1523 ton CO2e, which was a reduction at 32% compared to 2019, which was at 2253 ton CO2e (Wästbygg Annual Report 2020).

In the spring of 2021, the company prepared themselves for entering the Nasdaq stock exchange. As part of this, a preparation for using green finance was carried out. Hiring the external evaluator Cicero Shades of Green, that used the taxonomy, the company obtained the following results (Table 20.1):

Cicero found, summarizing the qualitative evaluation listed above, that 65% of the income, 66% of the operational costs (OPEX), and 100% of the investments (CAPEX) from Wästbyggs as a whole could be identified as sustainable according to the taxonomy.

In June 2021, Wästbygg received the Nasdaq Green Equity designation. This noticed share is dedicated to companies with more than 50% of the turnover occupied in sustainable activities and investments predominantly are done in sustainable activity. The turnover from fossil fuels has to be less than 5%.

The Cicero evaluation pointed to a few improvement points such as that "Wästbygg needs to identify physical climate risks and adaptation solutions for their activities. The company informs that they will develop an approach to climate the risk assessments in 2021" (Cicero Shades of Green, 2021).

By February 2022, the company had obtained a higher degree of sustainability as indicated in the table from the annual report of 2021. At the same time, the company

Table 20.1 Summary of Cicero's 2021 evaluation of Wästbygg

	Taxonomy criteria	Cicero evaluation of Wästbygg May 2021
1	Mitigation criteria	Green debt eligibility criteria are likely aligned for Swedish properties.
		Alignment on EPC B cannot be confirmed for properties in Norway, Denmark, and Finland.
		Likely aligned with criteria related to air-tightness and thermal integrity
		Not aligned to GWP requirement for current projects on buildings larger than 5000 m^2
2	Climate change adaptation	Likely partially aligned.
3	Sustainable use and protection of water and marine resources	Likely not aligned.
4	Transition to a circular economy (circular economy)	Likely aligned.
5	Pollution prevention and control	Likely aligned with self-developed properties. Alignment for contracted developments cannot be confirmed
6	Protection and restoration of biodiversity and ecosystems	Likely aligned with EIA requirement. Alignment toward construction on arable or forested land for existing properties cannot be confirmed

GWP global warming potential, *EPC* energy performance certificate, *EIA* Environmental Impact Assessment

announced that it considered all its activities covered by EU taxonomy and that they would publicly share their status later in the spring of 2022. However, due to the number of full-time employees, 350, they did not concern themselves as eligible of following the taxonomy.

To receive the Nasdaq Green Equity and carry out a Cicero screening of activities according to the taxonomy in spring and summer of 2021 makes Wästbygg a fore-runner, probably the first contractor to go this far in Sweden.

Case 2: Platzer

Platzer is a medium-sized real estate company operating in commercial real estate in the area of greater Gothenburg. Platzer manages and develops approximately 70 properties of just over 800,000 m^2 (www.platzer.se, 2022). Platzer got listed on the stock exchange in 2013. One of their profiled development projects concerns the developments of an old factory area (Gamlestadens Fabriker). The detailed plan for the district gained legal force in 2021 and enabled Platzer to start working on the area development, adding 300 homes and 68,000 m^2 offices, restaurants, and services (Platzer Annual Report 2021).

In 2018, Platzer ran a set of development activities that made them announce a focus on the UN goals, numbers 5, 7, 8, and 11. The following year this focus had been developed and fine-tuned into GRI (Global report initiative) index that Platzer commenced to comply with.

Table 20.2 Summary of Cicero's 2021 evaluation of Platzer

	Taxonomy criteria	Cicero evaluation of Platzer
	Acquisition and ownership of buildings (7.7)	
1	Mitigation criteria	Unable to determine alignment to energy efficiency criteria
		Likely aligned with energy management criteria
2	Climate change adaptation	Likely aligned
	Construction of new buildings (7.1)	
1	Technical screening	Likely aligned (70% of CAPEX)
		Likely aligned with criteria on air-tightness and thermal integrity
		Likely not fully aligned with criteria on GWP
2	Climate change adaptation	Likely aligned
3	Sustainable use and protection of water and marine resources	Likely not fully aligned.
4	Transition to a circular economy (circular economy)	Likely aligned with criteria on waste management
		Likely not aligned with criteria on circular economy
5	Pollution prevention and control	Likely aligned
6	Protection and restoration of biodiversity and ecosystems	Likely aligned

In 2021 Platzer commenced preparing themselves for entering the green finance area. In June 2021, Platzer was evaluated by Cicero Shades of Green, which made an evaluation vis-à-vis the EU taxonomy (Table 20.2).

The evaluation identifies two relevant EU taxonomy criteria areas: first acquisition and ownership of buildings, covering revenue and operating costs, and second construction of new buildings and renovation, covering investments. Cicero assesses that Platzer had no fully taxonomy-aligned turnover, OPEX, or CAPEX in 2020. Cicero estimates that it is "likely" that 70% of 2020 CAPEX would be aligned with the mitigation criteria for new construction (the second area of criteria). However, the assessment of the alignment is according to Cicero contingent on Platzer meeting their expected energy efficiency level. Cicero posits not to able to determine the alignment of Platzer's activities related to the technical mitigation thresholds for acquisition and ownership of buildings or renovation (the first area of criteria).

Platzer is evaluated to be "likely aligned" with the relevant DNSH criteria on climate change adaptation, but has, according to Cicero, gaps related to sustainable use and protection of water and marine resources and the transition to a circular economy. Cicero considers that Platzer does not currently fulfill all the minimum social safeguards of the EU taxonomy. See also the table.

Platzer receives the Nasdaq Green Equity designation in August 2021, as the second company in Sweden (Wästbygg being the first). By September Platzer could report that they had carried out their first emission of green bonds. Platzer has within the frames of their green bond program at 5000 million SEK, emitted one senior unsecured green bond. This bond is a variable loan with a max of 600 MSEK.

Case 3: COWI – A Proclaimed Forerunner

COWI is a broad-spectrum consulting engineering company with its main business areas being Denmark, Infrastructure, Norway, Sweden, Architecture in this ranking. The company also possesses extensive environmental competences. Notably the purchase of Arkitema in 2018 included 550 employees to the 6700 employees that already were in the company at the time. The growth of the two first areas (Denmark and infrastructure) have been continual over the last 10 years and involves huge material flows of what the greenhouse gas protocol would denote scope 3 emissions. Business area in Sweden involves around 1000 employees and has been characterized by a somewhat shaky development.

In 2019, COWI's management decided to aim at carbon neutrality in 2020 and to reduce actual direct and indirect CO2 emissions by more than 70% in 2030 compared to 2008. Carbon neutrality was, according to COWI, achieved in 2020 thanks to reduction efforts already made, reduced corporate travel due to COVID-19 and by acquiring CO2 credits (COWI Annual Report 2020). This neutrality does not encompass scope 3 emissions. In January 2022, a strategic CEO statement of COWI attracted attention. Within a few years, COWI announced the company will turn its back on everything to do with climate-damaging projects. The plan is, according to COWI, that they will restructure the business over the next 3–5 years. It will no longer have to deal with fossil energy and construction projects without a sustainable approach. The CEO also promised to leave behind a business area of natural gas. Challenged on their main business area "large bridges" the CEO made a distinction between sustainability and CO2 reductions (Andersen, 2022) and notes that COWI will still engage in making large bridges.

There are several reasons for why this strategy is not linked to the EU taxonomy. First COWI is not a stock exchange listed company; secondly, as a consulting engineering company they are not at present part of the taxonomy coverage. Nevertheless, COWI is a large company under scrutiny of Danish authorities together with 100 other large companies with seat in Denmark. Moreover, COWI could choose to follow the taxonomy, especially in the scope 3 area, where buildings they design would be placed. Buildings are developed above part of the taxonomy. A competitor on the Nordic market, AFRY, appears to have engaged for more profoundly in reporting CO2e emission. The announcement of the COWI CEO did immediately triggered questions from collaboration partners that became worried how COWI's move would impact on their own business.

Summarizing, although there is a risk that the CEO announcement will not be followed up by concrete actions, and results, we still consider that the announcement has placed COWI as a forerunner in sustainable transition of the building industry.

20.4 Analysis and Discussion

The taxonomy regulation is aiming at creating sustainability in companies' activities in the building and real estate sector. The initial legitimacy of the taxonomy is derived from the external experts enrolled to develop the regulation and the later reduced legitimacy is related to politicians and lobbyist impacting the developed criteria. From the process occurs, however, a set of passed legislations that have direct impact on the three cases presented. As we noted, the taxonomy is far from the only driver and/or direction giver in this field. Actually, numerous other dynamics, some enforcing, other contradictory, and other counter forces are active. One aspect of this is that the delegated acts and the adjacent taxonomies appear to come without specified consequences if not respected.

The three forerunners commence implementing enforced measurement, documentation, as well as CO2e reduction strategy, something that occurs partly in parallel with the overall process. There are, however, serious counter concerns: Cicero's evaluations of both Platzer and Wästbygg appear to indicate that the taxonomy is too easy to comply with. Moreover, the Nasdaq Green Equity designation only requests that the two companies obtained a mere 50% degree of sustainability scale. This means that for investing and financing the two companies will in the future be able to choose between green bonds and shares from the Nasdaq context and financing from banks that follow the taxonomy criteria. This implies that it is an easier path to follow to opt for Nasdaq financing when possible. In actual building projects when allying with other companies and clients it might be more likely to see banking financing compliant to the taxonomy.

Schütze and Stede (2021) find places in the taxonomy where it exhibits weaknesses and not clearly levers itself from present day standards. Certain parts of the economy are also held outside of the regulative frame. A similar pattern of possibility to low demands in the taxonomy applies for other real estate and consulting engineering companies, that have asked for a second opinion from Cicero Shades of Green during 2021 (Cicero, 2021). If this turns out to be a general tendency, the EU taxonomy risk entering a contradictory set of logics as EU will need tightening its demands if the Paris accord is to be realized, whereas equity financing will become even more attractive. Moreover, voluntary mechanisms (standards and certifications) that go further than the taxonomy are likely to delegitimize it as well as Nasdaq Green Equity. This also goes for COWI's strategy, which might prove highly profitable and possibly far less bureaucratic. However, we doubt whether COWI can continue avoiding reporting its direct and indirect emissions in the future.

As for COWI's development, one can note that the company is engaged in highly political infrastructure projects, which probably will need much stronger climate impact assessments in the future. Although COWI's role in the recent Danish governmental crises has not surfaced in the public, professionals in this area knows that it was COWI's tool of LCA evaluation of infrastructure that became questioned and led to the fall of a minister.

Both the character of the overall legislation and our three examples seem to suggest that there is a risk of approving something as sustainable that is neither state of the art nor scientifically grounded. This suggests that a changed form of greenwashing is occurring – from a simple form, where, for example, a green roof and electrical car charger might be called a sustainable house, to a governmentally approved annual reports and or climate declarations that still are misleading in terms of degree of sustainability.

Wästbygg, Platzer, and COWI all report very strong economic results for 2020 and 2021. It is thus from a position of strength that the companies engage with a higher priority of sustainability. It should be noted also that Platzer's financing need of new development project up to 50% degree is dependent on bank loans.

We find the legitimacy of the taxonomy and of the companies as a fluid issue that has to be addressed and re-established continually. The present dip in the EU taxonomy legitimacy because of the political entry of nature gas and nuclear power, might be substituted with a far broader interest in a number of sectors and countries for the possibilities of living up the criteria and the adjacent legitimacy gained in the public and in B2B relations. In a somewhat parallel manner, it is likely that not only COWIs but also Wästbygg and Platzers legitimacy might reduce. While Platzer and Wästbygg measure their development, they are able to maintain their legitimacy as long as they continue their efforts and demonstrate results. COWI has placed themselves in a far more vulnerable situation with their deep engagement in the climate issue as professionals and lack of measures published.

20.5 Conclusion

This paper raised the question of what the implications of the taxonomy as a regulation of finance and investments are, aiming at creating sustainability for companies in the building and real estate sector. We have presented the overall process of introducing the taxonomy and the three cases illustrating the changes the taxonomy can imply in the building sector.

The process of establishing the regulation and the adjacent taxonomy has occurred through continual tensions between the participating actors of state, science, and industry. This had meant that the content of the regulation has fluctuated along with the negotiations between politicians, industry lobbyists' interventions, and experts' panels. The consequence of these fluctuations is that the demands are reduced and the taxonomy loses its legitimacy as regulation for transitioning to sustainability.

We also presented and analyzed three cases viewed as possible forerunners of the impact of the taxonomy. The taxonomy appears to join other forces in a push for measurement and documentation of the company's impact on the climate, both direct and indirect. Especially the indirect effects are important in a project-based industry delivering larger complex products. Two of the companies engaged relatively recently in systematizing their measure and documentation of their climate impact and also established strategic goals of reduction. But it should also be noted that this extra effort of establishing documentation was carried within the 2 or 3 years where the companies have had an unusually good economy.

According to Scott, for the regulative pillar to change an institution (here sustainable building), one should establish rules and compliance mechanisms to frame and motivate the transition to climate change. However, its performativity can be questioned as our analysis shows that the taxonomy is not demanding enough to provide the needed push for living up to the Paris agreement and consequently risks becoming a publicly legitimized greenwashing.

References

Andersen, M. M. (2022). Cowi vil arbejde '100 % bæredygtigt': Er det så slut med at opføre store broer? https://ing.dk/artikel/cowi-vil-arbejde-100-baeredygtigt-saa-slut-med-at-opfoere-store-broer-253766

Ahmad, S. (2021). *The EU Taxonomy*. The Seven Pillars Institute. Downloaded from https://sevenpillarsinstitute.org/the-eu-taxonomy/

Cicero. (2021). *The Cicero shades of green homepage*. Accessed in February 2022.

Díez-Martín, F., Prado-Roman, C., & Blanco-González, A. (2013). Beyond legitimacy: Legitimacy types and organizational success. *Management Decision, 51*, 1954–1969.

EU. (2021). *FAQ: What is the EU Taxonomy and how will it work in practice?* Retrieved February 17, 2022 https://ec.europa.eu/info/sites/default/files/business_economy_euro/banking_and_finance/documents/sustainable-finance-taxonomy-faq_en.pdf

EU. (2022a). Reference to strategy package as of July 2022.

EU. (2022b). Commission Delegated Regulation (EU) …/… of 4.6.2021 reference to delegated act from June 2021.

EU. (2022c). *Members and Observers of the Platform on Sustainable Finance. Overview of plenary and subgroups*. European Commission.

European Commission. (2018). Action Plan: Financing Sustainable Growth, European Comission: Brussels

Hadfield, P., & Coenen, L. (2022). Contemporary financial capitalism and sustainability transitions in urban built environments. *Environmental Innovation and Societal Transitions, 42*(2022), 285–300.

Kern, A., Laguecir, A., & Leca, B. (2017). Behind smoke and mirrors: A political approach to decoupling. *Organization Studies, 39*, 543–564.

Lawrence, T., Suddaby, R., & Leca, B. (2011). Institutional work: Refocusing institutional studies of organization. *Journal of Management Inquiry, 20*, 52–58.

Mignerat, M., & Rivard, S. (2009). Positioning the institutional perspective in information systems research. *Journal of Information Technology, 24*, 369–391.

Pierson, P. (2004). *Politics in time: History, institutions, and social analysis*. Princeton University Press.

Ranta, V., Aarikka-Stenroos, L., Ritala, P., & Mäkinen, S. J. (2018). Exploring institutional drivers and barriers of the circular economy: A cross-regional comparison of China, the US, and Europe. *Resources, Conservation and Recycling, 135*, 70–82.

Schütze, F., & Stede, J. (2021). The EU sustainable finance taxonomy and its contribution to climate neutrality. *Journal of Sustainable Finance & Investment*. https://doi.org/10.1080/20430795.2021.2006129

Scott, W. R. (2010). Reflections: The past and future of research on institutions and institutional change. *Journal of Change Management, 10*, 5–21.

Scott, W. R. (2014). *Institutions and organizations*. Sage Publications.
Stål, H. I., & Corvellec, H. (2018). A decoupling perspective on circular business model implementation: Illustrations from Swedish apparel. *Journal of Cleaner Production, 171*, 630–643.
Suddaby, R., & Greenwood, R. (2005). Rhetorical strategies of legitimacy. *Administrative Science Quarterly, 50*, 35–67.
Suddaby, R., Bitektine, A., & Haack, P. (2017). Legitimacy. *Academy of Management Annals, 11*, 451–478.
Suchman, M. C. (1995). Managing legitimacy: Strategic and institutional approaches. *Academy of Management Review, 20*(3), 571–610.

Chapter 21
Circular Economy in the Nordic Real Estate and Construction Industry: A Policy Review

Rebecka Lundgren, Riikka Kyrö, and Stefan Olander

Abstract Circular economy (CE) has gained attention in the real estate and construction policy, practice, and research in the past few years. This study explores how two neighboring Nordic countries and EU member states, Sweden and Finland, are transitioning toward circularity in the real estate and construction industry context. The aim is reached through a review of CE policies on the national and industrial level, as well as a practice review on the city level. Common features for both countries include the promotion of circular business models, with digitalization seen as an enabler for circular economy. This indicates that the policies from both countries have moved beyond the so-called long loops and recycling focus. Still, much more focus should be placed on the short loops, such as sharing and access-based models, which are more efficient in reaching sustainable consumption targets.

Keywords Bioeconomy · Circular economy · Digitalization · Finland · Sweden

21.1 Introduction

The real estate and construction sector (RECI) is a major contributor to climate change and a key consumer of global resources (WEF, 2016), with 40% of extracted materials being in the built environment (Rees, 1999). The UN adopted 17 sustainable development goals (SDGs) in 2015. Due to the high environmental impacts of the RECI, there is an opportunity for the sector to contribute significantly toward meeting the SDGs. One way in which the sector can reduce its environmental impact

R. Lundgren (✉) · R. Kyrö
Real Estate Science, Faculty of Engineering at Lund University (LTH), Lund, Sweden
e-mail: rebecka.lundgren@lth.lu.se; riikka.kyro@lth.lu.se

S. Olander
Construction Management, Faculty of Engineering at Lund University (LTH), Lund, Sweden
e-mail: stefan.olander@construction.lth.se

G. Lindahl, S. C. Gottlieb (eds.), *SDGs in Construction Economics and Organization*, Springer Proceedings in Business and Economics,
https://doi.org/10.1007/978-3-031-25498-7_21

is through the adaptation of circular economy (CE) principles. CE, the idea of keeping resources in use indefinitely, has become increasingly relevant in popular as well as academic discourse. The two goals that are most relevant from the CE in the RECI perspective comprise SDG 11 *sustainable cities and communities* and SDG 12 *responsible consumption and production*. The EU declares to share the UN goals for a sustainable future, and the European Commission has even developed a tool with which to map how well EU policies respond to the different SDGs, with the most recent EU CE action plan relating to 14 out of the 17 SDGs (EC, 2022).

The CE concept has been gaining traction in the EU through the introduction of the Roadmap to a Resource Efficient Europe (EC, 2011), the Objectives of the Circular Economy Package (EC, 2015), the Resource Efficiency Scoreboard (EC, 2016), and A New Circular Economy Action Plan (EC, 2020). CE has also been increasing in popularity in several national governments around the world (Domenech & Bahn-Walkowiak, 2019), of which Finland and Sweden are two (Korhonen et al., 2018). However, CE policies have also received criticism due to their focus on recycling instead of reducing or reusing (Ranta et al., 2018). The New Circular Economy Action Plan (EC, 2020) only mentions five actions related to the RECI context, all of which are focused on new construction.

Similarly, in the RECI, research has mostly focused either on the design and construction of new buildings in a way that facilitates circularity in the future or on the recycling of building materials (e.g., Cruz Rios et al., 2021; Eberhardt et al., 2019). Kyrö (2020) notes that this recycling-focused approach goes against the well-known basic waste hierarchy of first reducing, then reusing, and only as a last resort, recycling. The efforts of policymakers should be put toward enabling *shorter loops* (refuse, reduce, resell/reuse, or repair) or even *medium loops* (refurbish, remanufacture, or repurpose), as opposed to the *longer loops* (recycling) (Reike et al., 2018). Research has suggested that the input side (short and medium loops) is more challenging from a policy perspective due to its connection to other policies and tends to be less measurable and non-mandatory than those policies concerning output (longer loops) (Domenech & Bahn-Walkowiak, 2019). Nonetheless, any CE policies aiming at reducing carbon emissions in the next decades should have a focal point on optimizing existing buildings as opposed to new construction.

Although the EU is highlighting CE with advanced policy frameworks, such as the New European Action Plan (EC, 2020), little is known on the implementation of the policies in member countries. The impact of the EU policy frameworks on national policies is inconsistent, and policies are fragmented (Domenech & Bahn-Walkowiak, 2019). Dytianquin et al. (2021) reviewed the application of CE in the construction and demolition sector in five EU countries, namely the Netherlands, Belgium, Germany, France, and Denmark, and found that in order for CE principles to translate to project outcomes, there needs to be political steering and thus highlighted the need for political steering in order to influence the industry toward a more circular economy. Giorgi et al. (2022) found that the implementation of CE principles in national policies in Belgium (Flanders), the Netherlands, United Kingdom, Denmark and Italy was fragmented. However, these previous policy reviews focus on countries that are very different in terms of, e.g., geographic location, regulatory environment, or demographics. Moreover, the United Kingdom is no longer an EU member state. It is therefore interesting to investigate any differences in national policies for two very similar EU countries, namely Sweden and Finland. Considering that they both fall under EU policy frameworks, the analysis will focus on any differences that might highlight a disparity between the two nations' CE frameworks as well as their level of maturity.

The purpose of the study is to explore how two EU countries, Sweden and Finland, are transitioning toward circularity in the RECI context. More specifically, we seek to find out whether and how CE policies in two very similar EU member states and neighboring Nordic countries differ. The aim is reached through a review of national and industry-level CE policies in the two countries. In order to deepen the policy comparison, one city from each country is selected and presented as a practical case example.

21.2 Research Design

This research follows the pragmatic research tradition and seeks to contribute to new knowledge that is context-dependent and useful in real life (Saunders et al., 2009). As the aim is to explore a novel phenomenon, a qualitative case study approach was deemed appropriate. Two very similar cases were selected for their expected high information content, based on an information-oriented sampling strategy (Flyvberg, 2006). The research is designed as an embedded case study with Finland and Sweden, RECI in the respective countries, and the cities of Tampere (FIN) and Malmö (SWE), are the cases. The countries were chosen due to geographical proximity and very similar regulatory and institutional environments and joint history. The industry was chosen due to its high environmental impact and expected gains from CE transition. It is worth noting that the key RECI players are active in both case countries. The selection of the cities was based on their similar size and national significance, as well as outspoken CE measures. Key cases characteristics are presented in Table 21.1.

Table 21.1 Case characteristics

Case	Country level	Industry level	City level
Finland	EU member since 1995 Population 5.53 Mio Sqm 337,030	25% of GDP 12% of workforce Largest companies: YIT, Skanska NCC	Tampere Population 226,696 Third largest city in Finland
Sweden	EU member since 1995 Population 10.35 Mio Sqm 450,295	16% of GDP 14% of workforce Largest companies: PEAB, Skanska, NCC	Malmö Population 344,166 Third largest city in Sweden

Table 21.2 Documents included in the document review

Document	Responsible	Country	Level
Circular Economy Action Plan (2017, 2019)	Ministry of Agriculture and Forestry, Ministry of Labor, Ministry of the Environment, and Finnish Innovation Fund Sitra	Finland	National
Seven goals for circularity in the RECI (2019)	Finnish Innovation Fund Sitra and Finnish Green Building Council	Finland	RECI
CE criteria for built environment projects (2019)	Finnish Innovation Fund Sitra and Finnish Green Building Council	Finland	RECI
The city of Tampere and the CE transition (2020)	City of Tampere	Finland	City
National Strategy for Circular Economy (2020)	Swedish Ministry of the Environment	Sweden	National
Roadmap for Fossil Free Construction Sector (2018)	Fossil Free Sweden	Sweden	RECI
Local roadmap city of Malmö (2020)	LFM30	Sweden	City

The data collection was based on retrieving existing documents from secondary sources. The data comprise policy documents such as the national action plans for CE from respective ministries, industry-level policies, as well as city-level documents. A full list of reviewed documents is presented in Table 21.2.

21.3 Theoretical Background

The Ellen MacArthur Foundation (2013) describes CE as restorative by intention, with value creation in every chain in the system and closing or slowing of material loops. CE has been gaining momentum since the late 1970s (Geissdoerfer et al., 2017). In 1970–1990s, the focus was on dealing with waste, and the concept of *reduce, reuse, and recycle* was gaining traction; however, the emphasis was very much on pollution limitation instead of waste reduction. There was a shift to input–output strategies in the 1990s–2010, and CE was starting to be viewed as an economic opportunity. This view is still in line with the views on CE today; however, more focus is being put on resource depletion and major sustainability challenges (Reike et al., 2018).

21.3.1 Circular Economy in the RECI

CE in the RECI is still a nascent research field, although there has been a constant stream of CE research since 2016 (e.g., Pomponi & Moncaster, 2017; Leising et al., 2018; Ness & Xing, 2017; Sanchez & Haas, 2018; Kyrö et al., 2019; Eberhardt et al., 2019; Domenech & Bahn-Walkowiak, 2019; Kyrö, 2020; Dytianquin et al., 2021; Cruz Rios et al., 2021; Lange, 2022). Outside of the academia, the Re-generate, Share, Optimize, Loop, Virtualize, Exchange (ReSOLVE) framework by the Elle McArthur foundation (EMF, McKinsey Center for Business and Environment, and SUN, 2015) was adapted to the built environment by Arup in 2016. Next, the CE elements from existing literature are grouped in accordance with the loop size presented by Reike et al. (2018), with an additional group for overarching elements. The key CE elements found in general CE and RECI specific research and practice are presented in Table 21.3.

Short Loops Decreasing demand would be the most efficient CE element in reducing the negative environmental impact from RECI (IVA, 2020). Digitalization has been highlighted in a number of studies as an enabler to CE and its elements (e.g., Antikainen et al., 2018; Agrawal et al., 2022). Another short loop aspect found in existing research is delivering access over ownership in order to satisfy users' needs without necessarily owning physical products (e.g., Kyrö, 2020; Brinkø et al., 2015). Industrial ecology focuses on industrial ecosystems, which aim to minimize

Table 21.3 Key CE elements organized by loop size

	CE element	Reference
Short loop	Decrease demand	IVA (2020)
	Digitalization	Antikainen et al. (2018); Agrawal et al. (2022)
	Space efficiency, shared spaces, access-based consumption	Bocken et al. (2014); Kyrö (2020); Brinkø et al. (2015); IVA (2020)
	Industry-level ecosystems	Korhonen et al. (2018)
Medium loop	Optimizing useful life	Leising et al. (2018); Langston et al. (2008)
	Adaptive reuse	Sanchez and Haas (2018); Conejos et al. (2015); Ellen MacArthur Foundation (2015); IVA (2020)
Long loop	Design for disassembly	Sanchez and Haas (2018); Cruz Rios et al. (2021); Eberhardt et al. (2019)
	Recycling of building materials	Domenech and Bahn-Walkowiak (2019)
	Bioeconomy	Chennouf et al. (2018); Lange (2022)
	Sustainable new construction	IVA (2020); Cruz Rios et al. (2021)
Overarching	Sustainable and circular business models	Bocken et al. (2014); Boons et al. (2013); Leising et al. (2018)
	Collaboration	Leising et al. (2018)
	Sustainable urban planning	Bott et al. (2019)
	Procurement	Ershadi et al. (2021)

raw materials from entering the system, much the same as CE short loops aim to do (Korhonen et al., 2018).

Medium Loops Leising et al. (2018) define CE for the built environment as optimizing the useful life of buildings. One way to do this is by adaptive reuse, or repurposing, of existing buildings, which is highly aligned to CE building principles (Sanchez & Haas, 2018). Prior to CE becoming mainstream in research, general sustainability research framed adaptive reuse as a superior alternative to new construction (Conejos et al., 2015; Douglas, 2006; Langston, 2008). Better use of existing infrastructure is needed for there to be the shift in RECI which is required to curb emissions (Langston, 2008).

Long Loops Current CE research in RECI is focused on new construction and design for disassembly. A key problem with focusing on new construction is that the building stock in developed nations renews itself slowly (Beccali et al., 2013; Conejos et al., 2014). Despite efforts in recycling efficiency, the demand for raw material has not decreased and, thus, the current measures are insufficient to reach resource efficiency targets (Domenech & Bahn-Walkowiak, 2019). The use of bio-based building materials is increasing around the world, and these aim to provide more sustainable raw materials (Chennouf et al., 2018). The link between CE and bio-based materials was recently made in Lange (2022).

Overarching CE Elements The development of sustainable business models is common in general sustainability research (e.g., Bocken et al., 2014; Boons et al., 2013) and to some extent in the built environment context (Leising et al., 2018). Leising et al. (2018) also highlight the importance of collaboration between stakeholders in order to deliver CE in the building sector. Sustainable procurement in a complex multidisciplinary industry such as RECI also calls for collaboration to achieve objectives (Ershadi et al., 2021). Further, sustainable neighborhoods as a result of sustainable urban planning are lifted in the architectural sector as having a positive impact on social, environmental, and economic sustainability (Bott et al., 2019).

21.4 Policy Review

Policies are recommended courses for action that do not have the binding nature of laws and regulations. The EC (2022) states that their policies are influenced by the UN SDGs, and EU policies are expected to impact the national policies in member states. Industry-level policies in a country should reflect the national and local policies, and ideally, these would be aligned (Fig. 21.1). That said, industry-level policy may also have an impact, particularly on local action plans.

Fig. 21.1 Connection between policy framework levels

21.4.1 *Finland*

21.4.1.1 National Circular Economy Policies

The first national Circular Economy Action Plan was published in Finland in 2017 as a joint initiative of the Ministry of Agriculture and Forestry, Ministry of Labor, Ministry of the Environment, and Finnish Innovation Fund Sitra. The focus of the action plan is on generating new business, promoting digitalization, and resource efficiency. Construction is mentioned in the action plan only as a waste stream, although there is a brief mention of a *construction ecosystem for CE* in connection with discussing potential circular innovation ecosystems. The second version of the action plan was published in 2019. The new version focuses on the potential for public–private collaboration. Bioeconomy is lifted as a parallel goal to CE. The action plan mentions the goal of developing specific CE criteria for the construction industry.

21.4.1.2 Circular Economy in the RECI

The development goal from the national action plan is implemented through a joint report by Finnish Innovation Fund Sitra and Finnish Green Building Council. The report outlines explicit CE criteria for built environment projects. The report uses different lifecycle phases to categorize the criteria. First, the criteria related to the planning phase include creating specific neighborhoods that follow circularity principles, as well as developing neighborhood-level energy systems. Second, the conveyance process of land plots should include a plan for circular economy. Individual building projects should follow circularity principles.

Promoting circularity in the procurement and tendering processes requires educating in circular issues, utilizing circular materials, and creating a material pass for circular materials. There should also be specific demands for planning and construction. Furthermore, the infrastructure building phase should consider the reuse of material and retaining the mass balance within a neighborhood. The building design phase should focus on preserving what is already made and recycled construction materials. The buildings should be viewed as material banks, with a plan for reuse and adaptation. Shared spaces are also included in the criteria for the design phase. The construction phase should utilize circular materials and include a material pass for circular materials. The report suggests a new process, "demolition review" prior any demolition, along with recycling and reusing the building materials. The use phase of the building should focus on the shared spaces, minimizing the need for adaptation, as well as use of recycled materials in fit-outs.

Another joint report by Finnish Innovation Fund Sitra and Finnish Green Building Council from 2019 outlines seven goals for circularity in the RECI. The seven goals are as follows. The industry should develop a joint goal to reach CE. The national legislation and industry guidelines should promote CE. The industry should adapt new ways of working that promote CE. Both infra and building construction should be based on lifecycle thinking. Urban planning should take into account the circularity principles. Furthermore, public procurement processes should promote circularity. Finally, all existing spaces should be in their most efficient use.

City of Tampere

The city of Tampere would like to become an "enabler and accelerator of the circular transition." A local report discusses real estate and construction as one of the sectors where CE should be implemented, along with infrastructure, waste management, and biological cycles. The three main focus areas are improving the vacancy rates of existing buildings, recycling construction waste, as well as new construction that follows circularity principles. Improving vacancy rates refers both to the direct involvement in city-owned facilities and to the role of the city in enabling building adaptations with, e.g., zoning and the permit processes. Recycling construction waste includes demolition practices and creating a market for recycled materials. Regarding new construction, the report mentions that the city is only in the beginning of the CE journey. One example of this is that the first urban design competition with a "circular building" is only planned to be launched in the year 2022.

21.4.2 Sweden

21.4.2.1 National Circular Economy Policies

The Swedish Ministry of the Environment launched a strategy for the shift toward a more circular economy. The main goal is here to create a society with effective use of resources in circular flows that contribute to the global sustainability goals expressed in Agenda 2030. The Swedish strategy suggests four focus areas to obtain the vision of a more circular economy: (1) Better product design and a more sustainable production process are essential for a higher circularity of materials and products. There is a need for a shift where more focus is on functions and services rather than products. The design of products is essential since it is in the design phase that most of the choices affecting the possibility of a higher degree of circularity are made. (2) Private and public consumption of product and services needs to be more sustainable with a high degree of circularity. A way to achieve this is to promote products with longer and more effective use. (3) Products and materials need to be recycled in toxic-free and circular processes. A more effective waste management is one success factor, where deposition of waste is a last resort when all possible circular alternatives are rejected. (4) There need to be drivers and incentives within societal, industrial, and commercial actors to take measures to promote circular business models. In a circular economy, products are to be produced with a longer life span and with the possibility of modernization and repairs.

21.4.2.2 Circular Economy in the RECI

The Ministry of the Environment points out that for RECI, a decreased environmental impact can be achieved through a more efficient use of facilities, as a lot of waste is generated in the construction and demolition phases of facilities. A higher degree of reuse and recycling in construction would have a great effect on the overall work toward a more circular economy. RECI in Sweden has developed a roadmap for a more sustainable industry. One of the goals in the roadmap is that in the year 2045, all materials in construction are circular, and zero waste is generated. This means a more effective use of buildings and other facilities for longer life spans, and that they function as material banks that can be reused after demolition. Flexible solutions are suggested where buildings and other facilities can change and be refined during their lives with the help of circular renovation processes. Digitalization is identified as one success factor to increase the efficiency of information processes and possibilities of a more effective use of resources through the whole supply chain.

City of Malmö

The municipality of Malmö has developed a local RECI roadmap for a climate neutral sector in the year 2030. The roadmap breaks down into six focus areas where three of them directly or indirectly relate to circular economy. Circular economy and resource efficiency strive to achieve 100% circularity for construction materials by 2030. The strategy is to promote a circular and bio-based construction by gradually increase the demands made in the municipalities' procurement of new buildings and other facilities. In addition, a reduction of waste in every part of the supply chain is one clear goal.

In the design and construction process, a lifecycle perspective is a starting point in order to achieve climate neutrality. The resource usage from a whole of life perspective should be optimized according to circular principles. These views are also suggested to be incorporated in the urban planning process and in the process concerning building permits.

The target of climate neutrality for building materials includes prioritizing circularity. In the procurement of new buildings and other facilities, increasing demands are about a higher degree of reused, recycled, and renewable materials promoting efficiency and climate neutrality in resource usage. Focus will be on those parts of the building that have the highest effect on the climate. The EPDs (Environmental Product Declarations) are used as a tool for gathering information about the circularity and climate effect of building materials.

Table 21.4 Key policy foci regarding CE elements on national, industry, and city level

	CE element	Finland			Sweden		
		National	RECI	City	National	RECI	City
Short loop	Decrease demand				x		
	Digitalization	x				x	
	Space efficiency, etc.		x	x		x	
	Industry-level ecosystems	x					
Medium loop	Optimizing useful life		x			x	x
	Adaptive reuse		x			x	
Long loop	Design for disassembly		x				
	Recycling	x	x	x	x	x	x
	Bioeconomy	x					x
	Sustainable new construction		x	x		x	x
Overarching	Sustainable business models	x			x		
	Collaboration	x	x		x		
	Sustainable urban planning		x	x			x
	Procurement		x				x

21.5 Cross-Case Analysis

Table 21.4 presents the finding of the policy reviews, on a national, industry, and city level, categorized using the theoretical background. The findings are further discussed in the following section.

21.6 Discussion

This study set out to identify key differences and gaps in Nordic circular economy strategies. Sweden has decided to lift UN's SDGs as a starting point for the national-level policy, which could be considered to restrict the policy to a very generic level. Sweden includes collaboration at a national level, while Finland is lifting public–private collaboration and industry ecosystems on the RECI level. These goals would fall under SDG 17 *Partnership for the goals*. Public–private partnerships are not commonly found in CE research. Finland continues to call for industry-level collaboration and even regulation in the form of industry standards. Taking circularity into account in planning and procurement processes is also included on the industry level, while in Sweden, they are only present in the city-level targets of Malmö.

Common features for both countries include the promotion of circular business models with digitalization seen as an enabler for CE. Circular business models have been well represented in existing research as well (e.g., Bocken et al., 2014). Digitalization as a circular strategy is present in the EU policy level (EC, 2020), the ReSOLVE framework (Ellen MacArthur Foundation, McKinsey Center for Business and Environment, and SUN, 2015; Arup, 2016), as well as existing literature (e.g., Antikainen et al., 2018; Agrawal et al., 2022).

Interestingly, both countries have included bioeconomy in their policies around CE, which is not typically considered a part of CE frameworks. However, the use of biomaterials is consistent with international trends (Chennouf et al., 2018) and recently linked to CE (Lange, 2022). Both Finland and Sweden lobbied strongly against the EU taxonomy due to the detrimental impact to forestry. Lifting bioeconomy solutions as part of the circular transition is aligned with this political activity.

Both countries include a large proportion of measures around recycling building materials. This aligns with earlier criticism from researchers concerning a focus on the materials level (e.g., Ranta et al., 2018; Pomponi & Moncaster, 2017; Kyrö, 2020). Circular new construction, another long loop measure, is included in policies in both countries. However, industry-level policies from the two countries also include the adaptive reuse of buildings, emphasized as a medium-loop circular measure in the literature (e.g., Ness & Xing, 2017; Sanchez & Haas, 2018).

It is noteworthy that sharing economy and access-over-ownership solutions are not the common reviewed policies, even though this is a short-loop measure, which is known to have a significant impact (e.g., Bocken et al., 2014; Kyrö, 2020; Brinkø et al., 2015; Ellen MacArthur Foundation, 2015). Still, elements of efficient use of space are included in the policies in both countries, and the Finnish industry-level policy even names shared spaces as one way to achieve space efficiency. It would be prudent to emphasize the short-loop measure of access-based consumption in any future CE policies.

Although the policies from both countries have moved further than the focus on recycling to shorter loop measures in recent years, this could further be enhanced for a more effective sustainability outcome. There appear to be high-impact CE elements missing or understated from the strategies, the most prominent being reducing demand, which is one of the most efficient ways to enable circularity. Perhaps, the absence is due to the implementation of shorter and medium loops being more challenging (Domenech & Bahn-Walkowiak, 2019). It is also evident that some things are included but are either branched out, and the main goal is not mentioned, e.g., efficient use of space and decreasing demand, or only the higher level is included, e.g., sustainable urban planning. The policy frameworks would benefit from having clear overarching aims. For example, sustainable new construction would be linked with actionable elements, such as design for disassembly, bio-based materials, and the use of reused and recycled materials. This would enable stakeholders to take appropriate action and knowing "why."

21.7 Conclusion

The cases of Sweden and Finland reveal an inconsistency in CE policies, which fall under the EU umbrella. Differences were mostly found in the level of which a CE element is lifted. Overarching CE elements such as collaboration and circular business models seem to be well covered and are expected to support reaching SDG 11 *sustainable cities and communities*. However, much more focus should be placed on short loops, such as sharing and access-based models, in order to reach SDG 12 *responsible consumption and production*.

Policy documents, unlike regulation, are non-binding recommendations for a course of action. One limitation of this study is the exclusion of existing economic incentives and legislation, which may work against the policies in both countries. We suggest that future research includes these elements to further investigate the CE maturity of RECI in Sweden and Finland. It would also be of interest to analyze the national policies against the EU policies in detail, in order to elicit where the disparities come from. An interesting further research avenue is to determine the reason for the focus on long loops in CE policies, research, and practice alike.

References

Agrawal, R., Wankhede, V. A., Kumar, A., Upadhyay, A., & Garza-Reyes, J. A. (2022). Nexus if circular economy and sustainable business performance in the era of digitalization. *Journal of Productivity and Performance Management, 71*(3), 748–774.

Antikainen, M., Uusitalo, T., & Kivikytö-Reponen, P. (2018). Digitalisation as an enabler of circular economy. 10th CIRP Conference on Industrial Product-Service Systems, IPS2 2018, 29-31 May 2018, Linköping, Sweden. *Procedia CIRP, 73*, 45–49.

Arup. (2016). *Circular economy in the built environment*. Retrieved from Arup website: https://www.arup.com/perspectives/publications/research/section/circular-economy-in-the-built-environment

Beccali, M., Cellura, M., Fontana, M., Longo, S., & Mistretta, M. (2013). Energy retrofit of a single-family house: Life cycle net energy saving and environmental benefits. *Renewable and Sustainable Energy Reviews, 27*, 283–293.

Bocken, N., Short, S., Rana, P., & Evans, S. (2014). A literature and practice review to develop sustainable business model archetypes. *Journal of Cleaner Production, 65*, 42–56.

Boons, F., Mantalvo, C., Quist, J., & Wagner, M. (2013). Sustainable business models and economic performance: An overview. *Journal of Cleaner Production, 45*, 1–8.

Bott, H., Grassl, G. C., & Anders, S. (2019). *Sustainable Urban Planning*. DETAIL.

Brinkø, R., Balslev Nielsen, S., & van Meel, J. (2015). Access over ownership – A typology of shared space. *Facilities, 33*(11/12), 736–751.

Chennouf, N., Agoudjil, B., Boudenne, A., Benzarti, K., & Bouras, F. (2018). Hygrothermal characterization of a new bio-based construction material: Concrete reinforced with date palm fibers. *Construction and Building Materials, 192*, 348–356.

Conejos, S., Langston, C., & Smith, J. (2014). Designing for better building adaptability: A comparison of adaptSTAR and ARP models. *Habitat International, 41*, 85–91.

Conejos, S., Langston, C., & Smith, J. (2015). Enhancing sustainability through designing for adaptive reuse from the outset: A comparison of adaptstar and adaptive reuse potential (ARP) models. *Facilities, 33*(9/10), 531–552.

Cruz Rios, F., Grau, D., & Bilec, M. (2021). Barriers and enablers to circular building design in the US: An empirical study. *Journal of Construction Engineering and Management, 147*(19), 04021117.

Domenech, T., & Bahn-Walkowiak, B. (2019). Transition towards a resource efficient circular economy in Europe: Policy lessons from the EU and the member states. *Ecological Economics, 155*, 7–19. https://doi.org/10.1016/j.ecolecon.2017.11.001

Douglas, J. (2006). *Building adaptation*. Butterworth-Heinemann.

Dytianquin, N., Gregersen-Hermans, J., Kalogeras, N., van Oorschot, J., & Ritzen, M. (2021). Circularity in selected EU countries: The case of construction and demolition. *IOP Conference Series: Earth and Environmental Science, 855*, 012017.

Eberhardt, L., Birgirsdóttir, H., & Brikved, M. (2019). Life cycle assessment of a Danish office building designed for disassembly. *Building Research & Information, 47*(6), 666–680.

Ellen MacArthur Foundation. (2013). *Towards the Circular Economy Vol. 2: Opportunities for the consumer goods sector*. Shared by Business. thirdlight.com. Accessed on 12-12-2021, 15:30.

Ellen MacArthur Foundation. (2015). *Delivering the circular economy: A toolkit for policymakers*. Shared by Business. thirdlight.com. Accessed on 12-12-2021, 14:24.

Ellen MacArthur Foundation, McKinsey Center for Business and Environment, and SUN. (2015). *Growth within: A circular economy vision for a competitive Europe*.

Ershadi, M., Jefferies, M., Davis, P., & Mojtahedi, M. (2021). Barriers to achieving sustainable construction project procurement in the private sector. *Cleaner Engineering and Technology, 3*. https://doi.org/10.1016/j.clet.2021.100125

European Commission. (2011). *Roadmap to a resource efficient Europe*. Communication from the Commission to the European Parliament, the Council, the European Economic and Social Committee and the Committee of the Regions. No. COM (2011) 571. Brussels, European Commission.

European Commission. (2015). *Closing the loop – An EU action plan for the circular economy*. Communication from the Commission to the European Parliament the Council, the European Economic and Social Committee and the Committee of the Regions. No. COM (2015) 614/2. Brussels, European Commission.

European Commission. (2016). *EU resource efficiency scoreboard 2015*. Prepared by Ricardo Energy & Environment. European Commission, DG Environment, Brussels.

European Commission. (2020). *The new circular economy action plan*.

European Commission. (2022). *Sustainable development goals*. https://ec.europa.eu/info/strategy/international-strategies/sustainable-development-goals_en. Accessed on 2022-02-08, 15:34.

Flyvberg, B. (2006). Five misunderstandings about case-study research. *Qualitative Inquiry, 12*(2), 219–245.

Geissdoerfer, M., Savaget, P., Bocken, N. M. P., & Hultink, E. J. (2017). The circular economy – A new sustainability paradigm? *Journal of Cleaner Production, 143*, 757–768. https://doi.org/10.1016/j.jclepro.2016.12.048

Giorgi, S., Lavagna, M., Wang, K., Osmani, M., & Liu, G. (2022). Drivers and barriers towards circular economy in the building sector: Stakeholder interviews and analysis of five European countries policies and practices. *Journal of Cleaner Production, 336*, 130395.

IVA. (2020). *Resurseffektiva lokaler i Sverige – Lokaldelning som norm*. En branscgrapport från IVA-projektet Resurseffektivitet och cirkulär ekonomi (ReCE).

Korhonen, J., Honkasalo, A., & Seppälä, J. (2018). Circular economy: The concept and its limitations. *Ecological Economics, 143*, 37–46.

Kyrö, R. (2020). *Share, Preserve, Adapt, Rethink – A focused framework for circular economy in the real estate sector*. Accepted for BEYOND2020 Sustainable Building Conference series, 2–4 November 2020, online conference.

Kyrö, R., Jylhä, T., & Peltokorpi, A. (2019). Embodying circularity through usable relocatable modular buildings. *Facilities, 37*(1/2), 75–90. https://doi.org/10.1108/F-12-2017-0129

Lange, L. (2022). Business models, including higher value products for the new circular, resource-efficient biobased industry. *Frontiers in Sustainability, 3*, 789435.

Langston, C. (2008). *The sustainability implications of building adaptive reuse*. Paper presented at the Chinese Research Institute of Construction Management (CRIOCM) International Symposium, Beijing, China, pp. 1–10.

Langston, C., Wong, F. K. W., Hui, E. C. M., & Shen, L. (2008). Strategic assessment of building adaptive reuse opportunities in Hong Kong. *Building and Environment, 43*, 1709–1718.

Leising, E., Quist, J., & Bocken, N. (2018). Circular economy in the building sector: Three cases and a collaboration tool. *Journal of Cleaner Production, 176*, 976–989.

Ness, D., & Xing, K. (2017). Toward a resource-efficient built environment. A literature review and conceptual model. *Journal of Industrial Ecology, 21*(3), 572–592.

Pomponi, F., & Moncaster, A. (2017). Circular economy for the built environment: A research framework. *Journal of Cleaner Production, 41*, 710–718. https://doi.org/10.1016/j.jclepro.2016.12.055

Ranta, V., Aarikka-Stenroos, L., Ritala, P., & Mäkinen, S. J. (2018). Exploring institutional drivers and barriers of the circular economy: A cross-regional comparison of China, the US, and Europe. *Resources, Conservation and Recycling, 135*, 70–82. https://doi.org/10.1016/j.resconrec.2017.08.017

Rees, W. E. (1999). The built environment and the ecosphere: a global perspective. *Building Research & Information, 27*(4/5), 206–220.

Reike, D., Vermeulen, W., & Witjes, S. (2018). The circular economy: New or refurbished as CE 3.0? – Exploring controversies in the conceptualization of the circular economy through a focus on history and resource value retention options. *Resources, Conservation & Recycling, 135*, 246–264.

Sanchez, B., & Haas, C. (2018). A novel selective disassembly sequence planning method for adaptive reuse of buildings. *Journal of Cleaner Production, 183*, 998–1010.

Saunders, M., Lewis, P., & Thornhill, A. (2009). *Research methods for business students fifth edition*. www.pearsoned.co.uk.

World Economic Forum (WEF). (2016). *Industry Agenda Shaping the Future of Construction A Breakthrough in Mindset and Technology Prepared in collaboration with The Boston Consulting Group.*

Part VI
Production and Procurement for Sustainable Development

Chapter 22
Conceptual Maturity Model for Strong Owner Capability Development in the Construction Industry

Ali El Chaabi, Nikolaos Anevlavis, and Christian Thuesen

Abstract Today, there is a known crisis in the construction industry worldwide where major projects and programs are experiencing substantial cost increases and long overruns. This is mainly due to the lack of skills and experience needed to set up successful projects or programs. This research study investigates how construction is being delivered and how owners' organizations can take up the leading role in shaping the construction industry. It builds on the concept of the strong owner organization concept, defined as an owner organization able to drive more effective and efficient project execution by having and developing capabilities within three dimensions: strategic, commercial, and governance capabilities. The study extends the three-dimensional capability map based on recent literature and interviews with industry practitioners including capabilities, critical success factors, industry challenges, and best practices. This forms the foundation for a conceptual maturity model that aims to help owner organizations within the construction industry assess their maturity level, to support them in performing better project planning and execution. Furthermore, the strong owner concepts and the maturity model will aid organizations in their sustainable transformational journey toward SDGs #9 Industry, Innovation, and Infrastructure, #11 Sustainable Cities and Communities, and #12 Responsible Consumption and Production, by connecting people, organizations, data, and technology.

Keywords Strong owner · Maturity model · Maturity assessment · Capability mapping · Sustainable development

A. El Chaabi (✉) · N. Anevlavis · C. Thuesen
Department of Management Engineering, Technical University of Denmark, Lyngby, Denmark
e-mail: chth@dtu.dk

G. Lindahl, S. C. Gottlieb (eds.), *SDGs in Construction Economics and Organization*, Springer Proceedings in Business and Economics,
https://doi.org/10.1007/978-3-031-25498-7_22

22.1 Introduction

Constructions play an essential role at the core of economic and social development. They provide the foundation for almost all modern economic activities, constitute an essential financial sector in their unique way, and contribute significantly to increasing living standards and quality of life (Barrie et al., 2006). In addition, the industry is critical for delivering societies' green transformations, which relies on massive investments in large-scale building and infrastructure projects. According to several sources, construction projects are delivered behind schedule, beyond the cost estimate, and fail to meet the public's expectations (Changali et al., 2015; Roehrich, 2020). Traditional research on mega projects explains this by strategic misrepresentation (Flyvbjerg et al., 2002), where benefits of projects are overestimated while budgets are underestimated and suggest concepts such as reference class forecasting (Flyvbjerg, 2008) as remedial strategies. While these are acknowledged as essential contributions, they fail to fundamentally address the underlying causes of increasing complexity and uncertainty (Love et al., 2012). More than ever, there is a significant need to improve construction delivery (Flyvbjerg, 2008; Love et al., 2012). According to (SNC-Lavalin Atkins, 2021), there are a lot of reasons why the best-laid plans fail. However, they usually lead back to a few common problems. First is the lack of capabilities, skills, and experiences needed to set up successful projects and programs. The second reason is the slow decision-making process with inaccurate baselines and time, scope, and budget expectations. The third reason is poor, and often out-of-date, information flows from projects (SNC-Lavalin Atkins, 2021). Considering the data available in the industry, the construction industry is behind others in terms of modernizing the way it manages and delivers significant programs. Inspired by other industries' successful transformation, the construction industry recognizes that they must change to achieve better performance of projects and programs. This paper investigates how construction is delivered and how it can be transformed by letting the owner organizations shape the market based on the capabilities needed for better project delivery. The strong owner concept motivates and supports the organizations to be ambidextrous (Turner et al., 2015). First, organizations can identify what capabilities are needed for their development, and second, what is the level of maturity required for each capability dimension to explore new markets and take advantage of existing and potential opportunities. However, despite "constructions" importance to society, and the vital role of owner organizations in them, the current practices of construction delivery are not living up to expectations. Thus, this study investigates the following research question.

How can owner organizations in the construction industry assess their capability maturity to improve their performance?

In answering this question, a conceptual maturity model is developed based on three-dimensional capability framework. This is aimed to help organizations within the industry to assess their maturity level, which is intended to aid them in road mapping their sustainable transformational journey by connecting people, organizations, data, and technology. The empirical context for the research is the Danish construction industry.

The paper starts with a literature review covering academic and industry background about organizational capabilities in the construction industry, then a research methodology describing the approach of the research study, followed by presenting major findings, and final part for discussion and conclusion.

22.2 Literature Review

The literature review examines existing knowledge and best practices from academia and industry regarding the organizational capabilities that cover the strong owner concept. The topic is broad, so the authors focused mainly on the academic and industrial literature directly or indirectly related to the Danish market segment. This is not to say that best practices worldwide have not been taken into account. The objectives were to systematically describe the characteristics of the strong owner concept, create a capability map of strong owners, and develop a maturity model that the proprietary organization can use in the construction industry to assess the maturity of their current capabilities. According to Winter (2003), the organizational capabilities on the broader concept were defined as the organizational routines. Inspired by a fundamental conceptual framework (Winch & Leiringer, 2016), the capabilities needed for an organization to expand its organizational competencies could be allocated in between three dimensions of capabilities: strategical, commercial, and governance. The criteria or so-called critical success factors define the elements needed for these capabilities to exist and mature inside the organization. It is also important to determine each dimension presented. First, the strategic dimension included capabilities defined, controlled, and owned by the owner organization (Winch & Leiringer, 2016). These capabilities were activities that are the full responsibility of the owner organization as they relate to the business and corporate strategy of the organization. Second, the commercial dimension referred to the capabilities an owner organization needs to interface with its suppliers. According to NAO (2010), owners' commercial capabilities have a crucial role in delivering assets, whereas many organizations, especially public organizations, lack these skills. Third, the governance dimension depicted an owner organization's capabilities to interface successfully with a temporary project organization (Winch & Leiringer, 2016). While governance frameworks were introduced more than 20 years ago in the United Kingdom and Norway, in Denmark, however, they are not yet firmly established (Biesenthal et al., 2018). By reviewing the capabilities from NAO (2010), Winch and Leiringer (2016), and Bang et al. (2016), Table 22.1 summarizes the capabilities needed for an owner organization.

Furthermore, based on the literature, the industry also seems to lack maturity assessment tools so that strong owners can evaluate the performance of their ability, as it was almost impossible to find articles in this field. Foremost, it is crucial for the

Table 22.1 Summary of capabilities derived from the literature review (Iteration 1)

Strategic	Commercial	Governance
Capital raising	Packaging	Assurance
Stakeholder management	Procurement	Project coordination
Project portfolio management	Relational	Asset management
ESG		

organization to map capabilities needed for optimal project execution; therefore, the capability map was an essential tool. One of the most widely accepted systems to increase the performance of the organization is the so-called Enterprise Architecture Management mechanism (Matthes et al., 2008). The architecture management mechanism is an essential mechanism that supports the management to steer and drive the organization in the desired direction. This mechanism can be applied to various aspects of the organization, such as the level of the organization of the business (Buckl et al., 2010). It is worth noting that a common Enterprise Architecture Management mechanism is the capability map used in this research. Moreover, according to Battista and Schiraldi (2013), a capability maturity model refers to a conceptual model that assesses an organization's level of maturity by considering a well-defined set of structured levels that show how effectively behaviors, methods, and processes can achieve the desired results reliably and sustainably. This indicated that a maturity model can be very effective in supporting an organization in understanding the current level of organizational capabilities and helping it to create the appropriate strategy to increase its performance. The Capability Maturity Model (CMM) is a development model that emerged in the early 1990s due to a study of data collected from organizations that had contractual agreements with the United States Department of Defense (Paulk, 2009). In recent years, capability maturity models from the perspective of the project management research field have become increasingly attractive (Fengyong & Renhui, 2007). Inspired by the existing maturity models and considering the level of organizational competencies that an organization needs for the efficient and effective execution of projects, the capability maturity model was created for the concept of a strong owner organization. The maturity model was created, based on the existing maturity model approach, and inspired by Sowden et al. (2010), Caracchi et al. (2014), Liu et al. (2018), and Wang et al. (2020).

22.3 Research Methodology

The research behind the paper adopted a Design Science Research (DSR) methodology. The intention of using DSR was to design a maturity assessment tool that industry practitioners could use. One of the essential characteristics of DSR is that it uses an iterative approach in a close exchange of data and information between the academic, industry literature, and industry practitioners (Gregor & Hevner,

2013). Therefore, DSR is well considered among researchers as an appropriate and practical approach, where the paradigm strives for close interaction between the relevant knowledge base and the practical research environment (Hevner et al., 2004).

The developed methodological structure based on the DSR approach can be seen in Fig. 22.1. This was created after having identified the problem definition by our research question. The model consists of five phases.

In Phase 1, the authors abductively conducted the literature review to find existing knowledge from academia and industry. Next, the research focused on finding organizational competencies, enablers, critical success factors, and maturity models related directly or indirectly to the topic of owner organizations in the construction industry. At this step, the industry reports from well-known multinationals and reputable organizations were considered highly important from the owners' or clients' side.

In Phase 2, a first version of the strong owner capability map was established by summarizing all the owner capabilities categorized in the three dimensions: Strategic, Commercial, and Governance. In addition to that, a prototype maturity model was developed based on the capability map. The approach used when developing the capability map's first design and the maturity model was based on deductive reasoning. It is worth noting that the data analysis followed the basic principles for qualitative data analyses to make full use of the value of the data. One of the critical concepts of qualitative data analysis is that it must be methodical and adhere to the qualitative content analysis method (Mayring, 2004). The

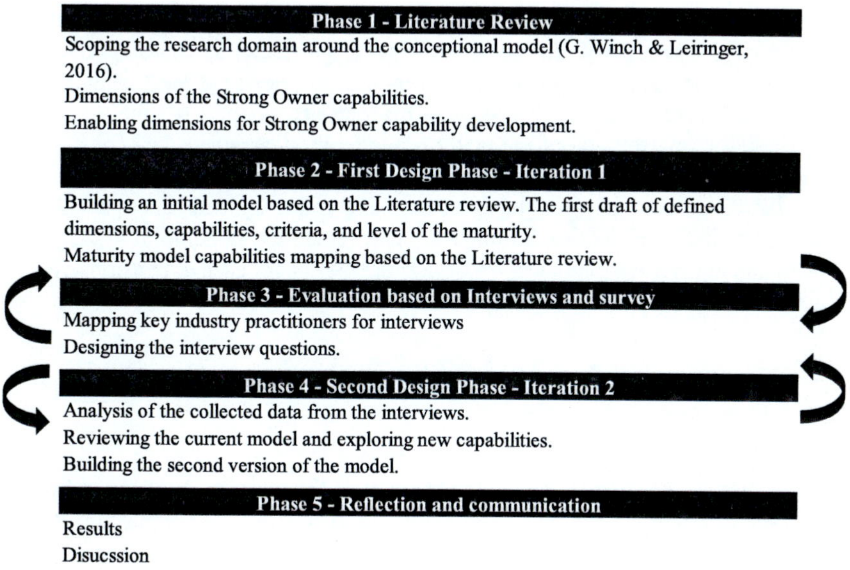

Fig. 22.1 Methodological approach

methodology and tool used to map out the capabilities are described and explained in Phase 4.

In Phase 3, the methodology behind the tools used for data collection was described and explained. Primary data were collected directly from the subject–matter expert—stakeholders and organizations operating either in the public or private sector in the construction industry. The pattern of choosing the organizations was focused on the fact that they should primarily operate in Denmark and act as owners, clients, or consultants. In the analysis, nine interviews were conducted, including a Danish Railway Client, Building Construction Client, Danish Regional Client, Danish Municipality, Danish Owners Association, and an International Pharmaceutical Company. It is worth noting that organizations that partially operate in these industries were considered to extract best practices around the industry. The primary data, as mentioned, consisted of nine semistructured interviews (Brinkmann & Kvale, 2018) organized with ten industry practitioners from seven different organizations. The interviews followed a qualitative data collection strategy; predetermined but open-ended questions were defined and presented in a guide format for informants. The interviews aimed to extract data about the existing capabilities, critical success factors, and maturity levels the organizations employ.

In Phase 4, the methodology of the analysis for the gathered data from practitioners from the industry was addressed. Preferably, a well-detailed thematic analysis was performed by employing the scientific software Atlas.ti to create a solid overview and flexibility in working with the data. To do so, deductive and inductive codes were defined. The deductive codes focused on validating the existing data from the literature review and interviews. In contrast, the inductive codes were used to explore new knowledge by sieving the acquired data and finding meanings and patterns of the interviews. Finally, the methodology's steps method was used to analyze the data collected from the interviews (Miles & Huberman, 1994). As a result, the second version of the maturity model was built.

In Phase 5, the final version of the maturity model was described. The conceptual model was mainly divided into five sheets. Foremost, a summary sheet provides a holistic overview of the maturity assessment, and a detailed sheet provides a more detailed scoring in each dimension and capability. In addition to that, the model included a separate sheet for each capability under each dimension, where users could assess themselves given the information provided.

22.4 Findings

In this section of the chapter, the findings from the literature (Phase 1) and industry interviews (Phase 3) are presented and discussed. As the core of the purpose statement for this research was related to exploring strong owner capabilities, their relative critical success factor (CSF), and then building a maturity assessment tool for strong owner capability development, the results were presented per capability, in accordance with the developed structure of the Owner Capability Maturity Model (OCMM) capability dimensions. The findings presented were utilized to define the maturity elements and maturity levels definitions of each capability in the OCMM framework.

The first dimension to be discussed is the strategic dimension. Strategic capabilities are defined, controlled, and owned by the owner organization (Winch & Leiringer, 2016). According to the analyzed data, there are mainly two drivers for successful capital raising. These CSFs are strategic planning and political negotiation. First, strategic planning is a process in which organizational leaders determine their vision for the future and identify their goals and objectives for the organization. This may include planning the project pipeline of an organization. One of the first significant findings was that organizations in the construction industry have difficulty planning their needs for the future. According to a Director at the Danish Owners Association, this was a significant challenge. Long-term strategic planning helps owner organizations shape and better define their strategic direction. Thus, the owner must be capable of making short-term planning for projects and doing long-term strategic planning. Several industry practitioners emphasized the incapability of long-term strategic planning in the interviews held. One last finding in the capital raising capability indicated that the private owner organizations could raise more capital for their investment projects from the public if they could contribute to the environmental and social sustainability goals, where projects could contribute at the local or national level. This was an excellent incentive for a recognized international manufacturing company. Regarding stakeholder management, one of the first findings regarding stakeholder management was that the owner needs to ensure a transparent communication process and see project requirements from different perspectives of all the stakeholders involved. One of the main challenges found is that some owner organizations struggle to provide reliable and accurate project information to the stakeholders. It was found that it is extremely important for the owner organization to try and maintain a good balance between the benefits and costs each decision has on the project stakeholders. This is considered the main bottleneck in construction projects as it is tough to satisfy all stakeholders on the same level.

A major challenge found in the industry, in general, was the complexity behind defining the proper requirements of the projects, from the idea or initiation phase to the construction or execution phase of the project. In addition, the challenge of designing a constructible building by matching the design and construction requirements was found to be a challenge. Another main finding was the emphasis on the importance of defining the scope of the projects in the portfolio and having the ability to plan and select projects driving the owner organization's strategic direction. Regarding ESG, one of the aspects of environmental sustainability that could be included is implementing green engineering solutions in the construction projects, whether it is in the design or construction phase of the project. Moreover, environmental sustainability could also be implemented within the owner organization's supply chain. For example, a program manager at International Pharmaceutical company mentioned their strategic direction in implementing what they call "Green Procurement Strategy."

The second dimension to be discussed is the commercial dimension. Commercial capabilities refer to the organizational competencies needed for an owner organization to interface with its suppliers. The owner organization's commercial capabilities have an essential role in collaborating between the organization and its suppliers. One of the major findings was that an owner organization needs to align with project stakeholders about the project's needs. This way, the owner organization can efficiently and effectively identify the project's needs. In that way, the organization can also maneuver when work packages need to be broken down into smaller ones or combined into larger ones to lure suppliers into bidding. One more finding was that the organization needs to create a contracting strategy for every project in its portfolio. To do so, the people responsible for the project can collaborate, including the legal department and the project leaders, to discuss the scope, purpose, strategy, the desirable outcome, risks, success factors criteria, and relevant information about the execution of the project. Moreover, nowadays, the organizations mainly work together with one supplier who can supply almost everything. In that way, the risk is allocated to this supplier, so the money as well. Regarding relational capability, the first important finding was that the organization needs to value the different cultural aspects of its organization and its suppliers. Furthermore, it highlights that trust is critical for collaboration between the organization and its suppliers to succeed in projects.

Globalization is a commercial capability induced from the inductive coding analysis performed on the interviews with industry practitioners. Globalization entails that the owner organization must collaborate with international suppliers besides the national suppliers. In this way, the organization can ensure better prices, better quality, new knowledge, and new skills. The findings presented are based on the analyzed data from the industry interviews. One main CSF for this important owner capability was identified. International collaboration CSF defines the cooperation between the owner organization and the international suppliers.

The third dimension to be discussed is the governance dimension. Based on the literature review, governance capabilities are mainly those organizational competencies required for an organization to become a strong owner organization to interface successfully with a temporary project organization. One of the first major findings regarding assurance is that the project's progress is reported to the board at meetings on the project milestones. However, a major challenge identified where the project continues and moves to the next phase even though the prerequired phase is not completed yet. That indicates a great risk, but as a section manager argued, stopping the project is even more significant. Another finding indicates that it is essential for the organization to effectively document the ownership definition, the roles of the project participants, and the organization's structure. One of the first findings regarding project coordination is that it seems that there is a certain competency that is missing in the industry in general. This competency was named "engineering management." Engineering management is a discipline that is set to cross-coordinate between different functions in an organization, such as engineering and finance. A final finding concerning project coordination was that data-driven project monitoring and controlling would improve transparency between all project stakeholders. According to the Director of Danish Owners Association, digitization of project logging would be an excellent next step in the industry. Regarding asset management, an important finding was that the organization should consider the differences between the design and the actual execution phases—buildability. Also, another finding indicates that the use of modern technologies such as 3D coordination and building information models helps tackle many of the problems. Another significant finding determines that it is essential for the organization to obtain accurate data regarding asset integration. Another important finding, especially for property owners, is the fact that procedures should be put in place regarding the delivery of the project on how to integrate the asset into the existing business operation.

The following Tables 22.2, 22.3, and 22.4 summarizes the findings from the literature and the interviews.

The study further identified enablers that could be leveraged for the owner organizations to develop the strong owner capabilities, and these enablers include systems engineering, project organizing, leadership, digitalization, and strategic partnerships. First, systems engineering is a field of engineering management that focuses on complex systems in the design, integration, and management of these systems and their life cycle (Roehrich, 2020). Second, project organizing is about how the project organizational setup is functioning (ICE & ICG, 2021). Third, leadership is a role and activity that the owner should possess in administrating the delivery of construction projects (Kortantamer, 2019; Kedri & Unsplash, 2021). Forth, digitalization is a powerful tool for dialogue with decision-makers and partners, but they also achieve increased learning by gaining insight into their performance (Bang et al., 2016). Fifth, strategic partnerships between the organizations create a repetition effect that would affect their collaboration in many different ways (Gottlieb et al., 2020). It is worth noting that the enablers and critical success factors relating to the dimensions were discussed in the findings and conclusions section.

After analyzing the data gathered from literature and the industry interviews, maturity definitions of each level were interpreted as visualized in Fig. 22.2. The developed assessment tool included a detailed interpretation and description of each

Table 22.2 Strategic dimension—capability definitions

Strategic dimension	
Capital raising	Organization's ability to raise the capital to fund investment projects
Stakeholder management	The process of maintaining the relationships with the involved project parties
Project portfolio management	Owner's organization ability to manage "its" projects, which shares scarce resources by selecting its project, mission, and strategy
ESG	Environmental, Social, and Corporate Governance evaluation of a firm's collective conscientiousness for social and environmental factors
Knowledge management	Ability of the owner organization to manage all the knowledge at the organizational level

Table 22.3 Commercial dimension—capability definition

Commercial dimension	
Packaging	Organization's ability to make a contracting strategy through which the work breakdown structure (WBS) is packaged into market-friendly clusters of work
Procurement	The ability to identify and select suppliers by understanding the supply market through the bidding processes, negotiation skills, contract management, and setting procurement plans
Authority and legal processing	Owner's involvement in facilitating tender regulation and authority approvals for projects
Relational	Owner's interaction with the supply chain mainly through innovative supplier management and working relationship between commercial parties and interorganization
Globalization	Owner's ability and flexibility to coordinate its needs with local and global suppliers

Table 22.4 Governance dimension—capability definition

Governance dimension	
Assurance	Owner's internal activities to inform the executive management about the progress of a project. This is mainly happening via communication, ownership definition, a well organizational structure and complying to the rules
Project coordination	Addresses the capabilities needed internally in the project organization to secure the ongoing progress of the project's life cycle
Asset management	Addresses the capabilities required for an organization to integrate and manage the final asset into existing operations

Level 1 Initial: CSFs unpredictable, poorly controlled and reactive	Initial level of maturity where the organization shows in general poor maturity in the defined critical success factor of each capability, co-related to the three capability dimensions. The overall score in the OCMM assessment tool (0 - 20\%)
Level 2 Managed: CSFs characterized for projects is often reactive	At this level, the organization will show a fair and managed performance in the defined critical success factor of each capability, co-related to the three capability dimensions. However, it still shows major weaknesses in many CSF. The overall score in the OCMM assessment tool (20 - 40\%)
Level 3 Defined: CSFs characterized for the organization and is proactive	At this level, the organization will show a good performance in each capability's defined critical success factor, co-related to the three capability dimensions. The recognition of the CSF in the owner organization would have significantly raised with the application of enabling dimensions, with some recognized weaknesses that the organization would be working on. The overall score in the OCMM assessment tool (40 - 60\%)
Level 4 Quantitatively managed: CSFs measured and controlled	At this level of maturity, the capabilities defined are generally very well applied and existent according to the CSF under each of them. Minor weaknesses would still be present. However, collaborative cooperation, technical performance, cross- organizational performance adjustment, and continuous development optimization are very well established in the organization. The OCMM value is gradually reflected in the final performance of the projects in the organization. The overall score in the OCMM
Level 5 Optimizing: Focus on processes and capability improvement	At this final level of maturity, the owner organization would be considered strong in performance in most of the defined CSF's. The organization's management would apply recognizing all the capabilities defined in the capability map and would have an excellent programs to continue to optimize all the mechanisms in both the projects and the organization in general. The overall score in the OCMM assessment tool (80 - 100\%)

Fig. 22.2 Maturity levels definition in the developed maturity assessment tool

capability at each maturity level. Specifically, the descriptions entailed defined critical success factors, characteristics, challenges, and maturity elements an owner organization would have at each level. The capability map built a base for the maturity model, consisting of five different maturity levels. These levels evolve with a step-by-step approach, with the lowest level being Level 1 to the highest at Level 5. These levels were linked with the performance of the organization, and the dimensions referred to strategic, commercial, and governance. The capabilities gather from the spectrum of the literature and interviews with public and private organizations. These capabilities were the central focus of this model. The evaluation criteria referred to the critical success factors, where a capability was divided into smaller capability factors. Moreover, the Success Factor Criteria matrix described in detail the full description for every success factor criterion at the five maturity levels. Finally, the Capabilities Development matrix depicted all the capabilities found and needed for an organization to increase its organizational performance at project execution.

22.5 Discussion and Conclusion

Today, the owner organizations make the earliest decisions in the construction process. Thus, they play a central role in shaping the building and construction projects that significantly affect the society. The owner has a key position in ensuring that all interests are taken into consideration and that the parties in construction work together to meet the project's objectives. To succeed, it requires owners to navigate the board of directors and ask the right questions to consultants and contractors to better manage projects, processes, and actors through everyone involved in the project. The model developed was tested by the Danish Client Organization Building Construction to evaluate the performance of the maturity level of its capabilities. It is worth noting that the reliability was examined to clarify whether the framework developed and the findings presented in this research are valid and reliable. The assessment tool developed is an excel-based tool with owned copyrights for the authors.

The model developed has significant implications for academia and industry. Impact on practice focuses mainly on the ability of strong organization owners to assess their capabilities' performance and maturity. In addition, it would be very convenient to compare the necessary organizational development goals with these tools. Owners' organizations could also discover the capabilities used by other organizations in the same industry to understand the capabilities required for growth. Last but not least, the models could serve the purpose of identifying the green development agenda in line with the UN Sustainable Development Goals.

Shedding light on the implication on academia can be seen how these models contribute further to the findings of Winch and Leiringer (2016). The latest research and literature about strong owner capabilities have mainly focused on the needed capabilities at owner organizations to execute projects more successfully. It has also been revolving around the fundamental industry problems tackled in this research study. However, not much research has been done on maturity models for the construction industry's owner capability development. This is where this research provides its main contribution to academia. The maturity model developed and the area researched entailed many of the recent works done in the area of owner capabilities, development of capabilities for better project execution, and empirical data from the industry practitioners, which only enhanced the credibility of the results and outcome of this work.

Moreover, this research study has set a baseline and direction for future research. One of the main topics that could follow this research is the study of the enabling dimensions. Future research should focus on using these enablers in the developed maturity model, or in other words, the "how" of the application of these enablers in theory and practice. These could be certain practices that owner organizations must do or develop at each level of maturity in each capability. Furthermore, a review of more enablers could be substantial to develop this model further.

Acknowledgments The authors would like to express their deepest thanks to all the organizations and companies that have been part of this research study so far. Their input has been valuable in developing a more industry-based and practical model that is not only based on academic theories.

References

Bang, H., Busch, A. K., Simonsen, R., & Ullum, H. (2016). *Professionelle bygherrers kompetenceudvikling strategier for Øget vaerdiskabelse*. Retrieved from www.bygherreforeningen.dk

Barrie, S., Pierre-Alain, S., & Michel, A. (2006). *Infrastructure to 2030*. OECD. https://doi.org/10.1787/9789264023994-en

Battista, C., & Schiraldi, M. M. (2013). The logistic maturity model: Application to a fashion company. *International Journal of Engineering Business Management, 5*, 5–29.

Biesenthal, C., Clegg, S., Mahalingam, A., & Sankaran, S. (2018). Applying institutional theories to managing megaprojects. *International Journal of Project Management, 36*, 43–54. https://doi.org/10.1016/j.ijproman.2017.06.006

Brinkmann, S., & Kvale, S. (2018). *Doing interviews* (Vol. 2). Sage.

Caracchi, S., Sriram, P. K., Semini, M., & Strandhagen, J. O. (2014). Capability maturity model integrated for ship design and construction. In *IFIP international conference on advances in production management systems* (pp. 296–303). Springer.

Changali, S., Mohammad, A., & Nieuwland, M. V. (2015). *The construction productivity imperative*. McKinsey and Company.

Fengyong, Z., & Renhui, L. (2007). Study on framework of construction project management maturity model. In *2007 international conference on service systems and service management* (pp. 1–5). IEEE.

Flyvbjerg, B. (2008). Curbing optimism bias and strategic misrepresentation in planning: Reference class forecasting in practice. *European Planning Studies, 16*, 3–21. https://doi.org/10.1080/09654310701747936

Flyvbjerg, B., Holm, M., & Buhl, S. (2002). Underestimating costs in public worksprojects: Error or lie? *Journal of the American Planning Association, 68*, 279–295. https://doi.org/10.1080/01944360208976273

Gottlieb, S. C., Thuesen, C., Frederiksen, N., & Berg, J. B. (2020). *Strategic partnerships in Copenhagen municipality: Experiences and results 2017–2019*. Department of Construction, City and Environment (BUILD), Aalborg University.

Gregor, S., & Hevner, A. R. (2013). Positioning and presenting design science research for maximum impact. *MIS Quarterly, 37*, 337–355.

Hevner, A. R., March, S. T., Park, J., & Ram, S. (2004). Design science in information systems research. *MIS Quarterly, 28*, 75–105.

ICE & ICG. (2021). *Project 13*.

Kedri, J., & Unsplash. (2021). *Project of the future*. Outcomes Report, May 2021.

Love, P., Edwards, D., & Irani, Z. (2012). Moving beyond optimism bias and strategic misrepresentation: An explanation for social infrastructure project cost overruns. *IEEE Transactions on Engineering Management, 59*, 560–571. https://doi.org/10.1109/TEM.2011.2163628

Mayring, P. (2004). Qualitative content analysis. *A Companion to Qualitative Research, 1*(2), 159–176.

Miles, M. B., & Huberman, A. M. (1994). *Qualitative data analysis: An expanded sourcebook*. Sage.

Paulk, M. C. (2009). A history of the capability maturity model for software. *ASQ Software Quality Professional, 12*(1), 5–19.

Roehrich, J. (2020). *A systems approach to infrastructure delivery*. Institution of Civil Engineers.

SNC-Lavalin Atkins. (2021). *PMO evolution: Using data to improve delivery*. SNC-Lavalin Group.

Sowden, R., Hinley, D., & Clarke, S. (2010). *Portfolio, programme and project management maturity model (p3m3®): Introduction and guide to p3m3®*. OGC.

Turner, N., Maylor, H., & Swart, J. (2015). Ambidexterity in projects: An intellectual capital perspective. *International Journal of Project Management, 33*, 177–188. https://doi.org/10.1016/j.ijproman.2014.05.002

Wang, G., Liu, H., Li, H., Luo, X., & Liu, J. (2020). A building project-based industrialized construction maturity model involving organizational enablers: A multi-case study in China. *Sustainability, 12*(10), 4029.

Winch, G., & Leiringer, R. (2016). Owner project capabilities for infrastructure development: A review and development of the "strong owner" concept. *International Journal of Project Management, 34*, 271–281. Retrieved from https://linkinghub.elsevier.com/retrieve/pii/S0263786315000356, https://doi.org/10.1016/j.ijproman.2015.02.002

Winter, S. G. (2003). Understanding dynamic capabilities. *Strategic Management Journal, 24*(10), 991–995.

NAO (2010) Assurance for high risk projects,National Audit Office (NAO), London

Matthes, F., Buckl, S., Leitel, J., & Schweda, C. M. (2008). Enterprise architecture management tool survey 2008. München: Technical University München.

Sabine, Buckl Thomas, Dierl Florian, Matthes Christian M., Schweda Frank, Harmsen Erik, Proper Frank, Schalkwijk Joseph, Barjis Sietse, Overbeek (2010) Practice-Driven Research on Enterprise Transformation Second Working Conference PRET 2010 Delft The Netherlands November 11 2010. Proceedings Building Blocks for Enterprise Architecture Management Solutions Springer Berlin Heidelberg Berlin Heidelberg 17-46

Liu, K., Su, Y., & Zhang, S. (2018). Evaluating supplier management maturity in prefabricated construction project-survey analysis in China. Sustainability, 10(9), 3046.

Kortantamer, D. (2019) Governing major transformation portfolios in practice: illustrations from the UK central government International Journal of Managing Projects in Business 12(3) 755-784 10.1108/IJMPB-09-2018-0174

Chapter 23
Experiences from Implementing a Collaborative Project Delivery Method

Atle Engebø, Julian Andreas Moen Rygh, Ole Jonny Klakegg, Jardar Lohne, and Ola Lædre

Abstract During the last decades, a change has happened in which the project delivery methods have been adapted to the increasingly complex projects. Essential elements are the focus on early involvement of key actors, shared decision-making, risk, reward, etc. Although more projects are carried out in this manner, there are still unanswered questions about the effects and the project participants' experiences. This chapter reports on a case project in Trondheim, Norway. Document studies, eight in-depth interviews, and observation of eight integrated concurrent engineering sessions have been conducted. The contractor was involved early in the project. Thus, the delivery method acquired collaboration among the contractor, client, architect, and relevant consultants to find the optimal solutions. The study revealed several lessons for future collaborative projects. First, the client should set a more explicit financial framework, potentially leading to more favourable working conditions and time savings. In addition, elements such as shared goals and objectives should be updated and maintained throughout the project. Finally, cultural elements are less tangible and concrete and therefore seem less attended to in these projects. As a result, relationship building seems to be downgraded in times of uncertainty and time pressure. This underpins the importance of facilitating relationship building and other cultural elements through the project's delivery method.

Keywords Project delivery methods · Partnering · Collaboration · Early contractor involvement · Pre-construction phase

A. Engebø (✉) · O. J. Klakegg · J. Lohne · O. Lædre
Department of Civil and Environmental Engineering, Norwegian University of Science and Technology (NTNU), Trondheim, Norway
e-mail: atle.engebo@ntnu.no; ole.jonny.klakegg@ntnu.no; jardar.lohne@ntnu.no; ola.laedre@ntnu.no

J. A. M. Rygh
Veidekke, Trondheim, Norway

G. Lindahl, S. C. Gottlieb (eds.), *SDGs in Construction Economics and Organization*, Springer Proceedings in Business and Economics,
https://doi.org/10.1007/978-3-031-25498-7_23

23.1 Introduction

For a long time, the construction industry has been exposed to criticism associated with low productivity measurements and poor performance (Davis & Love, 2011). A contributing factor is the global trend of increasingly large and complex projects. Since the early 2000s, the size of construction projects has increased to a particular degree (Hu et al., 2015). Also, more complex buildings and infrastructure emerge due to the "green shift" and technological advancements. To achieve more sustainable projects often include innovative and complex solutions. Gunhan (2019) points out that this has gained foothold in the industry.

To meet complex projects' demands, both the public and private sectors focus on developing and implementing new project delivery methods (Molenaar et al., 2009; Scheepbouwer & Humphries, 2011; Touran et al., 2011). This means designing a better system for organising design, production, operation, and maintenance activities, including making choices related to the structure for decision-making, contract form, contracting process, and more (Miller, 1997; Miller et al., 2000). However, the design of project delivery methods is a complicated exercise, which often leads to significant variations and challenges associated with finding a practice that supports and provides the necessary room for manoeuvre within the chosen model (Klakegg, 2017). Particularly, attention to the relational aspects of project delivery has been given emphasis in the last 20 years, and a shift towards collaboration-focused methods has happened (Bayramoglu, 2001; Dewulf & Kadefors, 2012; Stene et al., 2016). These methods have been given attention in the literature but are still in constant development.

The paper addresses how the construction economics and organisation research field can contribute to sustainable development by reporting project delivery methods. Project delivery methods have been identified as critical for successfully delivering sustainable construction projects (Mollaoglu-Korkmaz et al., 2013; Ahmad et al., 2019). Specifically, it reports the findings from a master thesis that examined the project delivery method and its effect in the pre-construction phase on a case project located in Trondheim, Norway. Thus, the purpose of this chapter is (1) to describe the project delivery method, (2) strengths and weaknesses observed, and thus (3) to suggest a potential improvement for future projects.

23.2 Literature Review

A construction project is a complex process that integrates interests of many stakeholders towards the eventual goal of realising a constructed facility (Brown et al., 2001; Ruparathna and Hewage, 2015). In the search for better performance, one route to consider is to work more collaboratively with other organisations (Hord, 1981; Harback et al., 1994; Latham, 1994). In the private marked, collaboration is attractive when performing an activity with a partner is greater to performing it either internally, in a market transaction or through merger (Porter, 1985). In project-based industries, these dynamics will be slightly different. Here, the key to success lies in the ability to attract and accomplish projects and in which a prerequisite will be that several organisations work together (Brady and Davies, 2004; Turner et al., 2009; Söderlund, 2011).

23.2.1 Project Delivery Methods in the Context of Sustainable Development

The United Nations sustainable development goal number 9 is to build resilient infrastructure, promote inclusive and sustainable industrialisation, and foster innovation. However, sustainable construction projects are also complex, and the industry is beginning to realise that such projects have additional requirements in their overall development process (Ahmad et al., 2019). For example, the delivery process for sustainable construction projects is more complex than traditional construction and requires more planning as the technical requirements are higher and more stakeholders are involved (Swarup et al., 2011). Thus, such projects increase the need for team interaction, increase the demand for specialised system expertise, and introduce new technologies. Therefore, adjustments in delivery processes are necessary to help facilitate, manage, and improve the effectiveness of these project team transactions (Korkmaz et al., 2011).

Particularly, the design process has been given increased attention and the importance in collaboration between disciplines to establish the building's desired function and create optimal solutions (Magent et al., 2009). Thus, a key aspect for a collaborative project delivery method would be to facilitate team integration to improve delivery practices (Mollaoglu-Korkmaz et al., 2013). The core strategy seems to be that of project owner integrating contractors and other major contributors into the project. By involving the contractor during the design phase, the project could optimise the construction processes in regard to sustainability (Son et al., 2011).

23.2.2 Deconstructing Collaborative Project Delivery Method into Contractual, Organisational, and Cultural Elements

Collaborative project delivery methods go under many names such as Alliancing, Partnering, IPD, or ECI. What has been seen is that many of the various methods share many similarities in the elements (referred to as hard- and soft elements, partnering mechanism, etc.) that are implemented. For example, Nyström (2005) found that the concept is understood by looking for a network of overlapping similarities and that two components were always included in the descriptions, trust and mutual understanding. Through commitment to mutual project objectives, collaborative problem-solving and a joint governance structure, these project delivery methods pursue collaborative relationships emphasising trust to pursue improved performance (Børve et al., 2017).

Categorising the elements found in collaborative project delivery methods is meaningful for organising, analysing, and creating a foundation for further comparison. We choose to categorise into three categories: contractual, organisational, and cultural (Wøien et al., 2016; Simonsen et al., 2019; Engebø et al., 2020). Contractual elements, or hard elements, are those that are directly regulated by the contract or have their basis in the procurement process (Yeung et al., 2007). Furthermore, in our categorisation, organisational elements are mechanisms that affect the organisation and the process surrounding the project – but not regulated in the contract. However, they are more tangible than cultural elements, making them easier to identify (Wøien et al., 2016; Falch et al., 2020). Collaborative project delivery methods are sometimes referred to as soft because they depend on elements beyond the contractual or organisational ones to ensure good collaboration. Therefore, selecting the proper organisational elements is vital for success. The cultural elements, or soft, are those that contribute to the relationship between the people in the project (Yeung et al., 2007). The cultural elements represent the project delivery method's mechanism that is designated towards the "soft side" – project culture (Pollack, 2007; Turner, 2021). The result is that, in practice, the client has a comprehensive list of contractual, organisational, and cultural elements to choose from.

23.3 Methodology

This study followed a qualitative methodology to study the selected case project. The study has followed Yin (2018)'s approach to case studies. Yin (2018) points out that case studies are preferred when the desire is to research a current event or a collection of events, particularly complex events where researchers cannot quickly isolate variables. Also, the procedure is beneficial in cases where there is adequate prior knowledge of a topic (Samset, 2014). The case study was chosen due to its approach to handling a wide range of data: documents, interviews, direct observations, and objects (Yin, 2018).

For data collection in the case study, several data collection methods have been used: observations, interviews, and document studies. First, the observation method was used to gather empirical data. Observations were made of the design sessions. The observations were carried out as an "observer-as-a-participant." In such a role, the observation is known to those involved (Saunders et al., 2016). Due to the uncertain situation surrounding Covid-19, the observations were mainly done digitally (of eight observations, one was conducted physically). Due to the uncertain situation surrounding Covid-19, most of the design sessions happened digitally. Thus, of eight observations, one was conducted physically and the remaining digitally. When physically present, informal conversations and questions in breaks/lunch contribute to understanding. It was also easier to interpret the participants. The change to digital observations posed some alternations of approach. The meetings were easier to observe as discussions were more "on the topic" and "straight to the point." However, some of the informal interactions among participants in physically present meetings were lost. Additionally, the observations were also a convenient means for making initial contact with key people and understanding the project. For example, the first observations were made in advance of the interviews serving as a reasonable contribution to preparing the interview guide.

The documents studied were acquired with permission from the contractor. Document studies are beneficial in case studies that explore detailed descriptions of a simple phenomenon, an organisation, or event (Bowen, 2009). Although documents can be a rich data source, it is essential to look at them critically. Documents should not be treated as completely precise, accurate, or a complete picture of what happened. In this study, the document study has mainly been used to supplement the primary data (interviews and observations) and establish an overall assertion of the project. About 100 pages of documentation were examined, including project documents such as organisational charts, responsibility matrices, extracts from contracts, and progress plans. The document study was conducted in two parts. The first part was to establish an overview of the project and work up enough knowledge regarding the project before observations and interviews. The second part was carried out to confirm, supplement, or reject information from the data collected from the observations and interviews. Any discrepancy between the data collection information and documents was returned to the project for clarification.

Interviews were conducted with key personnel associated with the case project. The interviews were semi-structured, meaning that they followed a pre-defined interview guide. The recruitment of interview objects was based, among other things, on recommendations from contact persons in the project and observations. A total of eight interviews were conducted. Two interviews were conducted physically, while the rest were conducted digitally. The client's project manager was interviewed, while from the consultants and subcontractors, we interviewed a technical coordinator, project manager (constructor), and project manager (technical). From the contractor, the project manager, site manager, design manager, and project procurer were interviewed. All interviews were recorded after the approval of the interviewee. After transcription, the minutes were sent to the interviewee for approval. The transcript was also used as part of the filtering and analysis of the data.

The data obtained may be understood as data composed of events, i.e. process data. The analysis focusses on understanding the phenomena by providing a "vicarious experience" of a real setting (Langley, 1999). Thus, an essential part of the analysis was to see the data in the context of the contractual, organisational, and cultural elements that functioned as thematic categories/units of analysis and in the light of the research questions (experiences – potential for improvement).

The study is limited to one in-depth case project investigating the pre-project phase or phase 1 of the so-called two-stage method (elaborated in Sect. 23.4.2). The case project was chosen because it represents a complex project with a collaborative delivery method. In addition, the data collection was conducted by a master's student who had gained access to the project through an internship the prior year. The study emphasised a descriptive approach describing the delivery method, identifying strengths and weaknesses, and presenting the potential for improvement. Furthermore, the study is limited to a qualitative case study approach in terms of methodology.

23.4 Results and Discussion

23.4.1 The Case: Trondheim Central Station

The project was studied throughout the collaboration phase in autumn of 2020 and spring of 2021, i.e. from the contractor's entrance into the project until the contractor handed over the final project description and associated target price. The project is currently in the phase of detailed design. The Trondheim Central Station project is set to expand the current Central Station and the surrounding area giving the city a new public transport hub where railways, ferries, and buses are linked (see Table 23.1). The new terminal building will be completed in 2025 and aims to become Norway's most modern public transport hub. At the same time, as a new terminal is being built, 13,000 m^2 of office space, approx. 5000 m^2 of commercial space, and over 200 new homes will be built. The project is part of Trondheim

Table 23.1 Project characteristics

Project type	Terminal building with office, commercial, and housing spaces
Client type	Public: joint ownership between rail authorities and the municipality
Clients maximum price	200MNOK. Phase 1 developed using a target cost process
Project delivery method	Two-stage delivery: Phase 1: joint project development; Phase 2: design-build contract for the delivery

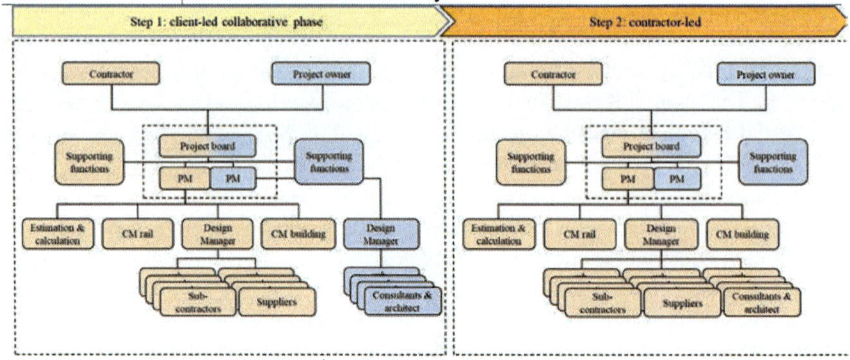

municipality's plan for green urban development. It follows up the municipality's goal of densifying areas close to the city centre so that more people can walk cycle and travel by public transport in everyday life.

23.4.2 The Delivery Method

The project is carried out as a two-step design-build. After winning a tender, the contractor became involved in the preliminary project. The client had already involved an architect and consultants, and the contractor then intended to take over these contractual conditions for phase 2. Phase 1 started in mid-2020, consisting of concept and target price development with collaboration. Phase 2 consists of detailed design and delivery. In this project, the organisation is defined as how the client and the contractor's overall organisation was structured. The project was divided into two phases. Phase 1 is a preliminary project with collaboration that leads to a fully developed target price. The project then moves to phase 2, detailed design and construction.

The first period after the contractor's involvement was initially supposed to be client-controlled before the contractor was to take over the responsibility of the contractual and procedural matter. The strategic aim of the client for the initial part of the collaboration phase was to involve the main contractor first in a consultant role for then to transfer the management responsibilities after a habituation period. Thus, making the first part client-led before the contractor took responsibility for the contractual and procedural matter. Nevertheless, it became clear to the project group relatively early in the collaboration phase that it would be appropriate for the contractor to take over control. The following argument for this organisational change was revealed through the interviews: (1) the contractor had a system and delivery method for managing the collaboration phase. This was considered a competence the project group could not fully use as long as the contractor had a consulting role. (2) The contractor believed that the project group was neither adequately nor well-staffed and desired to bring in more expertise. The client also believed it would be best for the contractor to take responsibility for the new contract conditions. Therefore, the contractor also took over existing contractual relationships with consultants and went into a more management role for the collaboration phase. The main contractor proceeded to bring subcontractors and suppliers into the project, which led to several replacements in the project group as the main contractor did not extend the existing consultancy contracts. The contractor conducted interviews focusing on experiences with collaboration and aimed at identifying those who best understood the task and what it takes to achieve good collaboration. Coincidently, most of the subcontractors and suppliers who came in had collaborated with the contractor before. Table 23.2 shows the contractual, organisational, and cultural elements emphasised in this particular project. The findings do not necessarily have to be representative of other collaborative projects.

Table 23.2 Identified elements in this case

Category	Identified elements in this case
Contractual elements	Joint project organisation Target cost development in phase 1 Design-build contract in phase 2 Early involvement of contractor Early involvement of subcontractor/supplier Function description Open book Conflict management procedure
Organisational elements	IT tools and BIM Integrated concurrent engineering Scrumsprints Savings list
Cultural elements	Motivation Common objectives Collaborative culture Relationship building Competence composition

23.4.3 The Delivery Processes

The target price was developed through five levels using continuous calculation and uncertainty analyses. The first level, target price 0, was a very rough calculation based on historical figures, with a low level of detail, which led to a high degree of uncertainty. Target prices 1 and 2 included more and more qualitative assessment, making the foundation for the investment decision. The target price 3 was a status assessment intended to optimise the project. Finally, target price 4 was a detailed assessment in which the project team clarified the most significant known uncertainties and risk factors. The development- and subsequent submission of the Target price (4) marked the end of the collaboration phase, as it was the final offer to the client. Accepting the Target price (4) allows the transition to phase 2 – detailed design and implementation, see Fig. 23.1.

The team worked on elapsed time. All actors registered the estimated hour budget to complete their respective work tasks in the preliminary project. The project team mainly reported positive experiences from this way of working. The target price development at different levels gave consecutive numbers to control the design from an early stage. There were relatively short horizons between target prices, leading to a common focus and collaborative climate towards reaching the relevant target prices.

On the other hand, the client reported that, from their perspective, some subject areas were perceived as rushed to finalise the target price, leading to inaccurate pricing. Several project participants expressed frustration about the client's lack of an explicit budget framework and thought this could lead to better working conditions and time savings. For example, when the group delivered the target price 2, the client deemed it too high. The actor thought they could have adapted the building to

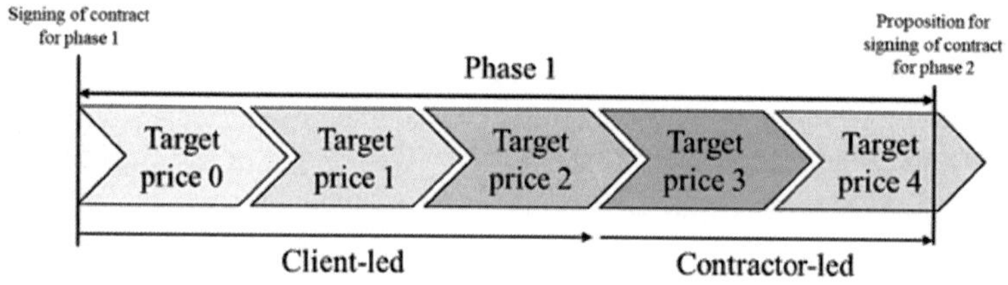

Fig. 23.1 The target cost process as planned by the project group

the client's wishes better if they had stipulated a better owner budget. This could have made the work on the preliminary project easier, saved time, and may be bypassed the need for a savings list. In this project, the client consisted of several constellations with their interests and budget (railway, ferry, bus, office rental, and shop rental). This ownership composition meant that the target price had to be distributed over different budgets. As stated by the client early in the project, this is special and challenging, and it may require more work than what the planned target price development process had taken into account.

Seen from the outside, it may look as if the contractors had expectations of a target price development process with a set client budget (the client's maximum price) where they should work to get the most within the fixed maximum price. In contrast, the client probably expected a target value process where the group would focus to a greater extent on creating the most significant possible value for money within a more uncertain budget. This discussion came to the surface when one constellation of owners experienced losing its most prominent tenant and thus had to reconsider its participation in the project. As a result, the project concept had to be significantly adjusted, which was frustrating for delivering entities. On the other hand, the project group gradually became better at working under uncertain conditions and better equipped to cope with such changes. It is also pointed out that there is a potential in improving the understanding between contractor and client when reviewing the uncertainty analysis. The delivering entities believe that it is often difficult for clients to familiarise themselves with these uncertainties and follow the review. A contractor representative acknowledged that something had to be done to make the uncertainties more understandable for the client. Bettering this part of the target cost process may improve the contractor's and the client's understanding, particularly by developing a more thorough description of the procedure for specific costs, associated risk, and added uncertainty when prices are collected from suppliers.

23.4.4 The Project Culture

The cultural elements are perceived as the most challenging to concretise, especially compared to contractual and organisational ones. The interviewees were clear on the importance of cultural aspects, such as project culture, relationship building, but struggled to articulate how these elements were associated with the project delivery method. When asked about examples, some said it was particularly challenging during the transition from client-led to contractor-led as many replacements and adjustments happened. In the project, when the contractor assumed the managing responsibilities, they simultaneously imposed their system for collaboration based on their experience with collaborative project delivery. Therefore, it has been a clear procedural responsibility to implement the collaborative elements on the contractor's part. In contrast, the client has actively participated in the process and learned a lot from it. It is also important to emphasise that the contractual and organisational elements do not guarantee an improved culture of collaboration.

A contractor representative expressed that such a collaboration phase required a lot from each participant. In addition, the actors involved, except the contractor, had not much experience with this kind of delivery method. An interviewee noted that when the contractor entered the project with their way of organising such processes, it collided somewhat with the existing consultant organisation structured in slightly different ways – especially considering meeting activities, etc. An interviewee stated: "I think it can be challenging for project participants and the management if culture and process are not agreed upon when new actors come in." For example, no new common goal preparation was made after the new project group was established (the contractor's consultants and suppliers), resulting in some feeling that the objectives were not well enough anchored in the project group. An interviewee believes that it could have been beneficial if the project participants had spent more time getting to know each other better.

Another aspect, Covid-19, resulted in the project group mainly interacting through digital collaboration. As a result, the project team lost essential elements such as social interactions during lunch, and coffee breaks, which are necessary for building and maintaining a good culture of collaboration. Another interviewee is aware that they should have been better at arranging team building activities within the possible framework (work context). It may have been challenging for the consultants to be transferred to the contractor, given they were used to working with the client. At the same time, the contractor obtained consultants they had a good past experience with. Some interviewees reflected that the project group had possibly relied too much on previous relationships, which could negatively affect over time since relationship building should happen continuously throughout the project. Some interviewees said they experienced that relationship- and teambuilding was downgraded in uncertainty and time pressure periods.

Some mentioned that too much competence and too many participants involved led to slowness in communication and confusing areas of responsibility. For example, there were up to ten architects involved at most. In addition, the various architects had different professional knowledge and areas of responsibility, which led to challenging communication and uncertainty associated with who should be involved in multiple decisions. This meant that the contractor eventually saw himself forced to reduce the number of architects in the project. However, interviewees also stated that the composition of professional competence was more than good enough in the early phase, which was somewhat contradictory to the statements from the contractor.

23.4.5 Suggested Improvements

The study uncovered some strengths and weaknesses of the project delivery method based on data obtained from the interviewees, observations, and considerations.

The transition from client-controlled to contractor-controlled collaboration gave the contractor time to get acquainted with the project organisation before deciding which actors/participants should be included going forth. However, finding the balance between making changes to the project group early on and maintaining continuity in the project organisation was challenging. As the client puts it, this intended overlap period was a somewhat strategic choice. It gave them room for manoeuvre to make changes along the way. As they mention: "We started a large organisation, but we were not sure that this was the right way to do it". Although the transitory period allowed the contractor to adapt the project to their wishes, much disruption could hamper continuity and reset relationships (Baiden et al., 2006). It also seems necessary to make changes in the project organisation relatively early, not negatively affecting the project culture. Naoum (2003) notes that it is particularly advantageous to ensure continuity by maintaining an equal project organisation from phase 1 to phase 2.

A challenge mentioned with such a model is associated with the contract form for phase 2, the intended design-build contract. First, this transition could potentially lead to the company boundaries becoming clearer and losing some of the collaborative culture developed through phase 1. Second, by exclusively including more subcontractors and suppliers in phase 1, there will be an uncertainty associated with the degree of utilisation of the market (in regard to pricing). The target cost development took place in levels, which was intended to create a smooth workflow and open up for mid-term assessments at the various levels. As outlined by Lahdenperä (2010), the possibility of making changes throughout the project is reduced, at the same time as the change costs increase. Therefore, it is essential for such a model to maximise the significance and potential in the collaboration phase (the period from the involvement of the supplier link to the determination of the target price). Several interviewees thought this allowed them to control the design early, which would be financially and time-efficient for the project. For example, as discussed in the previous section, the contractors expressed the need for an explicit maximum budget frame from the client's side. This is supported by Johansen et al. (2021), who believe that an unrealistic economic starting point often leads to the next target price being too high. Ambiguity and sudden changes to the budget may lead to cost-cutting instead of improvements and innovation if the client and the delivering entities share a common understanding of the process they conduct.

One can understand why the client is "playing on the defence" as projects tend to cost more than initial estimates (Andersen et al., 2016). This way, they get to probe the terrain and grasp whether the costs will be lower than they had imagined. As the project will progress into a design-build contract after the collaborative phase, the incentives for the delivering entities to get the target price substantially below the maximum client price are relatively low as that will affect their scope of work and potential margins. Also, another risk from the client perspective is that the delivering entities will withhold proposing cost-cutting solutions until after signing the design-build contract (then proceed with said solutions in the proceeding detailed-design phase). On the other hand, as stated by Lahdenperä (2010), the client benefits from the diminished risk of a cost over-run due to the incorporated "cost insurance" feature.

The client's idea that the delivering entities are withholding innovations and solutions has its founding in a principal-agent perspective where the actors are expected to act in their own most suitable interests. From a collaborative viewpoint, this does not seem to consider that the project was developed by people who want to perform well for this particular project. As Davis and Love (2011) point out, trust is vital for creating a joint project affiliation. Several interviewees highlighted such elements associated with the project culture aspect. Individual motivational factors are highlighted, such as being involved in a complex project with size and significance for Trondheim. In addition, in the collaboration phase, the participants can influence the building to a degree impossible at any later stage of the project ("*leaving a mark*"). In addition, for a comparative small marked that the Trondheim region is, project success (even in just the collaboration stage) could give merits that can be drawn upon for both individuals and organisations for ensuring future work.

23.5 Conclusion

This chapter explored the project delivery method and its effect in the preconstruction phase on a case project located in Trondheim, Norway. In particular, we sought to (1) describe the project delivery method and (2) the strengths and weaknesses observed, and (3) suggest a potential improvement for future projects.

In recent years, the two-stage collaborative project delivery method has increased in popularity in Norway, especially in public projects with characteristics making them a more complex endeavour. The study identified 18 collaborative elements in the case project, of which nine are classified as contractual, four organisational, and five cultural. In the context of the UN sustainable development goal, project delivery methods will play a role in building resilient infrastructure. The study shows that early involvement of the delivering entities in a so-called collaboration phase is central to these collaborative project delivery methods. The collaborative methods contribute towards fostering innovation as the emphasis on team interaction and focus on involving all the needed competence at once seem to be suitable for facilitating the planning, engineering, and involvement necessary for these projects, mainly since the delivery process for sustainability requires more planning as the technical requirements are higher and more stakeholders are involved. After all, such projects require far more competence, commitment, and collaboration to achieve goals beyond the traditional cost, time, and quality parameters. As for improvement, it seems that cultural elements get the least attention. The project participants express on a general basis the importance of cultural elements, but the abstraction of the concept makes it hard to make satisfactory considerations regarding them. This substantiates the significance of the project management focusing on the cultural aspects in the pre-construction phase.

Additional research on a better understanding of and improving the cultural aspects of collaborative project delivery methods is needed. In addition, this chapter focuses on phase 1, i.e. the collaboration and design, so it would have been worthwhile to also look at the implementation phase. In addition, there is a need for creating a body of knowledge on these types of projects allowing for systematic evaluation.

References

Ahmad, T., Aibinu, A. A., & Stephan, A. (2019). Managing green building development – A review of current state of research and future directions. *Building and Environment, 155*, 83–104.

Andersen, B., Samset, K., & Welde, M. (2016). Low estimates – High stakes: Underestimation of costs at the front-end of projects. *International Journal of Managing Projects in Business, 9*, 171–193.

Baiden, B. K., Price, A. D. F., & Dainty, A. R. J. (2006). The extent of team integration within construction projects. *International Journal of Project Management, 24*, 13–23.

Bayramoglu, S. (2001). Partnering in construction: Improvement through integration and collaboration. *Leadership and Management in Engineering, 1*, 39–43.

Børve, S., Rolstadås, A., Andersen, B., & Aarseth, W. (2017). Defining project partnering. *International Journal of Managing Projects in Business, 10*, 666–699.

Bowen, G. A. (2009). Document analysis as a qualitative research method. *Qualitative Research Journal, 9*, 27–40.

Brady, T., & Davies, A. (2004). Building project capabilities: From exploratory to exploitative learning. *Organization Studies, 25*, 1601–1621.

Brown, D. C., Ashleigh, M. J., Riley, M. J., & Shaw, R. D. (2001). New project procurement process. *Journal of Management in Engineering, 17*, 192–201.

Davis, P., & Love, P. (2011). Alliance contracting: Adding value through relationship development. *Engineering, Construction and Architectural Management, 18*, 444–461.

Dewulf, G., & Kadefors, A. (2012). Collaboration in public construction—Contractual incentives, partnering schemes and trust. *Engineering Project Organization Journal, 2*, 240–250.

Engebø, A., Klakegg, O. J., Lohne, J., & Lædre, O. (2020). A collaborative project delivery method for design of a high-performance building. *International Journal of Managing Projects in Business, 13*, 1141–1165.

Falch, M. R., Engebø, A., & Lædre, O. (2020). Effects of partnering elements: An exploratory case study.

Gunhan, S. (2019). Analyzing sustainable building construction project delivery practices: Builders' perspective. *Practice Periodical on Structural Design and Construction, 24*, 05018003.

Harback, H. F., Basham, D. L., & Buhts, R. E. (1994). Partnering paradigm. *Journal of Management in Engineering, 10*, 23–27.

Hord, S. M. (1981). Working together: Cooperation or collaboration? *Texas Univ., Austin. Research and Development Center for Teacher Education.*

Hu, Y., Chan, A. P., Le, Y., & Jin, R.-Z. (2015). From construction megaproject management to complex project management: Bibliographic analysis. *Journal of Management in Engineering, 31*, 04014052.

Johansen, A., Engbo, A., Torp, O., & Kalsaas, B. T. (2021). Development of target cost – By the owner or together with contractors – Target value design. *Procedia Computer Science, 181*, 1171–1178.

Klakegg, O. J. (2017, September 5–8). Project delivery models – Situational or fixed design? *2017 12th international scientific and technical conference on computer sciences and information technologies (CSIT)*, pp. 202–206.

Korkmaz, S., Riley, D., & Horman, M. (2011). Assessing project delivery for sustainable, high-performance buildings through mixed methods. *Architectural Engineering and Design Management, 7*, 266–274.

Lahdenperä, P. (2010). Conceptualizing a two-stage target-cost arrangement for competitive cooperation. *Construction Management and Economics, 28*, 783–796.

Langley, A. (1999). Strategies for theorizing from process data. *Academy of Management Review, 24*, 691–710.

Latham, M. (1994). *Contracting the team: Joint review of procurement and contractual arrangements in UK construction*. DDP Services.

Magent, C. S., Korkmaz, S., Klotz, L. E., & Riley, D. R. (2009). A design process evaluation method for sustainable buildings. *Architectural Engineering and Design Management, 5*, 62–74.

Miller, J. B. (1997). Engineering systems integration for civil infrastructure projects. *Journal of Management in Engineering, 13*, 61–69.

Miller, J. B., Garvin, M. J., Ibbs, C. W., & Mahoney, S. E. (2000). Toward a new paradigm: Simultaneous use of multiple project delivery methods. *Journal of Management in Engineering, 16*, 58–67.

Molenaar, K., Sobin, N., Gransberg, D., McCuen, T., Korkmaz, S., & Horman, M. (2009). *Sustainable, high performance projects and project delivery methods: A state-of-practice report* (p. 27). Charles Pankow Foundation.

Mollaoglu-Korkmaz, S., Swarup, L., & Riley, D. (2013). Delivering sustainable, high-performance buildings: Influence of project delivery methods on integration and project outcomes. *Journal of Management in Engineering, 29*, 71–78.

Naoum, S. (2003). An overview into the concept of partnering. *International Journal of Project Management, 21*, 71–76.

Nyström, J. (2005). The definition of partnering as a Wittgenstein family-resemblance concept. *Construction Management and Economics, 23*, 473–481.

Pollack, J. (2007). The changing paradigms of project management. *International Journal of Project Management, 25*, 266–274.

Porter, M. E. (1985). Technology and competitive advantage. *Journal of Business Strategy, 5*, 60.

Ruparathna, R., & Hewage, K. (2015). Review of contemporary construction procurement practices. *Journal of Management in Engineering, 31*, 04014038.

Samset, K. F. (2014). *Prosjekt i tidligfasen: valg av konsept*. Fagbokforl.

Saunders, M., Lewis, P., & Thornhill, A. (2016). *Research methods for business students*. Pearson education.

Scheepbouwer, E., & Humphries, A. B. (2011). Transition in adopting project delivery method with early contractor involvement. *Transportation Research Record, 2228*, 44–50.

Simonsen, S. H. F., Skoglund, M. H., Engebø, A., Varegg, B. E., & Lædre, O. (2019, July 3). Effects of IPD in Norway – A case study of the Tønsberg Project. *Proc. 27th annual conference of the international group for lean construction (IGLC)*, Dublin, Ireland, pp. 251–262.

Söderlund, J. (2011). Theoretical foundations of project management. In *The Oxford handbook of project management*. Oxford University Press.

Son, H., Kim, C., Chong, W. K., & Chou, J. S. (2011). Implementing sustainable development in the construction industry: Constructors' perspectives in the US and Korea. *Sustainable Development, 19*, 337–347.

Stene, T. M., Lædre, O., & Andersen, B. S. (2016). Samspill i gjennomføring av byggeprosjekter-Former for samspill og effekter.

Swarup, L., Korkmaz, S., & Riley, D. (2011). Project delivery metrics for sustainable, high-performance buildings. *Journal of Construction Engineering and Management, 137*, 1043–1051.

Touran, A., Gransberg, D. D., Molenaar, K. R., & Ghavamifar, K. (2011). Selection of project delivery method in transit: Drivers and objectives. *Journal of Management in Engineering, 27*, 21–27.

Turner, R. (2021). Forty years of organizational behaviour research in project management. *International Journal of Project Management, 40*, 9.

Turner, J. R., Ledwith, A., & Kelly, J. (2009). Project management in small to medium-sized enterprises: A comparison between firms by size and industry. *International Journal of Managing Projects in Business, 2*, 282.

Wøien, J., Hosseini, A., Klakegg, O. J., Lædre, O., & Lohne, J. (2016). Partnering elements' importance for success in the Norwegian construction industry. *Energy Procedia, 96*, 229–240.

Yeung, J. F. Y., Chan, A. P. C., & Chan, D. W. M. (2007). The definition of alliancing in construction as a Wittgenstein family-resemblance concept. *International Journal of Project Management, 25*, 219–231.

Yin, R. K. (2018). *Case study research and applications: Design and methods*. SAGE.

Chapter 24
Price Tendering in the South African Consulting Engineering Industry

Neil Govender, Samuel Laryea, and Ron Watermeyer

Abstract Competitive tendering has led to consulting engineering organisations reducing their fees. However, discounted or low fees have often been cited as one of the primary reasons for the declining quality of professional services in the South African construction industry. Since competitive tendering is a key process in the provision of consulting engineering services, this paper examined the impact of competitive tendering on pricing and services provided by consulting engineers. A survey with open-ended questions was conducted with ten experienced professionals in the consulting engineering industry. This study found that fees are still being determined using fee guidelines despite their publication being suspended in 2016. Participants in this study suggested that clients should stipulate minimum fees in tendered projects as a mechanism to address unsustainably low fees in the industry. Low levels of professional fees cause consulting engineering organisations to reduce services wherever possible and only spend time commensurate with the level of fees on projects. Reduced services may have downstream impacts on project outcomes during and post construction. Poor project outcomes adversely affect the provision of services related to several of the Sustainable Development Goals (SDGs). Therefore, clients in South Africa should assess the implications of competitive tendering and determine if the current approach is appropriate to ensure successful outcomes on infrastructure projects. This will assist in the attainment of the SDGs.

Keywords Tendering · Professional Services · Consulting Engineering & Pricing

N. Govender (✉) · S. Laryea · R. Watermeyer
School of Construction Economics and Management, University of the Witwatersrand, Johannesburg, South Africa
e-mail: 699591@students.wits.ac.za; Samuel.Laryea@wits.ac.za; watermeyer@ioptions.co.za

G. Lindahl, S. C. Gottlieb (eds.), *SDGs in Construction Economics and Organization*, Springer Proceedings in Business and Economics,
https://doi.org/10.1007/978-3-031-25498-7_24

24.1 Introduction

Since the introduction of competitive tendering, competition and a steep reduction of fees have been observed in professional service contracts in South Africa (CESA, 2021). Industry groups have asserted that competitive tendering has led to a high level of "discounting" causing unsustainably low fees that are detrimental to the quality of engineering services and the profession (Liebenberg & Wilson, 2011; Okonkwo & Wium, 2018; CESA, 2021). Professionals have indicated that within their organisations, delivering quality services is difficult especially if they are constrained by low fees (Okonkwo & Wium, 2018). Studies in the construction industry have tried to link low fees (arising from competitive tendering) with poor-quality professional services. Some of the studies mentioned that low fees result in declining quality (Andi & Minato, 2003; Gransberg et al., 2007; Ali & Au-Yong, 2021; Quapp & Holschemacher, 2021), while other studies reject this notion (Hoxley, 2000; Lam, 2012; Laryea et al., 2021). However, the construction literature does not examine how and to what extent pricing within consulting engineering organisations has been affected by competitive tendering (apart from mentioning that it has reduced), nor does it assert the extent to which professional services are impacted by competitive tendering.

Provision of inadequate or poor-quality consulting engineering services inevitably leads to poor project outcomes such as delays and cost implications on infrastructure projects (Love et al., 2006; Philips-Ryder et al., 2013). Poor project outcomes adversely affect the provision of services related to several of the Sustainable Development Goals (SDGs), namely, clean water and sanitation (SDG 6), health infrastructure (SDG 3), educational facilities (SDG 5) and reliable and affordable energy (SDG 7). Since competitive tendering is a key process in the provision of consulting engineering services, it is necessary to assess the impact of competitive tendering on pricing and services provided by consulting engineers, especially since these services are vital to achieve successful project outcomes which assist in attaining some of the SDGs.

Therefore, this paper first examines the effect of competitive tendering on how consulting engineering organisations price professional services. Second, this paper examines the extent to which consulting engineering services are impacted by competitive tendering. This paper contains important findings regarding the implications of competitive tendering and is relevant to all clients procuring consulting engineering services competitively.

24.2 Literature Review

The literature review was conducted in a semi-systematic manner and focused specifically on identifying literature that sought to address the objectives of this paper (i.e. (1) to examine how competitive tendering has impacted pricing of consulting engineering services and (2) to examine the impact of competitive tendering on the provision of consulting engineering services). Relevant literature was reviewed in three broad areas, namely,

- Competitive tendering processes to procure consulting engineering services
- Pricing consulting engineering services
- Implications of inappropriate fees on consulting engineering services

24.2.1 Competitive Tendering to Procure Consulting Engineering Services

Competitive tendering is used to procure consulting engineering services in various countries. Although, there are different approaches to implement competitive tendering, the main purpose is to ensure that clients obtain value for money during the procurement process (Kavanagh, 2016).

In the USA, many have criticised competitive tendering asserting that it is detrimental to the engineering profession and quality of professional services (Parks & McBride, 1987; ACEC, 2002). ACEC (2002) asserted that instead of procuring built environment consultants using competitive bidding, clients should rather utilise an approach that includes Quality-Based Selection (QBS). QBS ensures that a competent tenderer can be selected using objective evaluation criteria that are pertinent to the project (ACEC, 2002). Competence of built environment consultants was identified as one of the main factors responsible for project success during the procurement of consultants in the Swedish construction industry (Sporrong, 2011). Feldmann et al. (2008) posited that while it is important to appoint a competent tenderer that is best qualified for the job, tenderers should also be appointed at the "best" possible fee.

In the South African context, procurement of consulting engineering services for majority of public sector projects is conducted using competitive tendering. Competitive tendering in South Africa generally has a two-stage evaluation process that assesses quality and the financial offer, respectively. The quality component screens tenderers against a set of objective quality requirements relevant to the project. Tenderers that fail to achieve the minimum quality score are not considered further during the second evaluation stage. Tenderers that satisfy the quality criteria are evaluated during the second evaluation stage; however, scores for quality are not utilised further in the evaluation process. The second stage of the evaluation process assesses price and preference where price contributes 80% to the total score if the fee value is less than R50 million (approximately EUR 2.8 million), or 90% if the fee is greater than R50 million. The preference component contributes either 10% or 20% depending on the fee value. Preference relates directly to tenderers' Broad-Based Black Economic Empowerment (BBBEE) rating which is a measure of the racial and gender transformation within an organisation. In most cases, consulting engineering projects in South Africa are awarded based on the lowest fee tendered, with quality comprising a small component in the evaluation process.

24.2.2 Pricing Consulting Engineering Services

Professional fees are remuneration provided by clients to consulting engineers for professional services, that are usually executed as part of a professional services contract. The capital costs of construction projects are generally used as a benchmark for consulting engineering fees throughout the world. Publication of historical fees to guide clients and consultants on the suitability of professional fees is also a global norm. In the United Kingdom (UK), historical data (usually not more than 2 years old) from different types of construction projects are utilised to develop typical curves illustrating professional fees for various types of construction projects. A similar approach of publishing median fee curves is used in the USA.

Historically, in South Africa, professional fees for consulting engineers have been gazetted and regulated by the Engineering Council of South Africa (ECSA) through the publication of recommended fees (in the form of a fee guideline). The professional fee is guided by the capital value of the project; however, several factors, such as level of effort, risk, liability and complexity, may also influence the professional fee (ECSA, 2011).

In 2016, the Competition Commission ruled that the publication of fee guidelines restrict competition and result in indirect price fixing (CCSA, 2016). The finding by the Competition Commission in South Africa followed international best practice whereby many other countries moved away from utilising published/recommended fees to price professional services. Notwithstanding the Competition Commission's ruling, at the beginning of 2021, the ECSA began publishing fee guidelines again. Nonetheless, the ruling to suspend fee scales in the South African context is relatively recent when compared to events internationally.

Internationally the abolishment of fee scales and the introduction of competitive bidding facilitated increased competition and declining fees. In the USA, competitive bidding for professional services was introduced in the 1970s.While the phenomenon of abolishing fee scales began in Europe in the 1980s with the United Kingdom (UK) being one of the first countries to abolish fee scales. Thereafter, the rest of Europe gradually followed, after a report commissioned by the European Union (EU) entitled *Report on Competition in Professional Services* (2004) found that recommended prices negatively influenced competition. The report emphasised that professional fee scales of various professions (including architects, other built environment professionals and lawyers) were already abolished in regions, such as Finland, France and the UK. Germany was one of the last remaining countries in the EU opposed to abolishing fee scales for built environment professionals (particularly architects and engineers). However, in 2019, a court judgement by the Court of Justice of the European Union (CJEU) rejected contentions by the German government that fee scales protect consumers and safeguard quality (Laryea et al., 2021). The CJEU found that the use of fee scales in Germany was contradictory to EU law.

The abolishment of fee scales in South Africa has compelled consulting engineers to develop a system to price their services. Pricing of services is often imprecise and not an exact science (Sturts & Griffis, 2005). Many factors, such as project complexity, resources and level of effort, are considered when determining an appropriate professional fee. The competitive climate in the South African consulting engineering industry has necessitated fees to be reduced for consultants to stand a better chance of being appointed on projects. The downward pressure of competition on prices in the construction industry has been highlighted by several authors (Hoxley 2000; Sturts & Griffis, 2005; Quapp & Holschemacher, 2021; Laryea et al., 2021). However, the literature provides no in-depth examination of how and to what extent pricing within organisations is affected by competitive tendering. This is important, particularly in the South African context, where historically consulting engineers have been reliant on a fee scale to determine remuneration for their services. In the absence of fee scales, how are professional fees in the consulting engineering industry being determined considering the high level of competition on projects?

24.2.3 Implications of Inappropriate Fees on Consulting Engineering Services

Several studies in the international construction literature have asserted that low fees are problematic and adversely impact the quality of professional services in the construction industry (Andi & Minato, 2003; Gransberg et al., 2007; Ali & Au-Yong, 2021). Quapp and Holschemacher (2021) posited that an appropriate level of fee is required to maintain a level of quality for built environment professional services and ensure that constructed infrastructure is safe for public use. In the South African context, a handful of studies also suggested that low fees arising from competitive tendering negatively influence the quality of professional services in the consulting engineering industry (Liebenberg and Wilson, 2011; Malinda, 2017; Moos, 2017; Okonkwo & Wium, 2018; Akampurira & Windapo, 2018). On the other hand, there are a few studies that found that low fees arising from competitive tendering do not adversely affect the quality of professional services. The studies concluded that other factors apart from fees influence the quality of professional services (Hoxley, 2000; Lam, 2012; Laryea et al., 2021).

It is evident that most studies in the construction literature either accepted or rejected the notion that the level of professional fee may influence the quality of services rendered. However, studies on this topic were superficial and did not conduct in-depth examinations of how and to what extent services provided by consulting engineering organisations were impacted by the level of fees. This is an important topic that must be addressed since the quality of consulting engineering services has a direct impact on project outcomes.

24.3 Research Methodology

A preliminary investigation using a single-stage methodology that included a survey research strategy was conducted in this study. Surveys enable many participants to provide feedback in a timely and cost-effective manner (Fellows & Liu, 2015; Wilkinson & Birmingham, 2003). Surveys also allow researchers to influence the direction of information gathered in accordance with themes in a study (Wilkinson & Birmingham, 2003; Cohen et al., 2007; Saunders et al., 2019). An online survey containing 11 open-ended questions was designed. The survey was distributed to registered professionals in the South African consulting engineering industry in December 2021. Respondents needed to satisfy the following criteria to participate in this study:

- Hold a professional registration with the ECSA as either a Professional Engineer or Professional Engineering Technologist
- Be actively working in the consulting engineering industry
- Have a minimum of 10 years' experience in the consulting engineering sector

Questions in the survey addressed pricing and services provided by consulting engineering organisations. Open-ended questions were used to elicit rich in-depth feedback which was critical to develop a detailed understanding about how competitive tendering impacted pricing and services provided by consulting engineers. Qualitative data were collected from the surveys and analysed using thematic analysis. Categories/themes that were used during the development of questions in the survey and analysis of data emerged from the literature. The key themes are provided below:

- Factors considered when determining professional fees
- Pricing strategy adopted for competitive tendering
- Balancing competitive fees with providing quality services
- The impact of competitive tendering on professional services
- The impact of competitive tendering on training of staff
- The impact of competitive tendering on design optimisation and value engineering
- The impact of competitive tendering on the engineering profession
- Interventions to prevent unsustainably low professional fees

Since this was a preliminary investigation, the online survey was structured as a "one-way" data collection instrument whereby participants answered questions using an online form. Participants took an average of 10 minutes to complete the survey.

The survey was intended to provide a snapshot of the perceptions of consulting engineers in South Africa. However, the main disadvantage of using a "one-way" approach was the inability of the researchers to elicit additional information by asking follow-up questions. This may have inadvertently introduced bias into the study, especially in cases where built environment professionals were protecting their vested interests related to professional fees.

24.3.1 *Profile of Participants*

A total of 15 participants from the researchers' professional network in the South African consulting engineering industry were contacted to participate in this study. A response rate of 67% was achieved with ten participants responding to the survey. The participants all had more than 10 years' experience in consulting engineering with 80% of participants having more than 20 years' experience. All participants in this study were in management level positions in consulting engineering companies. Furthermore, all the participants were either professionally registered engineers or engineering technologists with the ECSA.

24.4 Results

24.4.1 *Factors Considered When Determining Professional Fees*

Participants in this study mentioned that a range of factors are considered when determining professional fees. The two common factors included project complexity and resources required on a project. Resources on a project are usually determined by the complexity and scope of works. In the engineering industry, the main resource used on projects are human resources (i.e. technical personnel that execute projects). Other factors that were mentioned included project location, current workload, the client and competition.

When determining the level of professional fee tendered, most of the participants mentioned that they still use the ECSA fee scale but combine it with a time and cost-based method. This involves calculating costs for resources based on the estimated time that they will spend on projects and their charge out rate. Only 30% of participants mentioned that they solely used the ECSA fee scale without checking it against the time and cost-based method.

24.4.2 *Pricing Strategy Adopted for Competitive Tendering*

Most of the participants mentioned that the fierce level of competition during tendering resulted in a "high" level of discounting of professional fees (compared to gazetted fees). Several participants mentioned that they looked for ways to reduce their price since most clients in South Africa appoint based on the lowest fee tendered. A "high" level of discounting is particularly prevalent on projects where there are many tenderers. However, one participant mentioned that their pricing strategy was not influenced by competitive tendering as they operated in a niche market. Another participant stated that pricing strategies were irrelevant if appointments were made via framework agreements.

Most of the participants (70%) mentioned that they changed their pricing strategy depending on the client where they were submitting tenders. Some factors influenced by the client that determined pricing strategies among consultants include project delays, change in scope, effort, frequency of client payments and whether the client is "informed" and technically competent. One of the participants mentioned that fees were generally higher for uninformed clients, to account for additional work and time that may be required on the project. Another participant mentioned that some private clients do not accept the fees gazetted by the ECSA; therefore, a more competitive strategy using a rates-based approach is utilised for those clients. A rates-based approach entails submitting hourly rates for resources on a tender and allocating those resources with their respective time allocation for tasks that are required to complete the scope of work. On the other hand, 30% of participants mentioned that their pricing strategy does not differ according to the type of client.

24.4.3 Balancing Competitive Fees with Providing Quality Services

All participants mentioned that regardless of the professional fee tendered, they try to render a quality service. This is achieved by undertaking due diligence, proper review and quality management prior to submission of deliverables. One of the participants mentioned that staff in their organisation were aware of low-profit margins on projects but still went the "extra mile" without additional remuneration, to ensure that good quality services were rendered. Another participant mentioned that time spent on projects by senior resources in their organisation was reduced to minimise project costs and ensure projects could be executed with lower budgets.

24.4.4 The Impact of Competitive Tendering on Professional Services

Participants in this study mentioned that time spent on tasks was minimised due to reduced fees as a result of competitive tendering. A few participants mentioned that the time spent on projects by senior staff was reduced and junior resources were provided opportunities for "day to day" execution of the project and other tasks previously done by senior resources. A few participants also mentioned that meetings were one of the services reduced to accommodate lower fees on projects. One of the participants mentioned detailed options analysis and life cycle costing were not undertaken if fees were "insufficient".

24.4.5 The Impact of Competitive Tendering on Training of Staff

There were contrasting views from participants on whether competitive tendering adversely impacted training of staff on projects. Some participants mentioned that training still needed to be undertaken despite the fee tendered and that the level of fees had no effect on training provided to staff. However, other participants mentioned that fee discounting had a big impact on training. Participants mentioned that because project budgets were reduced, there was often little room for training on projects. A few participants mentioned that because time allocated to activities were reduced, it was often difficult for training to occur. Another participant mentioned that junior resources were often forced to deal with situations that they were ill equipped to handle, since time spent on projects by senior resources was limited. On the other hand, one participant mentioned that despite difficulties to provide training on projects due to reduced fees, training was still undertaken and cross-subsidised with other more profitable projects.

24.4.6 The Impact of Competitive Tendering on Design Optimisation and Value Engineering

Most of the participants (80%) mentioned that discounted fees negatively influenced design optimisation and value engineering. Participants stated that when reduced fees were tendered, there was often no provision in the budget for design optimisation or value engineering, which resulted in standard designs/design principles being used. Another participant stated that instead of undertaking in-depth design optimisation, a technically feasible solution was usually sought on projects without exploring design options in detail. This approach takes cognisance of the reduced time that can be spent on design optimisation/value engineering due to a reduced fee. Another participant mentioned that on some projects, design optimisation/value engineering was part of the ongoing review process with the client and was not necessarily affected by the fee tendered.

24.4.7 The Impact of Competitive Tendering on the Engineering Profession

Majority of participants asserted that competitive tendering was detrimental to the engineering profession. Many of the participants expressed the view that to some extent, quality on projects was sacrificed because of competitive tendering. However, most of the participants (80%) also mentioned that a "higher" fee will not necessarily guarantee a better level of service. Two participants asserted that higher fees would allow more time to conduct thorough investigation and review; therefore, it will guarantee a better quality of service.

A few participants mentioned the need for value for money when procuring professional services; however, one participant stated that "value for money does not equate to the lowest cost". One of the participants opposing the view that competitive tendering is detrimental to the engineering industry asserted that Quality Cost-Based Selection is a global practice of procuring engineering services and ensures value for money.

24.4.8 Interventions to Prevent Unsustainably Low Professional Fees

A few participants mentioned that clients should consider setting minimum fees (or institute maximum discounts where fee scales are still used) on projects to address the issue of unsustainably low fees in the industry. Some participants mentioned that clients need to be more knowledgeable and understand "reasonable" costs for a particular service. Thereafter, they can assess risks associated with tenderers that have submitted prices which do not seem reasonable.

24.5 Discussion

This paper explored the impact of competitive tendering on pricing and services provided by consulting engineering organisations in South Africa. Literature on this topic asserted that competitive tendering has placed downward pressure on prices; however, an in-depth investigation on how and to what extent competitive tendering affects pricing of services in consulting engineering organisations was not conducted. This paper adds to the body of knowledge by explaining how consulting engineering organisations price their services during competitive tenders. This study supports the notion that there is downward pressure on prices due to competitive tendering.

24.5.1 Factors Considered When Determining Professional Fees

When determining fees, two main considerations by consulting engineering organisations include project complexity and resources required on projects. This study found that professional fee scales are still being utilised in South Africa to determine professional fees, despite their publication being suspended by the Competition

Commission of South Africa in 2016. However, many participants mentioned that fees derived from fee scales are checked using a time-based approach which is dependent on estimated time spent on projects by resources and hourly charge out rates.

24.5.2 Pricing Strategy Adopted for Competitive Tendering

Offering large discounts on gazetted fees was a common practice when tendering for projects and is intensified when there are many bidders tendering for the same project. Consulting engineering organisations also admitted to changing their pricing strategy according to the client for which they are tendering. Fees are increased when tendering for clients that lack technical knowledge or expertise. This is an important finding in the South African context, as many public sector clients lack technical expertise to roll out projects. Companies working for these clients may need to expend more effort to deliver projects while still working at a competitive fee. Therefore, experienced firms inevitably put a premium on their fees when tendering for these clients.

24.5.3 Impact of Fees on Consulting Engineering Services

Research literature on the effect of fees on the quality of professional services is also divided. Some papers found that "low" fees negatively influence the quality of professional services, while others rejected this notion. Nonetheless, the literature did not clearly explain how "low" fees affects the provision of services in the consulting engineering industry. Reduction of time spent on projects by resources is one of the common impacts on professional services due to low fees. There is also an emerging trend of allowing more junior resources to operate on projects and undertake tasks previously executed by senior resources. This may have implications on the quality of services provided; however, participants in this study mentioned that they try to deliver quality services regardless of the level of fees on a project. Assessing the quality of professional services was not specifically part of this study and is an important area that should be investigated in future studies on this topic. Design optimisation and value engineering are key services that are either reduced or completely omitted from projects. This has adverse implications on life cycle costs and value for money of the constructed project/asset. Other services that are deemed extra or "value adds" are also reduced or not undertaken by consulting engineering organisations to accommodate reduced fees.

Another implication of reduced fees is the difficulty to provide training on projects. Consultants recognised the importance of training and mentorship of staff; however, they indicated that training was more difficult when fees were reduced as additional time and resources cannot be expended on projects. Inadequate training of junior staff may have long-term implications on the consulting engineering industry and result in inadequate technical capacity being developed locally for future large-scale infrastructure projects in South Africa.

24.5.4 Interventions to Prevent Unsustainably Low Professional Fees

Majority of participants recognised the need for competition and value for money when procuring consulting engineering services but asserted that competitive tendering was detrimental to the consulting engineering industry particularly in relation to quality of services. This was a surprising finding since all the participants previously mentioned that they tried to ensure quality services regardless of the professional fee. Instating a stipulated minimum fee on projects was suggested by participants as a mechanism to prevent unsustainably low fees being tendered.

24.5.5 Relevance of the Study and Limitations

This paper contains information that is relevant to clients using competitive tendering to procure consulting engineering services. While competition may be beneficial and result in reduced fees being tendered, there may be implications on the service provided and value for money of the constructed asset in the long term.

The main limitation of the study was the small sample size and "one way" nature of the online survey which may have resulted in bias of professionals seeking to protect their vested interests. Future studies in this area should either try to increase the sample of participants to get a broader range of views or conduct in-depth interviews which will allow the interviewer to systematically interrogate feedback provided by participants.

24.6 Conclusion

Competitive tendering has compelled consulting engineering organisations to reduce their professional fees. Some of the major findings in this study are highlighted below:

- In the South African context, many participants are still reliant on fee scales published by the ECSA to determine professional fees on projects. However, two common factors considered when determining professional fees are project complexity and resources required on a project.

- The competitive nature of the consulting engineering industry in South Africa has led to fierce discounting of fees among consultants.
- All participants mentioned trying to render a quality service regardless of the professional fee tendered. One of the main impacts on projects with reduced fees is the use of junior resources to undertake most tasks, while senior resources tend to spend less time on projects.
- Most participants in this study mentioned that discounted fees negatively influence provision of design optimisation and value engineering services.
- Participants in this study asserted that competitive tendering was detrimental to the engineering profession with a few participants mentioning that clients should consider setting minimum fees (or institute maximum discounts where fee scales are still used) on projects to address unsustainably low fees in the industry.

Client bodies procuring consulting engineering services in South Africa should assess the implications of competitive tendering and determine if the current approach is appropriate to ensure successful project outcomes. This will assist in successfully attaining the SDGs.

24.7 Future Research Directions

Construction researchers in South Africa should consider the following areas for future research:

1. An in-depth study critically investigating the impact of fees and other factors on the quality of professional services.
2. An investigation of how professional councils in South Africa, such as the ECSA, set their professional fees. This should include determination of profit and overhead percentages used by the ECSA and comparison with international norms.
3. An investigation to fundamentally clarify how pricing of consulting engineering services should be conducted from "first principles". The investigation should address benchmarking of professional fees in the absence of professional fee scales.

References

Akampurira, E., & Windapo, A. (2018). Factors influencing the quality of design documentation on South African civil engineering projects. *Journal of the South African Institution of Civil Engineering, 60*(3), 41–48.

Ali, A. S., & Au-Yong, C. P. (2021). Risk factors in the refurbishment design process: The case of Malaysia. *International Journal of Building Pathology and Adaptation, 40*(4), 622–635.

American Consulting Engineers Council (ACEC). (2002). *Bidding is not the solution: Case studies in bidding*. American Consulting Engineers Council.

Andi, & Minato, T. (2003). Design documents quality in the Japanese construction industry: Factors influencing and impacts on construction process. *International Journal of Project Management, 21*(7), 537–546.

Cohen, L., Manion, L., & Morrison, K. (2007). Research methods in education sixth edition. London and NewYork, Routledge.

Commission of the European Communities. (2004). *Report on competition in professional services*. Commission of the European Communities.

Competition Commission South Africa (CCSA). (2016). *Competition commission South Africa annual report 2015/2016*. Competition Commission South Africa.

Consulting Engineers South Africa (CESA). (2021). *Bi-annual economic and capacity survey: January – June 2021*. Consulting Engineers South Africa.

Engineering Council of South Africa (ECSA). (2011). *Engineering Profession Act (46/2000): Guideline for services and processes for estimating fees for persons registered in terms of the act*. Gazette No. 34875. Board Notice 206. Engineering Council of South Africa.

Feldmann, M. L., Chrusciel, D., Pohlmann, A., Shelley, M. C., McCool, K., Morton, A. D., & Ahoy, C. K. (2008). Architectural and engineering fees from the public institutional perspective. *Journal of Management in Engineering, 24*(1), 2–11.

Fellows, R., & Liu, A. (2015). Research methods for construction. West Sussex, John Wiley and Sons.

Gransberg, D. D., Del Puerto, C. L., & Humphrey, D. (2007). Relating cost growth from the initial estimate to design fee for transportation projects. *Journal of Construction Engineering and Management, 133*(6), 404–408.

Hoxley, M. (2000). Are competitive fee tendering and construction professional service quality mutually exclusive? *Construction Management and Economics, 18*(5), 599–605.

Kavanagh, P. (2016). A case for negotiated performance-based contracting rather than competitive tendering in government public transport (bus) service procurement. *Research in Transportation Economics, 59*, 313–322.

Lam, T. Y. M. (2012). Economic perspective on outsourcing of property management services. *Property Management, 30*(4), 318–332.

Laryea, S., Watermeyer, R., & Govender, N. (2021). The influence of fees on the quality of professional services in South Africa. *Proceedings of the Institution of Civil Engineers – Management, Procurement and Law*. https://doi.org/10.1680/jmapl.20.00022

Liebenberg, J. J. E., & Wilson, S. J. (2011, September 11–14). *Tendering for professional services: Both sides of the coin*. Proceedings, 10th Conference on Asphalt Pavements for Southern Africa, KwaZulu-Natal, South Africa.

Love, P. E. D., Edwards, D. J., & Smith, J. (2006). Contract documentation and the incidence of rework in projects. *Architectural Engineering and Design Management, 1*(4), 247–259.

Malinda, M. J. K. (2017). *Quality of project documentation as a major risk source in infrastructure projects in South Africa.* MEng (Civil Engineering) Dissertation, Stellenbosch University.

Moos, M. F. (2017). *The impact of the current public sector procurement system on the consulting engineering industry.* MPhil (Civil Engineering) Dissertation, University of Cape Town.

Okonkwo, P. N., & Wium, J. (2018). Impact of discounted professional fees on the risk exposure of civil and structural engineering services consultants in South Africa. *Journal of the South African Institution of Civil Engineers, 60*(1), 10–20.

Parks, G. A., & McBride, R. R. (1987). Competitive bidding for engineering services. *Journal of Professional Issues in Engineering, 113*(1), 66–76.

Philips-Ryder, M., Zuo, J., & Jin, X. H. (2013). Evaluating document quality in construction projects – Subcontractors' perspective. *International Journal of Construction Management, 13*(3), 77–94.

Quapp, U., & Holschemacher, A. (2021). Appropriateness of fees for construction planning services of architects and civil engineers as a question of morality. *Journal of Legal Affairs and Dispute Resolution in Engineering and Construction, 13*(2), 04521004.

Saunders, M., Lewis, P. & Thornhill, A. (2019). *Research methods for business students eighth edition*. Harlow: Pearson Education Limited.

Sporrong, J. (2011). Criteria in consultant selection: Public procurement of architectural and engineering services. *Australasian Journal of Construction Economics and Building, 11*(4), 59–76.

Sturts, C. S., & Griffis, F. H. (2005). Pricing engineering services. *Journal of Management in Engineering, 21*(2), 56–62.

Wilkinson, D., & Birmingham, P. (2003). Using research instruments: a guide for researchers. London, Routledge Falmer.

Chapter 25
Managing Procurement Ability Over Time: A Study of Swedish Construction Clients

Stefan Olander and Henrik Szentes

Abstract There is a challenge in the construction sector to transfer knowledge from individuals and teams to the rest of the organisation. For a construction client organisation, one such challenge could be to transfer knowledge from procurement individuals and teams to the organisation as a whole. This study is focused on the procurement functions within construction client organisations, with the purpose of increasing the understanding of how these organisations manage their procurement ability over time. Based on an interview study, the strategies and actions taken within the different organisations represented are analysed and discussed. Empirical material was collected through a series of in-depth interviews and then analysed using a thematic approach and compared with previous research including the maturity model. Although there are examples of integration activities from the interviews, there is a lack of clear visible strategy of how to integrate procurement process in a learning perspective for both the procurement process in itself as well as for the whole construction project management process. More research is needed to increase the understanding of procurement as a value-adding process and how knowledge sharing and knowledge transfer can increase the learning concept in relation to procurement in construction.

Keywords Procurement · Buildings · Learning organisations · Construction clients

The original version of this chapter was revised. The correction to this chapter is available at https://doi.org/10.1007/978-3-031-25498-7_26

S. Olander (✉) · H. Szentes
Construction Management, Lund University, Lund, Sweden
e-mail: stefan.olander@construction.lth.se; henrik.szentes@construction.lth.se

G. Lindahl, S. C. Gottlieb (eds.), *SDGs in Construction Economics and Organization*, Springer Proceedings in Business and Economics,
https://doi.org/10.1007/978-3-031-25498-7_25

25.1 Introduction

Recent years, procurement strategies have gained increasingly more attention in construction management literature, including research recognising the need for analysing strategies over time. For instance, Boyd and Chinyio (2006) define construction procurement as the framework within which a construction project is acquired, or obtained, and that the client organisation is the main actor in the process setting the requisites for this framework. Thus, construction procurement can be viewed as the strategic process of how contracts for construction work is created, managed and fulfilled (Hughes et al., 2015). Hartmann et al. (2008) add that construction client organisations have an essential impact on implementing innovation and change in the construction sector, which is a necessity in order to meet goals concerning challenges, such as resource efficiency, sustainable development and climate change. In order to contribute to the UN Agenda 2030 global sustainability goals, the construction industry needs to design and build future buildings in accordance with the global goals. One of several ways to achieve this is to make conscious choices in the procurement processes. However, there is then a need for construction clients to reflect upon their procurement ability, and how to develop this ability over time.

Toumela-Pykkönen et al. (2015) highlight that the business environment for the construction sector is changing and that there is an increasing need to review procurement practices in order to achieve optimal value while minimising the use of resources. This is in line with UN target 12.2 sustainable management and use of natural resources and 7.3 double the improvement in energy efficiency. Furthermore, there is an ongoing shift to more partner-based approaches as well as more digital solutions to project delivery and project implementation (Ruparathna & Hewage, 2015). More partner-based approaches can help knowledge sharing and cooperation to better meet future demands in the construction industry, much in line with UN target 17.6 knowledge sharing and cooperation for access to science, technology and innovation. Thus, there is a need for client organisation to reflect upon their procurement ability also with respect to the changing business environment within the construction sector. Accordingly, Tassabehji and Moorhouse (2008) argue that there needs to be a shift in the procurement official's skill set. The traditional view of procurement as purely a set of administrative procedures needs to shift into a recognised skill set that adds value to the organisation, i.e. a construction client. However, the achievement of such a development, according to Tassabehji and Moorhouse (2008), requires a strong organisational support and a shift in how the role of procurement is viewed internally. Thus, there are several reasons for construction clients to address aspects related to competence as well as attitudes, in order to create a continuous learning about procurement within their organisation.

In his seminal work, Senge (1990) discussed learning organisations as those who develop their capacity in order to achieve desired results by nurturing a collective learning environment. On the one hand, the aspect of procurement abilities can be seen as a mean to achieve the learning aspirations within a construction client organisation, but on the other hand, it can also be viewed as an organisational unit in itself. Garvin (1993) defines a learning organisation as one that is skilled at creating, acquiring and transferring knowledge as well as at reflecting on new knowledge and insights in order to modify the organisations behaviour. Similarly, Hazlett et al. (2008) state that ongoing training on the subject of procurement is an effective method of sharing knowledge within an organisation. However, Hazlett et al. (2008) also argue that the key barriers for sharing procurement knowledge within an organisation are lack of time, resources and training. Senge (1990) proposes five components when addressing the concept of learning organisations: personal mastery, mental models, building a shared vision, team learning and systems thinking – described as the fifth discipline that joins the other four components together. The viewpoint being that without systems thinking there is not a sufficient base for learning. In addition, Goh and Richards (1997) specify that for organisational learning to be effective, structured managerial practices are a necessity, such as clarity of purpose and mission; leadership, commitment and empowerment; experimentation and rewards; transfer of knowledge; and teamwork and group problem solving.

When discussing learning in construction, the project-based nature of operations needs to be taken into account. Accordingly, Manley and Chen (2017) emphasise the importance of understanding the value of learning procedures for achieving an effective governance and performance. Furthermore, Chinowski et al. (2007) highlight that research in construction has often been focused on production aspects, whereas research on organisational development in construction is much scarcer. In their extensive study covering literature review, survey and case studies, they present and discuss several characteristics for learning organisations and argue that effective learning requires proactivity and that a reactive learning philosophy is not sufficient. There is thus an elevated challenge in the construction sector to transfer knowledge from individuals and teams to the rest of the organisation, which is also highlighted by Senaratne and Malewana (2011). For a construction client organisation, one such challenge could be to transfer knowledge from procurement individuals and teams to the organisation as a whole. Hence, there is a relevant need for construction client organisations to treat their procurement processes as a learning process.

This study is focused on the procurement functions within house-building client organisations, with the purpose of increasing the understanding of how these organisations manage their procurement ability over time. Based on an interview study, the strategies and actions taken within the different organisations represented are analysed and discussed, using the maturity model (Chinowski et al., 2007), as the main structure. The maturity model consists of five characteristics for a learning organisation – leadership, process and infrastructure, communication, education and culture. *Level 0* represents the starting point and *Level 5* represents maturity in all characteristics.

25.2 Empirical Study

25.2.1 Method

Empirical material was collected through a series of in-depth interviews, and then analysed using a thematic approach and compared with previous research including the maturity model by Chinowski et al. (2007).

25.2.1.1 Sample

The study comprises seven in-depth semi-structured interviews with client organisation representatives involved in house building in Sweden, as presented in Table 25.1. The sample was purposive (Silverman, 2006), with the aim to study client organisations ranging from small to large, private and public, and housing and commercial buildings.

Client organisation 1 Housing developer who owns both rental apartments and selling condominiums. They manage about 500,000 square metres of gross floor area and have a yearly turnover of about 350 million EURO.

Client organisation 2 A regional public client organisation responsible for managing and developing public buildings, such as hospitals and office buildings. They manage about one million square metres of gross floor area.

Client organisation 3 A local commercial real estate owner who develops and manages office buildings. They manage about 2.1 million square metres of gross floor area and have a yearly turnover of about 300 million EURO.

Client organisation 4 A local housing company that owns and develops student housing. They manage 200,000 square metres of gross floor area and have a yearly turnover of about 30 million EURO.

Client organisation 5 A nationwide housing developer who develops and sells condominiums. They have a yearly turnover of about 100 million EURO.

Client organisation 6 A publicly owned municipal housing company that owns rental apartments. They manage about 700,000 square metres of gross floor area and have a yearly turnover of about 100 million EURO.

Client organisation 7 A nationwide commercial real estate owner who develops and manages office buildings. They manage about 2.3 million square metres of gross floor area and have a yearly turnover of about 800 million EURO.

Table 25.1 Summary of studied client organisations

	Private	Public	Nationwide	Local	Housing	Commercial	Public facilities	Rental	Selling condaminiums	Gross floor area (m^2)	Turnover (EURO)
1	x		x		x			x	x	500,000	350,000,000
2		x		x			x	x		1,000,000	N/A
3	x			x		x		x		2,100,000	300,000,000
4	x			x	x			x		200,000	30,000,000
5	x		x		x				x	N/A	100,000,000
6		x		x	x			x		700,000	100,000,000
7	x		x			x		x		2,300,000	800,000,000

25.2.1.2 Interview Set-Up

To enable a combination of stringency and catching specifically interesting topics, each interview was recorded and transcribed, and field notes were also taken during interviews (Kvale, 1997; Alvesson, 2011). The interviews were based on the following main questions, although the in-depth characteristic of the interviews allowed flexibility and discussions:

- What is a successful procurement according to you?
- Which are the key factors that affect procurement ability over time?
- How is knowledge transferred between the procurement process and project execution?
- How is knowledge and experience from one procurement process transferred to other procurements/projects?
- What are your organisation's aims concerning the development of procurement ability over time?
- How is your organisation's procurement ability over time evaluated?
- Which are the main drivers for developing the procurement ability over time?

25.2.1.3 Analysis

To begin with, each interview question was analysed separately using a thematic approach to find interesting patterns, similarities and differences. The thematic approach led to some insights regarding specific actions taken by the different organisations. In addition, the model by Chinowski et al. (2007) was applied to better grasp the maturity level of procurement activities and to enable an analysis of characteristics that fit with each maturity level.

25.2.2 Findings

The initial thematic analyses of the answers to each interview questions entailed several interesting findings, as described in the coming subsections.

25.2.2.1 Successful Results of a Procurement

One main aspect raised by respondents concerning a successful procurement outcome is that it should result in a good business opportunity for all parties involved. However, there was also a narrow client-oriented perspective in the sense that a successful procurement is when all the requirements stated by the client are fulfilled in the end. Some requirements relate clearly to an economic perspective, such as price, while others stressed that parameters such as quality, environment, health and safety, and social aspects also are important evaluation criteria. Another view raised, is that a successful procurement process should result in a clear contract that can easily be understood by all project participants, especially the project manager who will carry out project implementation. A central issue is to clearly address and sort out unclarities and thus minimise resources used for contract interpretations during the project execution.

Moreover, the interviews also highlighted a distinction between a good procurement process, and a good procurement outcome. The result of a procurement outcome is in general described as more important than the procurement process as such, although the procurement process is also viewed as relevant. Good preparations from the client are essential to ensure that demands on project delivery is met. When project deliveries are insufficient, it is often the result of inadequate tender documents that failed to address relevant demands concerning, for example, user and client requirements, including operation and maintenance. Similarly, a prerequisite for a successful procurement process, raised by many respondents, was the ability to clearly describe the expected outcome of the construction project. Otherwise, there is a risk not only that the end result is not in line with the requirements but also that the procurement process becomes problematic and resource demanding.

A proactive approach to the procurement process was another issue raised as a success factor. However, the bottom line is often to obtain the best contractor as possible to the lowest possible price, as stated by one respondent. Thus, there is an issue regarding optimisation and best ration between price and quality of delivery. Partnering and collaborative approaches were mentioned as a potential way to increase the opportunities of obtaining best value. However, there were diverse opinions about the merits of partnering. Some respondents saw little or no use of engaging in partnering concepts, while others viewed partnering as a mean to get the contractors more engaged in sharing their experiences, and thus likely obtain a better outcome of the project. The ability to cooperate was by many viewed as a factor that increased the opportunity of both a successful procurement process and a successful project delivery. The focus on project outcomes was a clear common denominator with the respondents regardless of the contract form and whether it engaged partnering concepts or not.

There were also some thoughts raised about failed procurement processes, which one respondent described as a procurement process that does not result in a contract., e.g. due to lack of tenders and/or to expensive tenders. Another aspect mentioned was that tender documents were poorly executed in describing expected project outcomes, resulting in conflicts during construction on-site based on unclear agreements. Furthermore, this might even result in a finalised building that does not meet expectations and requirements. Poor preparations and insufficient client competence are issues that might negatively affect the execution of a construction procurement process.

25.2.2.2 Key Factors Affecting the Procurement Ability Over Time

The level of knowledge with the client project managers is very relevant since they are often the ones who gather and define the fundamental prerequisites for the expected outcome of the construction project, and thus lay out the framework for the procurement process to come. Although client project managers are crucial for each project, it was emphasised that competence and engagement in the procurement process within the client organisation in general are essential in order to develop the procurement ability over time. Knowledge management and follow-up of earlier project outcomes were regarded as an important process in order to develop the procurement ability. Not at least, when obtaining knowledge from a contractor perspective on how they have experienced the procurement process, project implementation and outcomes. Since construction on-site, often goes on for several years, it is equally important with feedback between contractor and client during this time in order to acknowledge aspects that might have been missed in the contract and in the tender documents

Property management and the future use of the finished building are another important aspect to consider when evaluating the procurement ability. Taking a user perspective early on increases the chance of achieving a functional building. However, there were only a few of the respondents that raised this as an important issue. One respondent, however, who was a developer of dwellings for open sale, raised the user perspective as very important since it affected marketing and sales.

25.2.2.3 Knowledge Transfer Between the Procurement Process and the Project Execution

There was a consensus among the respondents that, when possible, it is important to involve the coming construction project manager in the procurement process in general and when producing the tender documents in particular. There is often an ambition to keep the same construction project manager from the acquisition of land all the way to the finished construction project. This will ensure a continuity throughout the construction process and both specific and tacit knowledge about decisions taken and why. It is also important that there is a support by the procurement officials to the construction project manager in order to ensure that the different intentions of the procurement are clearly understood by both parties. For the business deal to be as good as possible there is a need for cooperation between the procurement officials and the construction project management.

25.2.2.4 Knowledge Transfer Between Procurement Processes of Different Projects

Some of the respondents, in the larger client organisations, highlight that they have a central procurement organisation that supports the procurement process when producing tender documents and managing contracts. They also support following up process about the outcomes of the procurement processes. There are also examples of internal training within the client organisation to share experiences and to develop procurement abilities. However, the knowledge transfer process is often informal and seldom documented. One respondent raised the notion that there probably is a need for more formal processes concerning the transfer of knowledge between procurement processes for different projects. Some respondents argued that there is an individual responsibility to learn and increase their own abilities.

Some respondent organisations work with standardised templates and documents to support the procurement process. However, there are drawbacks because of the variety of characteristics between different construction projects. The standardised templates and documents may be more problematic than helpful in the way that project-specific aspects may be overlooked or due to demands being stated that are not necessary or relevant, out of which both problems can have a cost driving effect. If standardised templates and documents are used, it is important that they are flexible and that they can easily be adapted to the circumstances of a specific project. Further, they need to undergo continuous revisions and improvements to stay relevant. Instead of standardised templates and documents, some respondents worked with standardised guidelines for their procurement process.

Other aspects affecting the ability to share knowledge and experiences between projects/procurements relate to organisational structure. The more centralised the organisation, the easier it is to facility share of knowledge, but, on the other hand, a more centralised management often means that the variety of projects within the scope is higher. One respondent raised that a distributed responsibility of authority to project managers may create engagement and ownership. Moreover, an extensive geographical distribution of the projects is often a hinder for sharing of knowledge. No matter the structure, it is important that communication channels and forums are defined.

Another important aspect is whether the client is obliged to use a public procurement act (PPA) or not. It was raised that the past performance of contractors was not possible for public clients to assess, according to PPA. Therefore, it is a challenge for public clients to build and nurture long-term relations with contractors and other suppliers. It was highlighted that it is a trade-off between competition and effective procurement.

25.2.2.5 Aims Concerning the Development of Procurement Ability Over Time

One respondent indicated that the goals in their procurement process were measured with different key performance indicators, in relation to minimizing and avoiding risks in the procurement process, and with a specific focus on sustainability. Some of the respondents from public clients raised the issue that goal concerning a correct process from a PPA perspective was a clear goal, both from the perspective of being a professional and respectful client and from the perspective of fulfilling public procurement legislation. Although there were no set indicators, there was an ambition from the public construction clients to find the balance between fulfilling public procurement regulations and achieving a god business deal.

Concerning the decision-making process when investing in a new facility, there needs to be a connection between the cost–benefit analysis and the requirements, technical as well as administrative, expressed in the tendering briefs. Practical experience was raised as a relevant aspect when developing the procurement ability over time, some respondents stated that there was a clear ambition to have the procurement abilities in-house and avoid outsourcing this to external parties. There was a variety of activities performed to manage and increase the knowledge in the procurement process, thus achieving a higher competence with project managers and procurement officials. However, none of the respondents had any measurable indicators to evaluate the knowledge management process.

One of the larger clients that have business nationwide stated the importance of unified procurement processes in the whole organisation. The purpose of these was that contractors should know how the client operates and manages the procurement process regardless of where the procurement takes place. Another topic raised was the concept of cooperation. One way to achieve this is to engage in various partnering concepts; however, the respondents raised that striving for cooperation was equally important in all projects, partnering or not. Further some raised the issue of long-term relationships, which increase the opportunity to promote more standardised procedures and production techniques, thus achieving better control and unity the delivery process.

25.2.2.6 Evaluation of the Procurement Ability Over Time

Most respondents recognise reoccurring problems in the procurement process and state that they try to identify these in order to update procurement-related documents and procedures. In addition, most respondents have procedures for exchanging experiences with their contractors. However, one respondent stated that the focus was strongly on the economic results and time management of the contract works when they evaluate their procurement ability, rather than the procurement process itself. Another respondent gave the view that to follow up a project budget is indirectly an evaluation of their procurement ability. Some respondents stated that they do not have any formalised evaluation about their procurement ability, although informal non-documented evaluation occur. Some respondents argued that procurement ability involves a process of learning by doing and that circumstances always changes. Something that worked well in one project may not work at all in another. Although some respondents raised the issue of education and developing the competence as important issues concerning developing their procurement ability.

25.2.2.7 Main Drivers for Developing the Procurement Ability Over Time

Some respondents raised the importance of obtaining the best value for money, i.e. to get what you pay for. The project results were claimed by many to be the main driver for how the procurement process is carried out. Mainly to keep the cost down without imposing undue economic pressure on the contractor. The contractor needs to be adequately compensated for demands made concerning time, quality and performance. However, another respondent argued that if a 5–10% saving can be made in the procurement of each project, this will be a lever effect for the results of the client organisation as a whole.

Another driver for developing the procurement ability was the challenge of setting clear demands for the construction project, not at least demands concerning sustainability and climate where functional demands need to be set concerning the buildings performance and the environmental impact of used materials, in relation to UN targets. The choice of delivery is crucial, for example, in a design-build contract too detailed demands controlling the delivery might hinder innovation opportunities, so that some of the advantages with the design-build scheme might be lost, whereas too flexible demands might miss the bar of achieving sustainable goals. A strategic perspective on procurement process with carefully thought through and competent choices throughout the entire tendering process. Competence and professionalism, both internally in the client organisation and externally in the supply chain, are important to maintain trustworthiness in the procurement processes.

25.3 Discussion

The five disciplines that Senge (1990) presented are personal mastery, mental models, building shared visions, team learning and systems thinking. When applying these to analyse the findings there is a clear emphasis on personal mastery, with a strong focus on procurement as an individual craftsmanship to master and understand procurement procedures and regulation in order to achieve a cost-effective procurement process. From a mental perspective, the thinking concerning the procurement is risk aversive, with a focus on decreasing cost and risk for the construction client organisation rather than focus on best value. This might not be wrong, since errors made in the procurement process can be costly. However, opportunities might be overlooked by not focusing on what creates value for the construction project and the client organisation. There is no clear evidence in the empirical material of organisations building shared visions about improving their procurement ability. There are examples of team learning, mainly from a local perspective within the procurement officials in close proximity to each other. Albeit the larger nationwide organisations exhibit examples of team learning on a higher organisational level between different procurement teams. However, systems thinking was not clearly expressed. Furthermore, the empirical material comprises no examples of learning on an industry level between different client organisations, and there are only a few examples of reasoning about how to ingrate the procurement process with other construction project activities.

To analyse the degree of systems thinking, the maturity levels described by Chinowski et al. (2007) were used, based on the following characteristics of a learning organisation:

- Leadership, championing the integration of new knowledge and taking proactive steps towards a shared vision. Leadership is clearly needed in the construction industry in general and within construction clients in particular, in order to meet the challenges that the UN sustainability goals address. In order to change the mindset, leaders are required to challenge and develop processes and infrastructure, communication, education and culture.
- Process and infrastructure, managing processes required to manage knowledge and implement new knowledge in an organisation
- Communication, knowledge sharing towards organisational improvement and eliminating barriers for communicating experiences
- Education, developing a systematic approach for training and dissemination of knowledge
- Culture. Receptiveness for new ideas and an openness for change

Based on the abovementioned characteristics of learning organisations and the typical attributes of the construction sector, Chinowski et al. (2007) define six maturity levels (0–5) in relation to learning.

At *level 0*, the organisation is at the beginning of transforming to a leaning organisation concept. At *level 1*, the organisation is establishing the leadership required to move towards a learning organisation concept. Focus is on the processes and infrastructure needed to implement knowledge sharing activities. At *level 2*, the organisation has completed their development of leadership, processes and infrastructure. The organisation is well on the way towards a clear focus on knowledge sharing and communication. At *level 3*, the organisation has completed organisation wide processes to support learning and promote a learning culture. Learning is viewed as an integral part of organisational activities. At *level 4*, the organisation has almost reached a full learning organisation maturity. There is a strong focus on learning on both an organisational as well as an individual level. At *level 5*, the organisation has reached full maturity as a learning organisation.

When we applied these levels of maturity to procurement processes within the studied client organisations, the following results emerge:

- The maturity level for leadership is mainly at level 1 or 0 since there is no clear evidence in this study of development of a shared vision within or between construction client organisations in connection to how they reflect upon their procurement process. An optimistic assessment is that one organisation is at level 2.
- From a process and infrastructure perspective, the maturity level can be assessed as 1 or possibly 2 in few organisations. There are examples of work being done with a focus on evaluating the procurement ability, because the responsibilities for this are on an individual level and sometimes on a community level within the local organisation. From this study, there are a few examples of developing processes and infrastructure on a higher organisational level.
- From an educational viewpoint, the maturity level is 1 because much focus is on individual training, not so much on reflections about how procurement processes affect the organisation as a whole. Procurement is mainly focused on processes

and infrastructure in order to avoid mistakes and minimise risks in relation to the client organisation in general and to the construction project in particular.

- From a communication perspective, the maturity level can be assessed as a 1 or 2. There are examples from some of the larger client organisations, particularly those that have projects nationwide, of organisational knowledge sharing. Procurement processes were evaluated over time, which resulted in organisational guidelines of how to conduct procurement processes. In the smaller client organisations, the individual focus was more apparent. Most evaluations made about the procurement process was on procedural aspect, and some even raised uniqueness as a problem. However, there was also expressed view that circumstances change between projects so it is more learning by doing, which in the model developed by Chinowski et al. (2007) would be an example of level 0.
- From a culture perspective, the maturity level can be assessed to 0 or may be 1. There is not much evidence on viewing procurement as a value adding process in the implementation of construction projects, although some respondents expressed best value as a driver for developing their procurement ability. However, the examples mentioned focused mainly on cost reduction and not about added value. Rather the main driver was on risk reduction by focusing on setting clear demands in tender documents. However, there are indications that there is a cultural change ongoing with a potential shift to level 2 and may be level 3.

25.4 Conclusion

This study represents an initial and preliminary reflection on how client organisations view and work with their procurement ability over time. The main conclusion is that procurement is mainly seen as a procedural activity which focuses on risk and cost reduction rather than potential in adding value to the organisation. This mindset might be a barrier when dealing with challenges concerning sustainability, environmental impact and climate change. A shift from cost reduction to a value perspective (where one important value in the future is to manage the UN sustainability goals and targets) is necessary. Here the procurement process is an important process in order to translate these values into demands in tender documents and contracts.

Furthermore, this study presents arguments that many client organisations view procurement as a support function rather than a core project activity. Thus, the procurement process is not fully integrated in the main construction project management processes. Although there are examples of integration from the interviews, there is a lack of clear visible strategy of how to integrate procurement process in a learning perspective for both the procurement process in itself as well as for the whole construction project management process. More research is needed to increase the understanding of how procurement can act as a value-adding process and how knowledge sharing and knowledge transfer can increase the learning concept in relation to procurement in construction. Further research may include a broad survey based on the maturity model by Chinowski et al. (2007) in order to achieve a more certain view of the situation in the construction sector in general, and also to better understand how an improved procurement process can contribute to the implementation of UN sustainable development goals.

References

Alvesson, M. (2011). *Interviews – Execution, interpretation, and reflexivity*. Malmö, Liber.

Boyd, D., & Chinyio, E. (2006). *Understanding the construction client*. Blackwell Publishing.

Chinowski, P., Molenaar, K., & Realph, A. (2007). Learning organizations in construction. *Journal of Management in Engineering, 23*(1), 27–34.

Garvin, D. A. (1993, July-August). Building a learning organization. *Harvard Business Review*, 78–91.

Goh, S., & Richards, G. (1997). Benchmarking the learning capability of organizations. *European Management Journal, 15*(5), 575–583.

Hartmann, A. A., Reymen, I. M. M. J., & van Oosterom, G. (2008). Factors constituting the innovation adoption environment of public clients. *Building Research & Information, 36*(5), 436–449.

Hazlett, S. A., McAdam, R., & Beggs, V. (2008). An exploratory study of knowledge flows: A case study of public sector procurement. *Total Quality Management, 19*(1–2), 57–66.

Hughes, W., Champion, R., & Murdoch, J. (2015). *Construction contracts- law and management* (5th ed.). Routledge.

Kvale, S. (1997). *The qualitative research interview*. Studentlitteratur.

Manley, K., & Chen, L. (2017). Collaborative learning to improve the governance and performance of infrastructure projects in the construction sector. *Journal of Management in Engineering, 33*(5), 1–14.

Ruparathna, R., & Hewage, K. N. (2015). Review of contemporary construction procurement practices. *Journal of Management in Engineering, 31*(3), 1–11.

Seneratne, S., & Malewana, C. (2011). Linking individual, team and organizational learning in construction project team setting. *Architectural Engineering and Design Management, 7*(1), 50–63.

Senge, P. (1990). *The fifth discipline: The art and practice of the learning organisation*. Doubleday.

Silverman, D. (2006). *Interpreting qualitative data* (3rd ed.). SAGE Publications.

Tassabehji, R., & Moorhouse, A. (2008). The changing role of procurement: Developing professional effectiveness. *Journal of Purchasing and Supply Management, 14*(1), 55–68.

Toumela-Pykkönen, M., Aaltonen, K., & Haapsalo, H. (2015). Procurement in real estate and construction sector (RECS) – Preliminary context-specific attributes. *Procedia Economics and Finance, 21*, 264–270.

Correction to: Managing Procurement Ability Over Time: A Study of Swedish Construction Clients

Stefan Olander and Henrik Szentes

Correction to:
Chapter 25 in: G. Lindahl, S. C. Gottlieb (eds.), *SDGs in Construction Economics and Organization*, Springer Proceedings in Business and Economics, https://doi.org/10.1007/978-3-031-25498-7_25

The book was inadvertently published with an incorrect chapter in the online version of the proceedings.

The chapter has been changed to Managing Procurement Ability Over Time: A Study of Swedish Construction Clients by Stefan Olander and Henrik Szentes.

The updated original version of this chapter can be found at
https://doi.org/10.1007/978-3-031-25498-7_25

G. Lindahl, S. C. Gottlieb (eds.), *SDGs in Construction Economics and Organization*,
Springer Proceedings in Business and Economics,
https://doi.org/10.1007/978-3-031-25498-7_26

Index

G. Lindahl, S. C. Gottlieb (eds.), *SDGs in Construction Economics and Organization*, Springer Proceedings in Business and Economics,
https://doi.org/10.1007/978-3-031-25498-7